HISTORY OF THE MARIANA ISLANDS

HISTORY OF THE MARIANA ISLANDS

LUIS DE MORALES, S.J. &
CHARLES LE GOBIEN, S.J.

Edited and commented by
Alexandre Coello de la Rosa

Prologue by
Joan-Pau Rubiés

UNIVERSITY OF GUAM
PRESS

Alexandre Coello de la Rosa (Barcelona, 1968) received his Ph.D. in History from SUNY at Stony Brook, USA (2001). He is editor-in-chief of the journal *Illes i Imperis* and is currently doing research at the Universitat Pompeu Fabra, Department of Humanities. He is a member of the research group *Imperis, Metròpolis i Societats Extraeuropees* (GRIMSE) and *Ethnographies, Cultural Encounters and Religious Missions in the Iberian World* (ECERM) at the same University, as well as an associate member of the Richard F. Taitano Micronesian Area Research Center (MARC) at the University of Guam. He specializes in colonial Latin American history, the ecclesiastical history of Peru and the Philippines, historical anthropology, and chronicles of the Indies. Some of his recent publications include *Jesuits at the Margins: Missions and Missionaries in the Marianas* (London & New York: Routledge, 2016) and *Elogio a la antropología histórica: enfoques, métodos y aplicaciones al estudio del poder y el colonialismo* (Zaragoza: Publicaciones de la Universidad de Zaragoza & UOC, 2016). E-mail: alex.coello@upf.edu

Originally published as *Histoire des isles Marianes.* Paris: Charles Le Gobien. 1700.

Copyright © Alexandre Coello de la Rosa
Prologue Copyright © Joan-Pau Rubiés
All rights reserved.
Published by University of Guam Press, Guam
First Edition, 2016
Second Edition, 2024

LIBRARY OF CONGRESS CATALOGING-IN-PUBLICATION DATA
History of the Mariana Islands/edited by Coello de la Rosa, Alexandre
Includes bibliographical references and index.
ISBN: 978-1-935198-09-3 (hardback)
ISBN: 978-1-935198-95-6 (paperback)
ISBN: 978-1-935198-96-3 (library ebook)
ISBN: 978-1-935198-98-7 (trade ebook)

University of Guam Press
R.F.T. Micronesian Area Research Center
UOG Station
Mangilao, Guam 96923
(671)735-2154
www.uog.edu/uogpress

This book is dedicated to Luz Helena Ramírez Hache, for all the years past and the years to come.

TABLE OF CONTENTS

ACKNOWLEDGEMENTS

Many people have contributed to bringing the new English version of this book to light, and my acknowledgements go to all of them. First, to my friend and colleague David Atienza de Frutos, of the University of Guam, and his wife Maruxa and their numerous family for the support and encouragement in this and other vital projects. Second, to Joe Quinata, Chief Program Officer of the Guam Preservation Trust; and to the Micronesian Area Research Center (MARC) and its director, Monique Storie, for publishing the *Historia de las islas Marianas* (that I edited earlier) there where it all started: in Guam. It was at the MARC in 2010 where Omaira Brunal-Perry, Marjorie G. Driver and John Peterson allowed me to consult the archive for the first time, giving me photocopies of maps, engravings and other documents that they have acquired over the years. And third, my friends and colleagues here and there, many of whom have been fundamental in my "discovery" of the so-called Pacific world. In Rome, Fathers Francisco de Borja Medina, S.J., Paolo Molinari, S.J. and Thomas McCoog, S.J., as well as the generous personnel of the Archivum Romanum Societatis Iesu (Roma), particularly Mauro Brunello and Nicoletta Basilotta, who helped me track down the documents related to the Jesuit missionaries destined to the Marianas archipelago. In Barcelona, Father Francesc Casanovas, S.J., director of the Arxiu Històric de la Companyia de Jesús a Catalunya, who gave me total access to the documents on the Philippines that contain the copies of Luis de Morales's *Historia de las islas Marianas*. In Dourados (Brazil), San Ignacio Velasco (Bolivia) and Santiago de Chile (Chile), I have shared memorable moments with friends and colleagues of the Jornadas Internacionales sobre las Misiones Jesuíticas, such as Maria Laura Salinas, Aliocha Maldavsky, Artur Henrique Barcelos, Carlos Paz, Eduardo Santos Neuman, Guillermo Wilde, Mercedes Avellaneda de Bocca, Lia Quarterli, Jaime Valenzuela Márquez, Arno Kern, Ignacio Telesca, Bartomeu Melià, Beatriz Vitar, and my beloved Nely Aparecida

Maciel (1970-2014).

The Universitat Pompeu Fabra (UPF) is a creative space where friends, students and professors share their projects and ideas, thus contributing to enhancing this particular book. Josep María Delgado and Josep María Fradera encouraged and supported me since the beginning to explore the colonial frontiers of the Pacific. With Daniele Cozzoli, Maite Ojeda, João Melo, Juan Carlos Garavaglia (1944-2017), Nicolás Kwiatkostki, David Ruiz Martínez, Manel Ollé, and Joan-Pau Rubiés, I have shared many enlightening conversations that have greatly enriched my own analyses. In the Universitat Autònoma de Barcelona (UAB), I also had the opportunity of sharing my first inquiries into the anthropology and history of the Mariana Islands with Doris Moreno, Verena Stolcke, Montserrat Ventura i Oller, Montserrat Clua i Fainé, Josep Lluís Mateo Dieste, Maite Ojeda Mata, Mónica Martínez Mauri, Joan Muela, Salvador Queralt, Pablo Domínguez, Alice van den Bogaert and Sabine Kradolfer. Xavier Baró (Universidad de Barcelona, UB) and Joan Pau Rubiés, Manel Ollé and João Vicente Carvalho de Melo Carreiro (UPF) share my interest in the Jesuits in Asia and Oceania. I thank Joan-Pau for his kindness in writing this book's prologue, and for his incisive commentary. In Madrid, María Dolores Elizalde Pérez-Grueso, of the Instituto de Historia (CSIC), showed great interest in my investigations on Spanish Micronesia. Her support and generosity were fundamental in my professional trajectory.

Last but not least, I would like to thank Yesenia Pumarada Cruz for her excellent translation; Ramón Alba, editor of Ediciones Polifemo, for publishing the Spanish edition; and again, the editors of UOG Press in partnership with NYU Press for publishing the English editions of this book, and thus making it available for the people of Guam today.

PROLOGUE:
APOLOGETICS AND ETHNOGRAPHY IN THE
HISTORY OF THE MARIANA ISLANDS
BY LUIS DE MORALES / CHARLES LE GOBIEN

Joan-Pau Rubiés
ICREA-Universitat Pompeu Fabra (Barcelona)

In 1700, the French Jesuit Charles Le Gobien (1653–1708) published a handsome book about one of the most remote spaces of the European colonial system, the Mariana Islands.[1] As recounted by Antonio Pigafetta, one of Le Gobien's sources, Ferdinand Magellan had "discovered" the islands during his crossing of the Pacific Ocean (before reaching what would eventually become the Philippines).[2] At first they had been named *Islas de los Ladrones* (Islands of Thieves) because of their natives' attempts to acquire, by unorthodox means, objects made of iron—a material that they coveted. The name Mariana Islands was imposed a century-and-a-half later in 1665, when a restless Jesuit missionary, Diego Luis de San Vitores, managed to persuade the Spanish Crown through the special intercession of Queen Regent Mariana of Austria, that it was the Monarchy's spiritual duty to Christianize the islands, which had apparently been abandoned to their fate because of their lack of economic significance. The islands were of course far from being abandoned, but the Spanish considered that their inhabitants were savages and that in order to bring Christianity and civilization to them, it was necessary to colonize the islands. In other words, they had to be conquered, whether peacefully or by force.

---

1. *Histoire des Isles MarianesSreligion chrestienne...* (Paris: Nicolas Pepie, 1700).

2. The word discovered is the correct one in this context because this is what Pigafetta used: "Circa de setanta legue alla detta via, in dodeci gradi di latitudine et 146 de longitudine, mercore a 6 de marzo **discopressemo** una ysola al maistrale picola, et due altre al garbino". *Il Primo Viaggio Intorno al Globo di Antonio Pigafetta, e le Sue Regole sull'Arte del Navigare*, a cura di Andrea da Mosto (Roma: Ministero della Pubblica Istruzione, 1894), p. 67.

What interested Le Gobien about the Marianas archipelago was pre-cisely the fact that it was a Jesuit mission in a poor and neglected region. As he emphasized in the book's letter of dedication to the Bishop of Ypres, Martin de Ratabon, the lack of resources in the islands demonstrated that the Society's motivations were purely evangelical. This, in turn, enhanced the contested legitimacy of the Society's missions elsewhere, such as in Japan, China, Tonkin and Cochinchina (Vietnam), or Canada. From the very beginning, the book's author presented Diego Luis de San Vitores as the key figure who, by means of his apostolic vision, was able to mobilize the support of the Governor of the Philippines and the Viceroy of New Spain.[3] The colonization of the Mariana Islands, and even the maintenance of the enormous missionary enterprise in the Philippines despite the vast distance, went beyond a purely economic logic. The attention paid by Le Gobien to the new mission of the Marianas was therefore part of the apologetics of the Society's missionary activity—a propaganda effort in which, precisely, Le Gobien was to specialize.

At the turn of the eighteenth century, the Jesuits were in a rather delicate position across Catholic Europe. Resentment from rival orders such as the Dominicans, Franciscans (including Capuchins) and Augustinians was long-standing, and throughout the seventeenth century, Jansenists and Gallicans in France joined the anti-Jesuit camp. The Chinese rites controversy, and a similar conflict that erupted at the end of the century regarding the Malabar rites in Southern India, fueled these attacks, bringing into question the legitimacy of the Jesuit method of cultural adaptation.[4] In parallel, Rome and Paris saw the emergence of de facto institutions such as the *Sacra Congregatio de Propaganda Fide* (Sacred Congregation for the Propagation of the Faith) and the *Société des Missions étrangères* (Society of Foreign Missions), which had been created outside the jurisdiction or *patronato* of the Spanish and Portuguese Crowns in order to encourage new missions in areas not under direct European politi-cal control. This was precisely the type of mission that had distinguished the Society of Jesus and had thrived since the mid-sixteenth century.

In line with the propaganda tradition of his order, Le Gobien, who

---

3. The Spanish colonization of the Philippines and the entire North Pacific depended on the logistical support from Mexico, because the Indian Ocean route belonged to the Portuguese Crown.

4. The rites controversy was provoked by the Jesuit attempt, inaugurated by Matteo Ricci, to integrate Confucian ethical teachings with Christianity. Critics questioedn that the ritual reverence showed towards ancestors in China, or towards Confucius himself, could be legitimately classified as merely civil ceremonies, and argued that they involved a form of religious worship, which would imply idolatry.

was Procurator of the Chinese mission in Paris, was committed to defending and publicizing the Jesuit missions by means of a direct intervention in the Republic of Letters. His collaboration with Louis le Comte in defending the Jesuit position regarding Confucianism in China was particularly notable, first in their *Lettre sur les progrez de la religion à la Chine* (1697), a rare work, and especially with the famous *Histoire de l'édit de l'empereur de la Chine* (1698). Both documents sought to counter the condemnation of the Jesuit method by Charles Maigrot (an Apostolic Vicar in Fujian, China) and by the Dominicans. In 1700, some of Le Gobien's propositions were censured by The Sorbonne.

Eventually, Le Gobien became one of the leading cultural figures of his time, launching in 1702 the famous series *Lettres édifiantes et curieuses*, whose very title revealed the blending of religious and enlightened values that characterized the work of that generation of French Jesuits. However, the defense of the non-idolatrous character of Confucian rites—and therefore, their compatibility with Christianity—was only one of the battlefronts faced by the Society. The Ignatian order was also often criticized for its preference for undertaking missions in wealthy and prosperous societies, such as those of Japan, Siam, and especially China, where its members would associate with the local elites and sometimes would even dedicate themselves to lucrative trading activities (notably, their intervention in the trade between Japan and the colony of Macao in China, until the Portuguese were expelled from the islands by the Shogun). The publication of a book about the "poor and abandoned" Mariana Islands allowed Le Gobien to provide evidence against this accusation, showing that the Jesuits were martyrs of the faith with a genuinely universal apostolic vocation.

The discovery and publication some years ago of the Spanish (Castilian) version of the *History of the Mariana Islands* — a manuscript found in the Provincial Archive of the Society of Jesus in Catalonia, and attributed to Father Luis de Morales — was a remarkable and unexpected event. Alexandre Coello de la Rosa, editor of this Castilian translation and a notable specialist in Jesuit missions in the Marianas, challenged the traditional attribution of the work to Charles Le Gobien. Coello noted that Father Luis de Morales, who was one of the members of the first missionary expedition to the Islands led by Luis de San Vitores, had written first-hand relations concerning the Marianas, and had also gathered information for many years with the hopes of writing a particular history of the mission.[5]

---

5. According to a document from May of 1689 included in Rodrigue Lévesque's monumental *History of Micronesia* (cited in the introduction by Alexandre Coello de la Rosa), Father Lorenzo Bustillo wrote a letter

Le Gobien cites Morales as one of his sources, but Alexandre Coello goes further by making available an original manuscript in Spanish that clearly identifies Morales as its author, and whose contents are practically identical to the book later published by Le Gobien. According to Coello's interpretation, Le Gobien would have been a plagiarist, albeit at a time when there was more flexibility concerning the notion of original authorship. In this respect, the Jesuits freely used each other's materials not as mere individuals seeking glory, but as men working together towards a higher aim. According to Coello, Morales was unable to complete his work when he was in Europe as the Society's procurator for the Philippines (1684–89), and continued to update the history during his long stay in New Spain (1690–1697), although he eventually returned to the Pacific islands without having published his text.[6] Le Gobien, on his part, was in France, a country central to the Republic of Letters and Europe's new great Catholic power, whose support had become essential for the Society's missionary activity. The final production of a book in French based on previous materials of a very similar nature but in Spanish would justify a new authorship, especially if both authors were committed to constructing an apology of the missionary order to which they both belonged.

Unfortunately, the manuscript found in Barcelona is not an original late seventeenth century document, nor an eighteenth century copy, but rather an unfinished version produced early in the twentieth century (Professor Coello de la Rosa has used Le Gobien's French text to complete the final chapters). It

---

from San Ignacio de Agadña (in the Marianas) to the new Procurator General of the Philippines, Antonio Matías Jaramillo, which explained that in 1687 Ambrosio Ortiz had asked from his post in Naples that the production of *relaciones* (accounts) about the Marianas should continue, "so that the *History of the Marianas* may be prosecuted, specifying individual people–which for this reason I have noted in the margins–and the lands". The relations would serve "as great guidance for the historian [who is] over there". These relations from 1685 had been sent to the Procurator General of the Indies in Madrid (it thus appears that the Jesuits in Naples, then part of the Spanish Monarchy, received their news about the Marianas from Madrid, and Madrid from Mexico). Bustillo added that "Father Luis de Morales says the same thing from Madrid, and he asks that nothing, however small, be left out of these relations, for back there everything makes a heavenly harmony" (*History of Micronesia. A Collection of Source Documents*, 20 vols. Québec, 1992–2001, IX, pp. 291–2). All of this implies that during the second half of the 1680s both the Procurator General of the West Indies, Diego Francisco Altamirano, and the former Procurator of the Philippines Province, Luis de Morales, had been promoting the arrival in Europe of new materials on the mission to the Marianas so that a history of the islands, which had already been started, could be completed. The identity of the "historian over there" who was to continue this work was not specified, but in all likelihood he was to continue the work begun by Francisco García, who died in 1685, and whose book of 1683 may be interpreted as the first installment of this *Historia*, enjoying great success at the court ("a heavenly harmony"). Not coincidentally, the same Ambrosio Ortiz who had written asking for the relations to continue was the Italian translator of García's work, and dedicated himself to updating its contents.

6. The bulk of the *History* covers the period of 1668 to 1683 (the year when Morales left the Philippine Islands to return to Europe as procurator). The tenth book, however, deals with a later period, from the start of the second CHamoru rebellion in 1684 to 1696.

is, moreover, written in modern Spanish. It is possible that the copyist, perhaps a Jesuit working on the history of the Philippines who continued his activities in Spain after the end of the colonial period in 1898 (which brings to mind the colossal work of Catalan Jesuit Pablo Pastells, or one of his collaborators), decided to modernize the old text as he transcribed it, thinking about a possible publication. However, it seems much more likely that the text found in the Provincial Archive of Catalonia is a direct translation of Le Gobien's book, produced in the early twentieth century. A number of observations, mainly philological, lead me to this conclusion:

1. The language is not the only modern element: the concepts themselves are modern. This is especially evident in the ethnographic section (book 2), where Le Gobien's phrase, "*les hommes peuvent pendre autant de femmes qu'il leur plaist,*" is translated as "polygamy is permitted." In a similar vein, the late seventeenth century concept *politesse* becomes "a degree of education," while *incivilité* is "lack of education". The three estates described by Le Gobien—the nobility, the people, and those of a middling condition—become three "classes," including the "middle class"; and while Le Gobien said of native housing that, "*on couche dans le premier appartement,*" and that, "*on mange dans le second,*" the Spanish version speaks of the bedroom and the dining room, as if describing a flat in Barcelona's *eixample*.

2. In numerous passages where the Castilian text is dubious or incoherent, Le Gobien offers a better text, suggesting that the manuscript provides a rushed and unrevised translation. For instance, the preface in Spanish reads, "I have put nothing of my own [in this history], only what I have found in their letters and relations…" (*"Nada he puesto de mi cosecha, y sólo sí lo que he hallado en sus cartas y relaciones…"*); whereas the French author told his readers that, *"Je n'y ay rien avancé que ce que j'ay trouvé dans les lettres et dans les relations de ces missionaires …"* which makes more sense. In a reference to Morales, the manuscript speaks of the missionary, "who came from Rome to look after the mission's affairs"; but in French he was a missionary *qui vint à Rome*, that is to say that he went to Rome, which is more logical. The CHamoru[7] woman who at home has no say (*"manda en absoluto"*), in reality rules without any limitations, *absolutement*. While there are countless other examples like these, I have not found a single passage in the Spanish text that is better in construction or content than what

---

7. The word CHamoru was initially used by some indigenous rights activists in the 1990s. Present-day indigenous inhabitants of the Marianas refer to themselves either as Chamorro or CHamoru. In the last years the latter term has gained popularity, especially in Guåhan, and for this new English edition we have adopted this usage.

was published by Le Gobien.[8]

3. The manuscript's notes reproduce Le Gobien's notes, but only selectively. That is to say, not all of them are included, and some references are incomplete.[9] It is hard to believe that Le Gobien, had he been a simple plagiarist, would have spent his time completing references and adding original historical notes. It is more likely that a modern translator, writing two centuries later, decided to keep only those notes that still seemed relevant.

4. The prefatory note that indicates that the author's sources came "from Rome, from Spain, and from the Low Countries," makes perfect geographical sense if the author was in France, which was Le Gobien's case.[10] Had the author been in New Spain, like Morales in the 1690s (after he left Spain), the sentence would still be coherent, but in reality the reports would have come to him directly from the Philippines, not Europe.[11]

5. The manuscript makes numerous references in the third person to the Spanish, and to Morales himself during his missionary activity in the Marianas between 1668 and 1671. However, this is not decisive evidence that the writer was not Morales himself: in historical writings (as opposed to letters or relations) it was perfectly possible to refer to oneself in the third person as a sort of rhetorical modesty, just as Julius Caesar had done in his *Commentaries*.

6. Finally, the Spanish version reproduces Le Gobien's prefatory note which, after offering a map drawn by Jesuit Alonso López, ends up also talking

8. Another example from the first book: Luis de San Vitores is described in the Spanish version as being "more distinguished by his virtue than by his birth, for he came from one of the most illustrious families of Burgos." In the French text, we are told that virtue brought San Vitores greater distinction than his birth, even though he was from one of the most illustrious families of Burgos (*"plus distingué par sa vertu que par sa naissance, quoy qu'il fust d'une des plus illustres maisons de Burgos"*). In another case, the manuscript says that navigation in the Pacific, "is not subject to bad weather and storms, for the wind is always favorable" (*"no es ocasionada [?] a contratiempo y tempestades, pues se tiene siempre el viento favorable."* This is an inelegant summary of the French text, where navigation, *"est douce et avisée; comme on a toujours vent arrière, on vogue tranquillement, sans estre exposé aux cops de mer et sans essuïer presque aucune tempeste."*

9. I will give three examples: the reference to Vitruvius when speaking of the domestication of fire as a key factor in the development of civilization (in the second ethnographic chapter) is incomplete in the Castilian manuscript. More remarkably, only in Le Gobien's text are we told that the governor of Manila in 1662 was Don Diego Salcedo. Le Gobien also makes explicit a reference to the gospel of St. Mathew in the memoir written by San Vitores to Philip IV in 1664 that one would expect in the original document.

10. The presence of the Low Countries in this list is explained by the important *relaciones* sent by the Belgian Jesuits Gerard Bouwens and Peter Coomans. Many of their letters, written in Latin, are still kept in Brussels, and were published by Rodrigue Lévesque in his *History of Micronesia* (see especially vol. V, 1995). See also Augusto V. de Viana, "Belgian Missionaries in 17th Century Marianas: the Role of Fr. Peter Coomans and Fr. Gerard Bouwens", *Philippiniana Sacra* XLVI, 136 (2011): 365–89.

11. For instance, after returning to Mexico from Europe in 1690, Morales prepared a summary of the news that reached him from the Philippines regarding the Marianas, noting especially his companions' numerous critiques against the incompetence of Governor Don Damián de Esplana, who had become fearful of the Chamorros after the 1684 rebellion (R. Lévesque, *History of Micronesia*, vol. IX, pp. 348–50).

about a brief memoir on the situation and extension of the islands that, "another missionary who came <from> [to] Rome in order to look after the mission's affairs left to one of his friends." The memoir is in fact attributed to Luis de Morales and it would be very peculiar for him to refer to himself in this way.

If we accept this analysis, why then did our early twentieth century hypothetical translator attribute the work to Luis de Morales, when in reality he had translated it directly from Le Gobien? It might be the case that he had previously seen (but not copied) a text written by Morales in Manila, which, however, nobody has described since. Or perhaps, knowing about the activities and writings of Morales as missionary and procurator, and seeing his name mentioned as one of the sources by Le Gobien, the translator concluded that what Le Gobien published was indeed plagiarized from a previous text by Morales, and decided to rectify this situation (perhaps driven by patriotic fervor) by identifying the Spanish Morales as the original author.

The fundamental fact is that although the manuscript we have is a translation from the French book, the narrative undoubtedly relies on the relations written by Morales and other Jesuits in the Marianas. As Alexandre Coello accurately notes, one of Le Gobien's sources was the life and martyrdom of San Vitores published by the fellow-Jesuit Francisco García in Madrid in 1683 (of which there was also an enlarged Italian version published in Naples in 1686).[12] García's text, on its part, was based on the annual letters sent to Spain through Mexico and the Philippines, which might have included some materials by Morales, who, as we stated earlier, two years later in 1685, was in Madrid as the Procurator for the Province of the Philippines.[13] García, a publicist of the

---

12. Francisco García S.J., *Vida y martyrio del venerable padre Diego de Sanvitores, de la Compañía de Jesús, primer apóstol de las islas Marianas, y sucessos de estas islas, desde el año de mil seiscientos y sesenta y ocho, asta el de mil seiscientos y ochenta y uno* (Madrid, 1683). In Italian, *Istoria della conversione alla nostra santa fede dell'Isole Mariane, dette prima de'Ladroni, nella vita, predicatione, e morte gloriosa per Christo del venerabile P. Diego Luigi di Sanvitores, e d'altri suoi Compagni della Compagnia di Giesù*, translated by Ambrosio Ortiz S.J. (Naples, 1686), with new sections that describe the 1684 uprising. I quote from this expanded edition. There is also a modern English translation: *The Life and Martyrdom of the Venerable Father Diego Luis de San Vitores First Apostle of the Mariana Islands, and Events of These Islands, from the Year Sixteen Hundred and Sixty-Eight, Through the Year Sixteen Hundred and Eighty-One*, ed. James A. McDonough S.J., et. al. (Guam, 2004).

13. If we compare the brief geographical description of the islands by Morales incorporated in the *Histoire des Isles Marianes*, with García's geographic data (Book III, chapter 1), we can see that Morales has more exact data, which suggests that García used a different, and probably earlier, document to write this section. Some years earlier, the Procurator of the Philippines Andrés de Ledesma had published a few accounts of San Vitores's mission, collecting materials about the period 1665–1672 to be used for his beatification cause. No doubt these were also used by García. Some of these documents are described by Lévesque in his *History of Micronesia*. See, for example, vol. V, pp. 478–509 (years 1668–1672); p. 510 (years 1669–72). The Royal Academy of History in Madrid has a copy of a summary, or *Historia Breve*, by García, which looks like a synthesis of the relations and letters that he was going to use for the history of the Marianas during the period 1672–1681 (R. Lévesque, *History of Micronesia*, vol. V, p. 513). By contrast, although missionary

Jesuit order at the court of Charles II in Madrid, was evidently assigned the task of writing a book with the purpose of advancing the beatification of San Vitores, whose life was interpreted as a re-enactment of that of the apostle to the Indies Francis Xavier, who had already been sanctified in 1622. García's hagiography of San Vitores became the first sketch of the history of the Mariana Islands, as the evident similarities between it and Le Gobien's book attest. Specifically, Le Gobien's most significant ethnographic chapter—the second "book" in the present text—closely follows García's account, either directly or through other Jesuit texts that copied from García's narrative, such as Gabriel de Aranda's *Vida y martirio* of father Sebastián de Monroy (Seville, 1690).[14]

It is not clear if Le Gobien had access to a complete text written by Morales in New Spain circa 1696, that is, before the latter embarked again towards the Philippines, as Coello de la Rosa suggests in his introduction; or if he wrote a synthesis of various narratives and reports produced by García, Morales, and other missionaries, and obtained from different locations in Europe, as the French writer states in his introduction. Le Gobien could also have received a package of materials from the General of the Society in Rome, Tirso González de Santalla (1687-1705), with the request to produce an integrated narrative for publication. What is, however, beyond any doubt, is that his narrative voice incorporates those of Morales, García, and perhaps other Spanish and Belgian Jesuits.

In any case, thanks to Alexandre Coello's efforts, readers of this version of the book will be able to enjoy a text layered with multiple significations. The tension between two seemingly contradictory themes is particularly notable: on the one hand, the hagiographical providentialism surrounding the history of the activities of the apostle Diego Luis de San Vitores and his companions; on the other, the scientific and sometimes even sympathetic tone of the descriptions of the way of life and customs of the supposedly savage islanders, the CHamoru people. The ethnographic chapter, in particular, is a small gem.

We are told that, isolated from the world and ignorant of its ways, the

---

Peter Coomans wrote in Latin an interesting *historica narratio* of the islands during the period 1667–1673 (R. Lévesque, *History of Micronesia*, vol. VI [1995], pp. 23–68), the different contents suggest that García did not use it. Nor was García able to use two further relations published by Morales after his arrival at the Court concerning the state and progress of the mission in Marianas in the years 1681–1684, which were based on the letters by Father Solórzano (R. Lévesque, *History of Micronesia*, vol. VIII [1996], pp. 66–75), for García had already published his work in 1683, when Morales was still in the Philippines.

14. Gabriel de Aranda S.J., *Vida y gloriosa muerte del venerable padre Sebastián de Monroy, religioso de la Compañía de Jesús, que murió dilatando la fe alanceado de los bárbaros de las Islas Marianas* (Sevilla, 1690), chapter 40, pp. 214–20.

CHamoru had no laws, but they rigorously held on to their customs, which allowed them to live pleasant and healthy lives, unperturbed by thefts or murders. They had no science, but they enjoyed beautiful music, poetry, and word games, and they considered themselves the world's wisest people, and indeed, the original stock of all mankind. Their natural liberty did not imply social equality: on the contrary, caste differences were very rigid and were maintained by strict rules of purity and a system of matrilineal succession (one that could be reminiscent of the *Nayar* caste in South India, also known to the Jesuits in the seventeenth century). Their wars were like ritualized encounters, with a great deal of shouting but little bloodshed, and a preference for ambushes. They combined the inconstancy of human passions with calculated dissimulation when taking revenge. Women were the heads of the household, and so much so that unmarried men preferred to live in common houses for single men (*uritaos*), enjoying sex with young women hired or bought from their fathers. They practiced no religious worship, but they entertained numerous beliefs about the spirits of the dead, which the missionaries dismissed as mere superstitions. The balance of this ethnographic discourse is striking; on the one hand, the CHamoru are described as enjoying a carefree and secure life, but on the other they are condemned for the darkness of their ignorance—in any case, something far more complex than the simple opposition between civilization and barbarism.

The justly famous speech pronounced by Hurao, a member of the native elite of Hagåtña in the island of Guam (or San Juan) who led the first CHamoru rebellion in 1671, speaks to this complexity. The rebellion was symptomatic of the catastrophic historical process through which the natives rejected the impositions of the Castilians, and found themselves caught in a destructive cycle of rebellion and repression which was, as a matter of fact, very much in keeping with what had taken place in many other areas of European colonization. The dynamic certainly was very far from the peaceful evangelization that Luis de San Vitores and his companions had initially sought: as Alexandre Coello reminds us, the combination of *entradas*, or military expeditions, and the impact of new diseases, had a devastating effect upon the indigenous population. However, the most notable element of Hurao's speech is not the analysis of the historical process, but rather its rhetorical elaboration as an example of the clash between civilization and barbarism, according to the conventions of humanist historiography as practiced by the classically trained Jesuit writers.

We know from the testimony of Francisco García and other sources that

Hurao was indeed a historical figure, but the arguments that Le Gobien puts in his mouth are very different from those recorded by the Spanish writer.[15] García says that those of Hagåtña "began to shout for their old liberty and their old impunity from criminal justice, which the foreigners were taking away from them (they called the foreigners tyrants for preventing their own tyranny)". Le Gobien, by contrast, with rhetorical equanimity, makes the speech sound like a well-thought-out argument against colonization.[16] This contrast is manifest in both tone and content. García presents a malicious and presumptuous Hurao who led the CHamoru people, but does not put any speech in his mouth, instead emphasizing that the CHamoru rejected the notion of justice brought by the Spanish, that is, the rule of law, "preferring instead the customary evil to the unknown good".[17]

In the *History of the Mariana Islands*, however, Hurao delivers a brilliant speech, invented according to the European writer's idea of what the barbarian might have said (following the example of classic historians such as Tacitus and Sallust). In it, Hurao defended the traditional CHamoru lifestyle, arguing that, "we do not need their help to live happily," and opposing the natural simplicity of the natives, embodied in their very nakedness, to the corruption of a civilizing process that in reality created artificial needs. It was not worth the while to lose their natural liberty in exchange for a few trinkets and the promise of a happy afterlife. The stories told by the missionaries were no less fictitious than the local fables. Hurao also questioned the purity of the

---

15. A letter by Fr. Francisco Solano, the mission's superior in Agadña (Hagåtña),, written in April of 1672 and addressed to Morales, Bustillo and Casanova, who had returned to Manila, offers what might be the first reference to Hurao as the leader of the 1671 rebellion—a rather vague reference, to be sure—and also to the death of San Vitores in 1672 (R. Lévesque, *History of Micronesia*, vol. V, p. 425).

16. The author could have been inspired by the invented speech that García put in the mouth of Aguarín, a *Chamorri* leader of a later rebellion in 1676 (F. García, *Istoria della conversione*, p. 499). The version of Aguarin's speech that appears in Le Gobien (*Histoire*, pp. 245–47) differs little from the speech written by García, who probably invented it. It emphasizes the theme of lost freedom, and especially insists on the Spanish intention to exterminate the Chamorros through various means, from baptism (with "poisoned water") to marrying their daughters to Spanish men, which, moreover, meant that the fathers lost the income from their service at an *uritao*. That is, more than rejecting civilization as superfluous, as Hurao did, Aguarín denounced the destructive character of the Spanish presence. The rhetorical elaboration of this same speech by Gabriel de Aranda in his life of Sebastián de Monroy (*Vida y martirio*, chapter 60, pp. 347–52), is another demonstration of the flexibility and liberty with which Jesuit writers managed their source materials. Besides suspecting the malefic nature of Christian rituals such as baptism, and lamenting the loss of control over their own children (a true *casus belli*), Aguarín declared that the material benefits from trade with the Spanish were scarce; he complained about the exploitation that Chamorros were subjected to, and the transformation of their traditional customs; he criticized the missionaries' intellectual arrogance, for they pretended to show them the way to heaven; and above all he pointed to the deadly soldiers that such teachers brought with them as a sign of their true intentions. Aranda also offered a counter-discourse in which, to hide their intentions, the rebels declared to the Spanish exactly the opposite of what they were thinking (pp. 352–4).

17. *Istoria della conversione*, p. 254.

Spaniards' intentions, especially the missionary rhetoric of selflessness, and ended up insinuating that the foreigners were responsible for bringing diseases, rats, and other pests.[18]

It should not come as a surprise that this sharp speech had a great impact in the European Enlightenment, and that it would be read by writers such as the antiquarian Charles de Brosses as something equivalent to Rousseau's ideas on the advantages of the natural life of the savages (in his *Discours sur l'origine et les fondements de l'inégalité parmi les hommes*). Gilbert Chinard detected its direct influence on Diderot. More recently, in an interesting article, Carlo Ginzburg has shifted the focus onto its origins, defending the influence exercised by Montaigne on the French Jesuits of Le Gobien's generation (of course, this influence would be much harder to argue if the speech's original author were in fact Luis de Morales).[19] Coello de la Rosa also highlights the iconic character of Hurao's speech in modern Guam.

Beyond recognizing its subsequent impact, I believe it is important to insist on the rhetorical character of Hurao's supposed speech. This was not a libertine speech, like Montaigne's essay on cannibals, or like the dialogues with the Canadian Huron, Adario, published in 1703 by the Baron de Lahontan; nor was it an argument against the colonizing mission. The speech must be interpreted in its broad narrative context, as a dramatic counterpoint to a fundamentally apologetic, and even hagiographic text. Hundreds of years earlier, Tacitus, the Roman senator, had had the British leader Calgacus utter a stirring but futile defense of liberty that criticized the "false name of empire" given to what, in reality, was awful slavery, only to see him immediately succumb to the Roman arms commanded by Agricola, the writer's admired father-in-law (*Agricola*, 30–32).

In a similar fashion, at the turn of the eighteenth century the Jesuit historian put a call to rebellion in the mouth of a native leader who was under the misapprehension that, since they were many and the Spaniards few, the rebels were bound to win. The barbarian's eloquence was a tragic counterpoint to the

18. Le Gobien, *Histoire*, pp. 139–46. The reader can find the full speech, with a more detailed commentary, in Alexandre Coello de la Rosa's introduction.

19. Denis Diderot, *Supplément au voyage de Bougainville*, ed. Gilbert Chinard (Paris, 1935), pp. 118–19. Carlo Ginzburg, "Alien voices. The Dialogic Element in Early Modern Jesuit Historiograpohy", *History, Rhetoric and Proof* (Hanover and London, 1999), pp. 71–91. As Ginzburg notes, the point is not to uncover a sort of Jesuit skepticism, but to identify the attempt of various French Jesuits to mobilize Montaigne (by then, a forbidden author) against libertinism and in support of natural reason, through intermediate authors such as Saint Réal. Ginzburg's argument is that the Jesuits' "dialogic monologue" also echoes the true Chamorro voice, particularly in the comment about rats and disease. The importance of Hurao's speech was also noted by Sergio Landucci in *I filosofi e i selvaggi 1580–1780* (Bari, 1972).

triumph of civilization, produced by civilized writers who were capable of feeling nostalgia (of a Stoic character) for a golden age of simplicity and innocence that was irrevocably lost. Neither Seneca nor Tacitus wanted to renounce their accumulated wealth and return to the genuinely savage lifestyle of the barbarian tribes: they would rather enjoy a "natural" retreat in a tranquil garden. Similarly, neither Montaigne nor Rousseu seemed to want to go beyond identifying and criticizing the moral corruption of civilized man. There was no turning back.

The Jesuits, ideologically more conservative than either of these philosophers, never doubted that Christianity and civilization should go together in a project that was invariably paternalistic. Men like Morales were, of course, critical of the greed of many Spanish commanders and colonists, who took advantage of the justification provided by the evangelizing mission in order to become wealthy at the expense of the natives.[20] However, this was a condemnation of the abuses of colonization of a Lascasian type, not a complete rejection of the imperial project. In fact, the history of the Marianas is a late example of the process through which, time and again, wherever missionaries went with their ideals of evangelization, provided colonization was possible (however slowly and haphazardly), there eventually followed armed men in search of material gain.

---

20. As procurator of the mission in Madrid, Morales not only stood out for demanding that the evangelical mission should continue after the native rebellion of 1684 (with arguments which apparently offended the Council of the Indies in 1687), but also for proposing that colonists be sent to the Marianas from Mexico (R. Lévesque, *History of Micronesia*, vol. IX, p. 362).

INTRODUCTION

1. An unfinished "history"?

In the section entitled "Filipinas" of the *Arxiu Històric de la Companyia de Jesús a Catalunya* (henceforth, AHCJC), there are two unfinished, twentieth century copies of an unpublished *History of the Mariana Islands (Historia de las islas Marianas)*, attributed to Jesuit priest Luis de Morales.[21] The first is constituted by ten notebooks of four-to-ten pages each, for a total of 150-numbered, double-sided pages. The second is a copy of the first in quarter-page letter size. Although it is a draft of the former, it has a different calligraphy. Both copies end abruptly towards the end of Book Seven, and neither of them is signed.[22] We do not know if an original text still exists, and if it is concealed somewhere in a repository in Mexico or the Philippines. However, we do know that two different copyists wrote these copies in Manila using modern Spanish, and that they were eventually sent to the Archive of the Jesuit Province of the Kingdom of Aragon (*Aragoniae*).[23]

---

21. Luis Martín de Morales (or Luis de Morales) was born in Tordesillas (Valladolid, Castile) on September 29, 1641. He joined the Society of Jesus on August 28, 1658. After taking his scholastic vows, he went to the East Indies as a missionary. He took his final vows and completed his formation in 1671. ARSI, "Primus Catalogus Anni Personarum Anni 1671." Philippinae Cat. Trien. 1649-1696, Vol. 2-II, f. 353v. See also Carlos Sommervogel, SJ, *Bibliothèque de la Compagnie de Jesus*, 12 tomes. Brussels – Paris: Oscar Schepens and Alphonse Picard, 1890-1900, Vol. V, 1894, p. 1283.

22. Luis de Morales, SJ, *Historia de las islas Marianas* (AHCJC, FIL HIS – 061, E.I, c-05/2/0, 149 ff. The second copy is in AHCJC, FIL HIS – 061, E.I, c-05/3/0). There is a stenciled 1970 copy entitled *Historia de las Islas Marianas, convertidas recientemente al cristianismo y de la muerte gloriosa de los primeros misioneros que en ellas predicaron la fe*, in the Hispanic Document Collection of the Micronesian Area Research Center (henceforth, MARC). In 2023 the entire Philippine collection of the Arxiu Històric de la Companyia de Jesús a Catalunya (AHCJC) was moved to *Archivo de España de la Compañía de Jesús en Alcalá de Henares* (AESI-A).

23. In 1863, the Jesuit Province of Spain was divided in two: the Province of Castile (*Castellanam*) and the Province of Aragon (*Aragoniae*). While the missions of Cuba, Puerto Rico, Fernando Poo, Colombia, Ecuador, Guatemala, and Portugal were under the jurisdiction of the first, the missions of Argentina, Chile and the Philippines were under the jurisdiction of the second. Nearly one-third of this province's 600 Jesuits worked in these missions at that time, under provincial Fermín Costa Colomer (1863–1867) (Manuel Revuelta González, S.J., *La Compañía de Jesús en la España Contemporánea, Vol. I. Supresión y reinstalación*

Even though it is a truncated *History*, it was not unknown to its author's confreres. On May 18, 1689, in San Ignacio de Agaña [Hagåtña, Agadña], Father Lorenzo Bustillo wrote a letter to General Procurator Antonio Matías Jaramillo in which he acknowledged that since May of 1685, they had been, "remitting quite detailed *Relaciones [reports, relations]* to the General Procurator of the Indies in Madrid,[24] so that he can continue his Marianas' *History*, explaining the particulars of persons (which is why I have written them on the margins of said *Relación*) and lands; because Father Ambrosio says that it will be a useful guide for the Historian who is over there.[25] And Father Luis de Morales says the same from Madrid, and that we should not omit anything from the *relaciones*, not even the smallest elements, because he is composing a heavenly harmony over there."[26]

From the beginning of his rule, French King Louis XIV (1643-1715) promoted himself as the Sun King and professed to rule by divine right. Not only did he dominate Europe but also favored the Society of Jesus, facing the opposition of Jansenits, *parlamentaires* and Enlightenment *philosophes*. He also pushed France's borders to their fartherst point, and with papal encouragement, the Society's missionary efforts expanded across Asia[27]. Jean Baptiste Colbert (1619-53)'s decision in the spring of 1669 to dispatch a large, well armed royal fleet to the Indian Ocean was part of his strategy to establish French economic power in Asia to the expense of the Dutch. As a result, between 1669 and 1672, the French King put political pressure on Dom Joao VI's son, the Portuguese regent Dom Pedro, to form an international alliance against the Dutch in Asia, but Lisboa declined the offer.[28] Spanish empire was in decay and European

(1868–1883), Madrid–Santander–Bilbao: Sal Terrae & Mensajero & Universidad Pontificia Comillas, 1984, pp. 13–23; 47). I do not know whether Father Pablo Pastells (1846–1932) had a role in the redaction or transfer of these documents—there is unfortunately no biography of this illustrious Jesuit.

24. As Martínez-Serna pointed out, "the post of court procurator to the Spanish court in Madrid was created in 1570 to "arrange in the court the affairs that concerned principally the Spanish provinces of the Order"–to serve as the primary mediator between the court of Spain and the Society of Jesus" (J. Gabriel Martínez-Serna, "Procurators and the Making of the Jesuits' Atlantic Network", in Bernard Bailyn & Patricia L. Denault (eds.), *Soundings in Atlantic History. Latent Structures and Intellectual Currents, 1500–1830*, London, England: Harvard University Press, 2009, p. 185).

25. He was probably referring to Father Francisco García, S.J., author of *Vida y martyrio del venerable padre Diego de Sanvitores...* (Madrid, 1683). I thank Joan-Pau Rubiés for the information.

26. Rodrigue Lévesque, *History of Micronesia. A Collection of Source Documents*, vol. 9, Québec: Lévesque Publications, 1997, pp. 291–92.

27. Eric Nelson, "The Historiography of the Pre-Suppression Jesuit Mission in France". [Available online] <https://referenceworks.brillonline.com/entries/jesuit-historiography-online/the-historiography-of-the-pre-suppression-jesuit-mission-in-france-COM_193392> [Consulted March 2, 2024].

28. Glen J. Ames, "An Elusive Partner": Portugal and Colbert's projected Asian alliance, 1669-1672," *Revista Portuguesa de História*, XXVIII, 1993, p. 38.

powers were fully aware of it.[29] The first treatises to divide the Spanish monarchy, according to Valladares, included the Philippines in the French treaties.[30] In addition, Frederik Vermote has recently pointed out that "the French Jesuits no longer accepted the exclusive connection between the Jesuit network and the French government".[31] This might explain why in 1685, the Sun King, sent six Jesuits to Siam [now Thailand], under the direction of the missionary and (unofficial) ambassador Father Guy Tachard, SJ (1651-1712).[32] Other French Jesuits were sent to southern India, cooperating with the Portuguese in the region of Madurai. Eventually these priests founded the mission of Karnataka circa 1700, carrying out successful conversions, even of Hindu Brahmin. Finally, the Jesuits took over the mission of Tonkin (in Vietnam), which had been founded in 1627 by Father Alexandre de Rhodes, SJ (1583-1660), but had been abandoned because of the persecution unleashed against Christians.[33] It was in this context that Father Charles Le Gobien, SJ (1653-1708) translated and published the *History* written by Father Morales.

Not much is known about this French Jesuit's life. Born in Saint-Malo (Brittany) in 1653, he entered the Society of Jesus on November 25, 1671. He taught rhetoric and humanities in various Jesuit schools in France—including at the College of Tours [34] and the Royal College of Alençon[35]—before transferring to Paris. Hoping to elicit interest in the mission, he published in 1697 the *Lettres sur les progres de la religion a la Chine* (Paris: Antoine Lambin). This text was followed a year later by the *Histoire de l'edit de l'Empereur de la Chine en faveur de la religion Chrétienne avec un éclaircissement sur les honneurs que les Chinois rendent a, Confucius et aux morts* (Paris: J. Anisson) [*History of the Chinese Emperor's Edict in favor of the Christian Religion, with a clarification on*

---

29. For a critical review of the concept of Iberian empires, see Christian Hauser and Horst Pietschmann, "Empire. The concept and its problems in the historiography on the Iberian empires in the Early Modern Age". *Culture & History Digital Journal*, vol. 3, nº 1, 2014, pp. 7-16.

30. The second treaty, which dates to 1698, was far from the original one. After the failure of Portugal and Colbert's projected Asian alliance, French interests drastically changed and the Philippines fell into the hands of the Duke of Bavaria's son together with the bulk of the Spanish empire (Rafael Valladares, *Castilla y Portugal en Asia* (1580-1680). *Declive imperial y adaptación*. Leuven/Louvain, Belgium: Leuven University Press, 2001, p. 94).

31. Frederik Vermote, "Travellers Lost and Redirected: Jesuit Networks and the Limits of European Exploration in Asia," *Itinerario*, 41, nº 3 (2017), p. 501.

32. *Diccionario Histórico de la Compañía de Jesús. Biográfico-Temático* (henceforth, DHCJ), dir. Charles E. O'Neill, SJ and Joaquín M. Domínguez, SJ. 4 vols. Madrid-Rome: IHSJ & Universidad Pontificia de Comillas, 2001, vol. IV, p. 3686.

33. "Francia. Antigua Compañía (siglos XVI-XVIII)" (*DHCJ*, vol. II, p. 1502).

34. Pierre Delattre, S.J., *Les établissements des Jésuites en France depuis quatre siècles*, tome IV, Enghien–Wetteren (Belgium): Institut Supérieur de Théologie & Imprimerie de Méester Frères, 1956, pp. 1436–1448.

35. P. Delattre, *Les établissements des Jésuites en France depuis quatre siècles*, tome I, 1949, pp. 118–136.

the honors paid by the Chinese to Confucious and the dead].

In this publication, Le Gobien defended the Jesuits' accommodating and tolerant attitude towards the so-called "Chinese rights," which had been criticized by the Franciscan and Dominican orders since their arrival in China in 1631, and had been condemned by the Sacred Congregation de Propaganda Fide in 1645.[36] Moreover, as part of his campaign to generate support for the Society's Christianization efforts across East Asia, Le Gobien published the *Histoire des Isles Marianes nouvellement convertes à la religion chrétienne* in Paris in 1700. Finally, sometime after the publication of this book, Father Le Gobien edited the *Lettres édifiantes et curieuses*, an anthology of letters sent by the French Jesuit missionaries to their provincial in Paris, with the purpose of of replying to the Jansenist attacks and "edifying" the European Jesuits.[37] This publication, which came out in French in order to reach the largest possible audience, was an instrument of propaganda that would continue throughout the eighteenth century.

Unlike the confreres whom he wrote about, Le Gobien never went to India, China, or the Mariana Islands. But like Federico Palomo has pointed out, writing (and publishing) was also missionary work, for it perpetuated and expanded evangelical labor through the written word.[38] Judging from his publications, this is precisely what Le Gobien did in the Professed House of Paris in 1698, when he was charged with writing the history of the Orient's Jesuit Missions.[39] There he met the Father Louis Daniel Le Comte (1655–1728), a

---

36. For a thorough study on the Congregation of Propaganda Fide, see Giovanni Pizzorusso, *Propaganda Fide. I. La Congregatione Pontificia e la Giurisdizione Sulle Missioni*. Rome: Edizione di Storia e Letteratura, 2022.

37. Guillermo Zermeño, "Entre el saber y la edificación: una relación inestable," *Introducción a las Cartas edificantes y curiosas de algunos misioneros jesuitas del siglo XVIII. Travesías, itinerarios, testimonios*. Mexico: Universidad Iberoamericana, 2006, pp. 17–57; Carlo Ginzburg, "Alien Voices: The Dialogic Element in Early Modern Jesuit Historiography," in *History, Rhetoric, and Proof. The Menace Stern Jerusalem Lectures*, Hannover & London: Brandeis University Press & Historical Society of Israel, 1999, pp. 71–72; Eva María St. Clair Segurado, "El obispo Palafox y la cuestión de los ritos chinos en el proceso de extinción de la Compañía de Jesús," *Studia Historica. Historia Moderna*, 22, 2000, pp. 145–70. Imbruglia, "A Peculiar Idea of Empire: Missions and Missionaries of the Society of Jesus in Early Modern History," in Marc André Bernier, Clorinda Donato, and Hans-Jürgen Lüsebrink (eds.), *Jesuit Accounts of the Colonial Americas. Intercultural Transfers, Intellectual Disputes, and Textualities*. Toronto: University of Toronto Press & UCLA Center for Seventeenth-and Eighteenth-Century Studies and the William Andrews Clark Memorial Library, 2017, 21-49. For more on Father C. Le Gobien, see Carlos Sommervogel, SJ, *Bibliothèque de la Compagnie de Jesus*, T. III, 1892, pp. 1512-1515; and the brief biography in *DHCJ*, vol. III, pp. 2303-2304.

38. Federico Palomo, "Misioneros, libros y cultura escrita en Portugal y España durante el siglo XVII," in Charlotte de Castelnau-L'Estoile, Marie-Lucie Copete, Aliocha Maldavsky, and Ines G. Županov (eds.), *Missions d'Évangélisation et Circulation des Savoirs, XVIe – XVIIIe siècle*, Madrid: Casa de Velázquez, 2011, pp. 146-47. For an interpretation of the role of Jesuit missions in the construction of a peculiar Spanish colonial empire, see Imbruglia, "A Peculiar Idea of Empire…," pp. 21-49.

39. Catalogus Primus Provinciae Franciae Anni 1700. Domus Profesesa Parisiensis (Archivum Romanum

missionary who had returned to France to take care of the Chinese mission's affairs, and who had been retained by King Louis XIV as the Duchess of Burgundy's confessor. In 1706, Father Le Gobien was named procurator of the Franco-Chinese Province,[40] but this task was cut short by his death in Paris on March 5, 1708, at the age of 55.[41]

A comparison between Father Morales's truncated *History* and Father Le Gobien's *Histoire* reveals that they coincide almost entirely, except for their footnotes. Indeed, Le Gobien says in the *Histoire des Isles Marianes* that he based his text on a text written by Morales, "*Memoire du Pere Louïs de Morales, Jesuie, touchant la situation, la distance et la grandeur des Isles Marianes,*" whose existence was relatively known at the time.[42] In 1752, Jacques Nicholas Bellin (1703–1772), a hydrographic engineer of the French Navy, made a map of the islands based on the maps drawn by Father Alonso López[43] and the cited Memoir of Father Luis de Morales (see Map 1, annexed in the *History's* foreword).

The hypothesis that guides this introduction is that Charles Le Gobien's *Histoire* cannot be attributed to him alone, for among the various sources that he used, he relied heavily on Father Morales's *Memoria*. Le Gobien's publication cannot be considered a palimpsest—in the sense of a rewritten or manipulated text—of Morales's work, but a collective product that formed part of the Society's propaganda efforts regarding its missionaries in Asia.[44] Therefore, I disagree with what Joan Pau Rubiés suggests in the prologue of this edition, which is that the copyists in Manila simply translated and modernized Le Gobien's 1700 text,

---

Societatis Iesu (henceforh, ARSI), France, Vol. 17, Cat. Trien. ff. (2) 35r; 195r).

40. *DHCJ,* vol. III, pp. 2303-2304.

41. A sample of the edifying letter that was written about Father Le Gobien can be read in Élesban de Guilhermy, S.J., *Ménologie de la Compagnie de Jésus. Assitance de France.* Vol. I, Paris: Typographie M. Schneider, 1892, pp. 324–25.

42. C. Le Gobien, *Histoire des Isles Marianes nouvellement converties à la religion chrétienne* (París: Nicolas Pepie, 1700), avertissement. See also C. Sommervogel, SJ, *Bibliothèque de la Compagnie de Jesus,* T. V, 1894, p. 1283

43. Alonso (or Alexius) López was born in Plasencia (Extremadura) on July 16, 1646. He joined the Society of Jesus on September 30, 1662. He was missionary to the Indians (*operario de indios*) in the Mariana Islands from 1672 until his death in 1675; he was also a cartographer (ARSI, "Primus Catalogus Anni Personarum Anni 1672." Philippinae Cat. Trien. 1649–1696, Vol 2-II, f. 363v). López wrote the Annual Letter that spanned the period from June 1674 to June 1675 (ARSI, "Primus Catalogus Anni Personarum Anni 1672." Philippinae Cat. Trien. 1649–1696, vol. 2-II, f. 363v). See also Francis X. Hezel, *From Conquest to Colonization. Spain in the Mariana Islands, 1690 to 1740,* Saipan, Mariana Islands, Division of Historic Preservation, 1989, p. 90.

44. One of the first colonial "palimpsests" is the *Diario* de Colón, rewritten by the Dominican friar Bartolomé de Las Casas (Antonello Gerbi, "Diario: autenticidad." *Cuadernos hispanoamericanos,* 512, 1993, pp. 11–14). See also José Rabasa, *Inventing A-m-e-r-i-c-a. Spanish Historiography and the Formation of Eurocentrism,* Norman: Oklahoma University Press, 1993.

and not one written by Morales.[45] In any case, the truth is that the procurators sent to Spain arrived with their arms full of memorials, *relaciones*, and letters on the Philippine and Mariana missions[46]. At the end of the seventeenth century, the 1670 and 1671 martyrology letters were used by Francisco de Florencia (Seville, 1673) and Francisco Garcia[47] to write the biographies of Fathers Luis de Medina (Madrid, 1673)[48] and Diego Luis de San Vitores (Madrid, 1683).[49] Le Gobien's reliance on this material was no exception. Proof of this is the content of his *Histoire's* second chapter, which reproduces entire paragraphs found in texts written by Morales and Garcia, which means that they, and not Le Gobien, are the authors of the chapter's original matrix and analysis.[50] This should not surprise us, for:

> ...this tendency towards uniformity, intended to facili-
> tate governance and decision-making at the highest level of the
> Society, meant that various ethnographic and historical texts that
> emanated from the missions were written by various persons (by
> several hands) and often without identifying an author.[51]

Many of the historians and chroniclers, especially in Spain, were men

---

45. See the Prologue by Joan Pau Rubiés in this book.

46. As Martinez-Serna remarks, "among the various types of reports sent by a provincial procurator to Rome were detailed catalogs of the province's personnel and material possessions; expense reports for his province, including individual accounts for each mission, residence, and college; and various types of religious commentary, some of which had detailed ethnographic information that was passed on to Jesuits in other provinces and assistancies. Provincial procurators also edited the reports before sending them to Rome, and such reports sometimes made their way into print, usually in *relations* of the various missionary activities of the order (J. G. Martínez-Serna, "Procurators and the Making of the Jesuits' Atlantic Network", p. 195).

47. Father Francisco García, S.J., (1641–1685) was a preacher in the popular missions, but he is best known as a hagiographer of Father General Francisco Borja (Alcala, 1671); St. Francis Xavier (Madrid, 1676); Fathers Luis de Medina (Madrid, 1673) and Diego Luis de San Vitores (Madrid, 1683) (*DHCJ*, vol. II, pp. 1572–1573).

48. Francisco García, S.J., *Relación de la vida del devotísimo hijo de María Santísima, y dichoso mártir padre Luis de Medina, de la Compañía de Jesús, que murió por Christo en las islas Marianas (llamadas antes de los Ladrones) con otro Compañero seglar, llamado Hipólito de la Cruz* (Madrid, s. n., 1673); Francisco de Florencia, S.J., *Ejemplar vida y dichosa muerte por Cristo del Fervoroso Padre Luis de Medina de la Compañía de Jesús*, Sevilla, Imprenta de Juan Francisco de Blas, 1673. For the martyrology letters used in these hagiographies, see Alexandre Coello de la Rosa, "Tres cartas martiriales de los misioneros jesuitas en las islas Marianas (1668–1686)." *Revista Española del Pacífico*, nos. 21–22/Años XVIII–XIX, 2008–2009, pp. 27–65. See also the letter written by Father Diego Luis de San Vitores in Hagåtña on May 14, 1671, on the exemplary life of Father Luis de Medina (Real Academia de la Historia (henceforth, RAH), Fondos Cortes, 567, 9-2676/13, ff. 1r-7v).

49. Francisco García, S.J., *Vida y martyrio del venerable padre Diego de Sanvitores...* (Madrid: Iván García Infanzón, 1683).

50. C. Le Gobien, *Histoire*, book II, pp. 53-54.

51. Charlotte de Castelnau-L'Estoile, Marie-Lucie Copete, Aliocha Maldavsky, Ines G. Županov [eds.], "Introduction", in *Missions d'Évangélisation et Circulation des Savoirs*, p. 9.

HISTOIRE

DES ISLES

MARIANES;

NOUVELLEMENT
converties à la Religion Chreſtienne;
& de la mort glorieuſe des premiers
Miſſionnaires qui y ont prêché la Foy.

Par le Pere CHARLES LE GOBIEN,
de la Compagnie de JESUS.

A PARIS,
Chez NICOLAS PEPIE, ruë S. Jacques, au
grand Saint Baſile, au deſſus de la
Fontaine de S. Severin.

M. DCC.

AVEC PRIVILEGE DU ROY.

Figure 1: La Histoire des isles Marianes; nouvellement converties à la Religion chrétienne; & e de la mort glorieuse des premiers missionnaires qui y ont prêché le Foy, del padre Charles Le Gobien, S.J.}

of letters who had never acted as missionaries.[52] In the case of the Mariana Islands those who were stationed there periodically sent letters and *relaciones* to their friends and relatives in Europe, usually written in French, Spanish, or Latin, in which they described in detail what took place in the islands. This correspondence formed the basis of the hagiographies written by Father Francisco Garcia, who never put his feet there, as well as Father Morales's *History of the Marianas*. Curiously enough, nobody knew of the existence of this truncated manuscript until now, even though historian Rodrigue Lévesque, in his encyclopedic *History of Micronesia: A Collection of Source Documents* (1997) writes in a footnote that there are two copies of this *History* in the AHCJC.[53] Said *History* synthesized the first-hand sources written by the missionaries who, like Father Morales, had gone with Father Diego Luis de San Vitores to the archipelago, such as the Jesuit missionaries' *Cartas Anuas* [Annual Letters]. However, Lévesque was not aware that Le Gobien's *Histoire* and Father Morales's *Memorias* were one and the same *History*.[54]

This edited translation and publication of the History of the Mariana Islands has two objectives. First, to make this interesting text available and intelligible for contemporary English-language readers, so that they may learn more about one of the Jesuits' most conflictive missions in Spanish Asia.[55] And second, to rescue and make known the figure of Father Luis de Morales as a significant missionary, procurator, writer, and Jesuit provincial. After all, Morales' descriptions and reflections regarding the Mariana mission —its people's customs and society and their insurrections and resistance; the relations between civil and spiritual authorities, etc.— and the missionaries' role in the expansion of Christianity and the colonial state, went beyond his personal perspective, and were indicative of his religious order's views. After all, the author uses of all sorts of primary sources, especially letters, reports, relaciones, and memorials written by the Jesuit missionaries stationed in the archipelago to write history.[56] In the

---

52. Asunción Lavrín, "La visión de la historia entre las órdenes mendicantes: el cronista religioso en la cultura urbana novohispana", in Francisco Javier Cervantes Bello (coord.), *Libros y lectores en las sociedades hispanas: España y Nueva España (siglos XVI-XVIII)*, Puebla, Benemérita Univesidad Autónoma de Puebla, 2016, p. 29.

53. Lévesque probably did not travel to Barcelona, but instead consulted the copies held in MARC.

54. R. Lévesque, *History of Micronesia*, vol. 9, 1997, p. 432.

55. To carry out this annotated edition using the method of ecdotics, it was necessary to return it to its original context (Ignacio Arellano, "Problemas en la edición y anotación de las crónicas de Indias," in I. Arellano and J. A. Rodríguez Garrido (eds.), *Edición y anotación de textos coloniales hispanoamericanos*, Madrid & Frankfurt: Iberoamericana & Vervuert, 1999, pp. 45-74).

56. I am aware of the ambiguity of the term "history" and its relation to other genres, such as hagiography, philosophy, and rhetoric. On this respect, see Chantal Grell, "Introduction," in Chantal Grell (dir.), *Les Historiographes en Europe de la fin du Moyen Âge à la Revolution*, Paris: Presses de l'Université Paris-Sorbonne,

end, the text serves as a window into the Society of Jesus' universal Christian project; that is, their belief in the establishment of a human community under the principle of a universal Church, and in that sense, their self-representation as a global force chosen by God to fight evil.[57]

In terms of its content, the book addresses the first stage of the Mariana Islands' conquest and colonization, from the arrival of Father San Vitores in 1668, to the conquest of the northern islands (Rota, Saipan, and Tinian), carried out by Sergeant Major Joseph de Quiroga y Losada in July of 1695. Most of the text uses a direct style, giving it the veracity of an eyewitness account.

Ethnographies written by Jesuit missionaries, such as this book's second chapter, captivated the European public, eager for information on the distant regions of Asia in a context of Western expansion across the Pacific.[58] Indeed, the *History* provides data on the social and political organization of the CHamoru or Marianas' natives: their economic system, customs, and religious practices. Its curious anecdotes and representations of everyday life and practices of CHamoru society reveal an internal logic between the beliefs and conduct of the Marianas' natives who first encountered the Europeans, which is perceived by the reader but seemingly missed by the writer, perhaps because of the latter's use of the language of Christianity to classify and describe "pagan" reality.

Beyond the classification or separation of Christians from pagans as the major category of human identity, the *History* combines a civil language that tries to systematize the information on the villages and regions that the Jesuits were trying to Christianize, as well as critiques on the "un-Christian" practices of Spanish officers and soldiers that negatively impacted the local population.[59] Let us not forget that in the eighteenth century the anthropological discourse as such did not exist, and ethnological analysis was instead incorporated in philosophical and historical discourses.[60]

2006, pp. 9-17; Norma Durán, Retórica de la santidad, Mexico: Iberoamericana, 2008, pp. 89-132.

57. On the idea of mission and universal history in the Society of Jesus, and particular in José de Acosta's work, see Imbruglia, "A Peculiar Idea of Empire...," 27-31.

58. Joan Pau Rubiés suggests that there was an underlying ethnological approach to Jesuit writing on Chamorro customs and social institutions (Joan Pau Rubiés, "The Early Spanish Contribution to Asian Ethnology in the sixteenth and seventeenth Centuries." *Renaissance Studies*, 17:3, 2003, pp. 418–48), but I disagree. Missionaries did not analyze these subjects through principles of diversity or cultural specificity, or to learn about other cultures: on the contrary, they did so trying to find the linkages that could facilitate bringing Christianity into Chamorro society.

59. For a reflection on the precursors of contemporary ethnology, see Rubiés, "New Worlds and Renaissance Ethnology." *History and Anthropology*, vol. 6, no. 2–3, Harwood Academic Publishers GmbH, 1993, pp. 170–71.

60. Michèle Duchet, Antropología e historia en el Siglo de las Luces. *Buffon, Voltaire, Rousseau, Helvecio, Diderot*, Madrid: Siglo XXI, 1971] 1975, pp. 11-21.

As part of their apostolic work, Ignacio de Loyola directed Jesuit missionaries to write letters that included reflections on four subjects found in their missions: the kings and the nobles, the common people, the Society, and the missionaries themselves. In her analysis on the dispute between two Jesuit missionaries in the Madurai mission (India), historian Inés G. Županov developed a typology of Jesuit missionaries' epistolary writing that corresponds with Loyola's four epistolary topics. According to her, missionaries':

> ...reflections on the common people became an *ethnographic description*. The encounters with the 'native' kings, priests, or nobles were recorded in *dramatic/theatrical* vignettes. The disputes and entanglements concerning the members of the Company were often couched in *dialogical/polemical* terms and repartees. Finally, the individual ambitions appeared most clearly in the rhetoric of *sainthood* and *utopianism*.[61]

It is not surprising that the four prescribed subjects of reflection are present in Father Morales's *History*. However, I believe that some of Županov's types or modes are more useful if they are framed differently in the book. First, the *theatrical* mode is best applied to the elements destined for edification; that is, hagiographies, hagiologies, or *vitae* of the first martyrs, particularly Diego Luis de San Vitores, Luis de Medina, and Sebastian de Monroy, for these followed formal literary exigencies. They all had promising beginnings (virtuous births, infancies, youths, and novitiates); a narrative/spiritual development (acts that revealed their doctrinal purity, abnegation, and heroic virtues following Christ's example, and the experience of miraculous recoveries, intercessions, or other such miracles until they reached the road of perfection), and an exalted ending (passion or death desired by the saint through an ascending scale of suffering and acts of penance). This type of history chose and emphasized the facts that could best exalt its religious brethren and secular readers. Edification meant "enseñar a través de los ejemplos de hombres que encarnaban las virtudes morales y espirituales que la iglesia se proponía inculcar en todos los cristianos".[62]

Hagiographers often used "dialogues" between the murdered Christians and their tormentors that reflected the latter's cruelty and hatred of the Gospel, demonstrating that their death met the martyrdom requisites set by Rome.[63]

---

61. Ines G. Županov, *Disputed Mission. Jesuit Experiments and Brahmanical Knowledge in Seventeenth-century India*, New York & Oxford: Oxford University Press, 1999, pp. 6–7.

62. A. Lavrín, "La visión de la historia entre las órdenes mendicantes", p. 30.

63. For an analysis on the dramatic and heroic dimension contained in Sebastiao Gonçalves' *Primeira*

The *utopian* mode that Županov ascribes to the individual missionaries' self-reflection applies in this case to the order's work: the Jesuit missionaries in the Marianas are represented as authentic "soldiers" that fight under the banner of Christ as providential messengers of God's divine grace.[64] The *ethnographic descriptions* that abound in the *History* refer to the information and analysis gathered on both the common people and their nobles and leaders, all of whom were the object of the missionaries' evangelization efforts. In this sense, missionaries not only wondered about the meaning of certain cultural practices; they also actively sought to change them, adapting them to the new religion and the new culture that it entailed, and thus adopting an active role in the social configuration of colonial society. The fourth type, the *dialogical/polemical* mode, characterizes the disputes or differences between the missionaries and the civil authorities more than those differences between the missionaries or vis-à-vis their superiors in Europe. The Marianas mission did not generate the theological, doctrinal, or ideological conflicts that some missions in India, China, or Japan did, but the *History* reveals that political and economic disputes vis-à-vis civil and ecclesiastical authorities did exist.

There is copious documentation on the Mariana Islands' colonial history, fundamentally written by the Jesuit missionaries stationed in the archipelago. This means that the ethno-historical reconstruction of political, social, and economic relations between Spaniards and CHamoru is based on a canonical documentation framed within European colonialism. It is a "history of the victors," in the words of Walter Benjamin, an apologetic narrative that responded to the hopes and interests of the dominant group.[65] However, Father Morales tried to produce a new historical document, one that avoided the colonizer vs. colonized dichotomy, where the "people without history" recognized the Spanish nation as a new Messiah. The implementation of the new Universal Christian order in the Marianas could not be built upon the mere demonization of the CHamoru as the barbaric enemies of the faith. As an eyewitness, Father Morales knew that Spaniards had also acted cruelly and with little regard for Christian virtue. His *History* was a first attempt to critically revise the process of conquest and evangelization of Micronesia.

Parte da História dos Religiosos da Companhia de Jesus e do que fizeram com a divina graça na conversão dos infieis a nossa sancta fee catholica nos reynos e provincias da India Oriental (1614), see Ines G. Županov, "The Prophetic and the Miraculous in Portuguese Asia: a Hagiographical View of Colonial Culture," in Sanjay Subrahmanyam (ed.), Sinners and Saints. *The Successors of Vasco da Gama*, New Delhi: Oxford University Press, 1995, pp. 135-161.

64. Županov, *Disputed Mission*, pp. 10–11; 16.

65. Walter Benjamin, *Angelus Novus*, Barcelona: La Gaya Ciencia, [1962] 1970.

2. Father Luis de Morales, missionary in the Mariana Islands (1668–1671)

Father Luis de Morales was born in the Castilian town of Tordesillas on September 29, 1641—the feast of St. Michael the Archangel. He joined the Society of Jesus at the age of seventeen, on the feast of St. Augustine [August 28, 1658]. After studying theology in Salamanca and taking his scholastic vows, he asked to be sent as a missionary to the Philippine Islands. His request was approved, and on March 23, 1662,[66] he left the port of Acapulco in an allegorical ship of the militant Church, along with fourteen other Jesuit missionaries, including fathers Diego Luis de San Vitores,[67] Tomás Cardeñoso,[68]

---

66. L. de Morales, *Historia*, chap. 1, f. 23; C. Le Gobien, *Histoire*, book 1, p. 38. See also Morales's "Resumen de los sucesos del primer año de la misión en estas islas Marianas," San Ignacio de Hagåtña, April 26, 1666 (ARSI, Philippine Historiae, 1663–1734, Vol 13, ff. 5r-8v); Manuel de Solórzano, S.J., "Descripción de las islas Marianas, costumbres de sus naturales. Una relación del estado en que se hallaban las misiones que había en ellas con el número de convertidos a Nuestra Santa Fe; varios acatamientos y persecuciones padecidas por los padres dimanado de una falsa semilla aparecida por cierto sangley Choco; y algunos casos maravillosos que Dios obro a favor de su causa y para confusión de la idolatría" (1683) (Biblioteca del Palacio Real, II/2866, f. 123r). Bustillo continued on to Manila, returning to the Marianas on June 4, 1675, aboard the San Telmo \ galleon (P. Murillo Velarde, *Historia de la provincia de Filipinas*, f. 336; Horacio de la Costa, S.J., *Jesuits in the Philippines, 1581–1768*, Cambridge, Massachusetts: Cambridge University Press, [1961] 1989, p. 456).

67. Diego Luis de San Vitores was born in the Castilian city of Burgos (Spain) on November 12, 1627. He joined the Society of Jesus on July 25, 1640. He was a graduate and professed of the four vows (2/12/1660) (ARSI, "Primus Catalogus Anni Personarum Anni 1696." Philippinae Cat. Trien. 1649–1671, Vol 2-II, f. 351r).

68. Brother Tomás Cardeñoso was born in Paredes de Nava (Palencia, Leon, Spain) in 1640. He joined the Society in 1664 and was a missionary to the Indians in the Marianas. The date of his death remains unknown (ARSI, "Primus Catalogus Anni Personarum Anni 1672." Philippinae Cat. Trien. 1649–1696, Vol. 2-II, f. 381v).

{Figure 2: Juan Carreño de Miranda (1614–85), *Doña Mariana de Austria, viuda* (1669). Museo del Prado (Madrid)}

Luis de Medina,[69] and Pedro de Casanova;[70] and Brother Lorenzo Bustillo.[71] This group would later found the mission in the islands formerly known as the Velas Latinas Archipelago, which they reached on June 16, 1668. The Jesuits had rechristened the archipelago the Mariana Islands after the Spanish Queen Regent Mariana of Austria (1634–1696)[72] had approved Father San Vitores's project to evangelize there [1665]. The name was meant to honor both her, and the Virgin Mary, to whom the islands were consecrated.

After their arrival, they set up their headquarters, so to speak, in the village of Hagåtña, in the island of Guam [Guåhan], the largest of the archipelago. According to the Cartas Anuas, they estimated Guåhan's native population to

---

69. Luis de Medina was born in Malaga (Andalusia, Spain) on February 3, 1638. On March 25, 1656, he joined the Society of Jesus. He studied in the Jesuit school of Malaga and was admitted into the Society despite his stuttering and a lame leg, because, as Father Visitador Francisco Franco stated, they wanted "to have a saint" (Xavier Baró Queralt, "Recibámosle para santo": las hagiografías sobre Luis de Medina (1637–1670), protomártir de las islas Marianas.", in Los Jesuitas: Religión, política y educación (siglos XVI–XVIII), eds. J. Martínez Millán, H. Pizarro Llorente and E. Jiménez Pablo (Madrid: Pontificia Universidad Comillas, 2012), pp. 1503–1522). Between 1659 and 1662, Medina studied philosophy and a year of theology in the college of Santa Catalina (Cordoba, Spain). In 1660, he coincided with Diego Luis de San Vitores, who was on his way to the missions in the Pacific. After finishing his theology studies at the College of San Pablo in Granada (1663), he went to Montilla, where he was ordained (August 1664). He was still in Montilla when he wrote to provincial Father Cristóbal Pérez asking to be sent to the missions (April 1666). Medina finally left for Mexico on July 11, 1667, and there he found Father San Vitores, who had arrived from the Philippines to organize the Marianas mission. By the end of March, Medina, San Vitores, and four more Jesuits sailed for the archipelago, arriving on June 16, 1668 (Alexandre Coello de la Rosa and Xavier Baró i Queralt (eds.), Luis de Medina, S.J. Protomártir de las islas Marianas (1637–1670) (Madrid: Sílex, 2014), pp. 13–14). Father Medina preached the Gospel in Guåhan as well as in the islands of Tinian and Saipan (Juan de Santa Cruz, "Vida y martirio del padre Luis de Medina," AHCJC, Carpeta "EI.b-9/5/4. "Mártires de la Provincia de Filipinas de la Compañía de Jesús. Manila, 12 de mayo de 1903," in folder EI/b-9/5/1-7. These documents are classified under the title "Martirios y varones ilustres"). On the life of Father Medina before his arrival at the Marianas, see P. Murillo Velarde, SJ, Historia de la provincia de Filipinas, ff. 303r-309r; DHCJ, vol. III, pp. 2602-2603; and Wenceslao Soto Artuñedo, "Celebraciones por las canonizaciones de jesuitas en el colegio de Málaga en la Edad Moderna." Revista de Historia Moderna. Anales de la Universidad de Alicante, 21, 2003, pp. 83–101.

70. Father Pedro de Casanova was born in Almeria (Andalusia) on August 26, 1641. He joined the Society of Jesus in 1658, and in 1672 took vows as a scholar (ARSI, "Primus Catalogus Anni Personarum Anni 1672." Philippinae Cat. Trien. 1649–1696, Vol. 2-II, f. 363r). The Real Academia de la Historia houses a letter that Father Casanova wrote to Father Guillén aboard the galleon on June 17, 1668, in which he describes his trip to Acapulco (RAH, Fondo Cortes 567, 9-2676/5). On March 12, 1689, Casanova left his post as rector of the Dapitan Residence ("Catalogus Brevis Personarum Provinciae Philippinarum Anno 1690." Philipp. Catal. Trien. Philipp. 4, f. 88v).

71. Lorenzo Bustillo was born in Burgos on August 10, 1642. He joined the Society on July 2, 1664, and was a formed spiritual coadjutor at the time of his arrival at the Marianas (15/8/1681) (ARSI, "Primus Catalogus Anni Personarum Anni 1684." Philippinae Cat. Trien. 1649–1696, Vol 2-II, f. 424r).

72. Spain's Queen Regent Mariana of Austria was widow of Philip V (1605–1665) and mother of Charles II (1661–1700). On June 4, 1665, Queen Regent Mariana wrote a Real Cédula supporting Father San Vitores's project and blessing his mission with a sum of 21.000 pesos for its maintenance and defense. With this gesture, the Queen sought to regain the divine protection that the Spanish empire seemed to have lost. See Antonio Astrain, S.J., Historia de la Compañía de Jesús en la Asistencia de España, Vol. VI (Nickel, Oliva, Noyelle, González, 1652–1705), Madrid: Razón y Fe, 1920, p. 811; H. de la Costa, Jesuits in the Philippines, p. 456; and Paul Carano & Pedro C. Sánchez, A Complete History of Guam, Rutland, Vermont & Tokyo, Japan: Charles E. Tuttle Company, 1964, p. 63.

be around twenty thousand, with thirteen thousand being baptized that very year. Father Morales had the honor of baptizing the first CHamoru: a girl, christened Mariana, "like a flower consecrated to Most Holy Mary."[73] Thus began the so-called second period of Marianas history (1668–1698), an extremely convulsive, albeit brief, stage in the islands' colonial development.[74]

A majority of the population was distributed across nearly 180 towns dispersed throughout the coast and the interior.[75] One of the highest-ranking CHamoru leaders, Kepuha (Hispanicized as Quipuha)[76], welcomed them cordially, and after allowing himself to be baptized and christened Juan, he authorized the missionaries to preach the Gospel and baptize the people in his territory.[77] Having won this leader's support, San Vitores sent Fathers Morales and Cardeñoso, along with various assistants and soldiers, to the islands of Tinian and Saipan.[78] On August 14, 1668, in Saipan, wary CHamoru leaders who mistrusted the priest's motives wounded Father Morales on one leg, and ended the lives of Sergeant Lorenzo Castellanos and his Tagalog translator, Gabriel de la Cruz, which precipitated the missionary's return to Guåhan (October, 1668).[79] He stayed there until he recovered from his wounds, but returned to

---

73. Murillo Velarde, transcribed in R. Lévesque, *History of Micronesia*, vol. 12, 1998, p. 47.

74. Luis Ibáñez y García, *Historia de las Islas Marianas con su derrotero, y de las Carolinas y Palaos, desde el descubrimiento por Magallanes en el año 1521, hasta nuestros días*, Granada, Imprenta y Librería de Paulino V. Sabatel, 1886, p. 35; and Paul Carano & Pedro Sánchez, cited in James B. Tueller, "Los chamorros de Guam y la colonización española: una tercera etapa, 1698 a 1747," in Mª Dolores Elizalde, Josep Mª Fradera & Luis Alonso (eds.), *Imperios y naciones en el Pacífico. Volumen II. La formación de una colonia: Filipinas*, Madrid: CSIC & AEEP, 2001, p. 385.

75. "Relación de las empresas y sucesos espirituales y temporales de las islas Marianas, que antes se llamaban Ladrones, desde que el año de sesenta y ocho se introdujo en ellas el Santo Evangelio por los Religiosos de la Compañía" (RAH, Fondo Cortes 567, 9-2676-8, f. 3v; ARSI, Philippine Historiae, 1663–1734, Tomo 13, f. 5r; Manuel de Solórzano, S.J., "Descripción de las islas Marianas…" (1683) (Biblioteca del Palacio Real, II/2866, f. 124r). See also Murillo Velarde, transcribed in R. Lévesque, *History of Micronesia*, vol. 12, 1998, p. 48; A. Astrain, S.J., *Historia de la Compañía de Jesús*, vol. VI, p. 812.

76. As Vicente M. Diaz says, "in the vernacular, the name 'Kepuha'–the seventeenth-century chief who is said to have welcomed the Catholic mission to Guam–is translated as "dare to overturn" (as in "dare to overturn a canoe")." Monsignor, Interview, January 13–18, 1991, Makati, Philippines" (Vicente M. Diaz, "Pious Sites: Chamorro Culture Between Spanish Catholicism and American Liberal Individualism," in Amy Kaplan & Donald E. Pease (eds.), *Cultures of United States Imperialism*, Durham & London: Duke University Press, 1993, p. 336).

77. ARSI, Philippine Historiae, 1663–1734, Tomo 13, ff. 5r-5v. See also Francisco García, pp. 47–50, cited in Soto Artuñedo, "Luis de Medina…"; P. Carano & P. C. Sánchez, *A Complete History of Guam*, p. 65; Don A. Farrell, *History of the Northern Mariana Islands*, Guam: Public School System of the Northern Mariana Islands, 1991, p. 153; and Omaira Brunal-Perry, "Los misioneros españoles en las Marianas," in Javier Galván Guijo (cord.), *Islas del Pacífico: el legado español*, Madrid: Ministerio de Educación y Cultura, 1998, p. 45.

78. L. de Morales, *Historia*, chap. 3, ff. 43–44; 56; C. Le Gobien, *Histoire*, book 3, p. 76.

79. ARSI, Philippine Historiae, 1663–1734, Tomo 13, f. 5v; L. de Morales, *Historia*, ff. 43; 53; C. Le Gobien, *Histoire*, book 3, p. 76. See also Augusto V. de Viana, "Filipino Natives in Seventeenth Century Marianas: Their role in the establishment of the Spanish mission in the islands." *Micronesian Journal of the Humanities and the Social Sciences*, vol. 3, nos. 1–2, 2004, p. 21.

Tinian, where he reestablished the mission. He was accompanied by Father San Vitores, who used the island as a platform from where to explore the northernmost islands, starting with Saipan.[80]

The *chamorris, maga'låhi* or noblemen who had attacked Father Morales were *matua*, members of the Chamorro society's highest-ranking group, in which prestige and social standing depended on their capacity to accumulate supernatural power or mana. Mana was embodied in the specific objects and symbols that they carried: bones, necklaces and belts made of conch, and other materials.[81] It was essential for the *chamorris* to retain this distinctive prerogative over the common people, the *mangatchang* (also spelled *manachang*), and since baptism was interpreted as a privileged means to access supernatural elements, the attackers wanted to prevent the missionaries from baptizing everybody. They did not understand why the missionaries wanted to evangelize and share the sacraments with the entire population of the islands, without regard for their rank.[82]

The Jesuits were determined to delimit a small grid-pattern town, following the Castilian model, and they chose the most appropriate place to edify a church, which was founded on February 2, 1669, with the name *Dulce Nombre de María* (Sweet Name of Mary). They also marked the places to build a school (which was abandoned in 1674), and the mission house that was to serve as the ecclesiastical and civil authority center.[83] With these ceremonies, they consecrated the mission's space, certifying the right and the duty of the Spanish Crown to propagate the Gospel and defend these territories under its protection from its heretic enemies.[84] After all, the Jesuits had come to stay.

---

80. Murillo Velarde, transcribed in R. Lévesque, *History of Micronesia*, vol. 12, 1998, p. 48. See also D. A. Farrell, *History of the Northern Mariana Islands*, p. 156.

81. Roger Keesing, *Rethinking Mana* (1984), cited in Carlos Mondragón Pérez-Grovas, "Reflexiones historiográficas en torno a las percepciones oceánicas durante los primeros encuentros entre europeos y melanesios en el Pacífico," in Miguel Luque Talaván & Marta M. Manchado López (coord.), *Un océano de intercambios: Hispanoasia (1521–1898). Un homenaje al profesor Leoncio Cabrero Fernández*, vol. I, Madrid: Agencia Española de Cooperación Internacional, 2008, p. 91. See also Laura Thompson, "The Native Culture of the Marianas Islands," Bernice P. Bishop Museum. Bulletin 185, Honolulu, Hawaii & New York, Kraus Reprint Co. [1945] 1971, p. 13.

82. According to Louis de Freycinet (*Voyage autour du monde*, 1829–1837), Chamorro society was divided into nobles (*matua*), semi-nobles or a "middle class" (*atchaot*) and the common people or lower class (*mangatchang*). Each strata was highly endogamous, and their contact with the others was highly restricted through various taboos and rites (cited in L. Thompson, "The Native Culture of the Marianas Islands," pp. 13–14).

83. Francisco García, S.J., *Relación de la vida del devotíssimo hijo de María Santíssima, y dicho Mártir Padre Luis de Medina, de la Compañía de Jesús que murió por Christo en las Islas Marianas (llamadas antes de los Ladrones) con otro Compañero seglar, llamado Hipólito de la Cruz*, Madrid, 1673, chap. IV, f. 61. See also Dirk A. Ballendorf, "From *Latte* to Concrete: Chamorro Influences in Construction on Guam, 1521 to Post-World War II." *Journal of the Pacific Society*, vol. 20, no. 1–2, 1997, pp. 21–26.

84. According to the *relación* written by provincial Father Ledesma, towards the end of 1668 three Dutch

Responding to the desires of provincial Miguel Solana (1668–1670), Fathers San Vitores and Morales visited the other large and highly populated islands of the archipelago, that is, Saipan (christened San Jose) and Tinian (christened Buenavista Mariana) to the north and Rota (also known as Serpana or Zarpana, and christened Santa Ana) to the southwest. They were accompanied by the *Armada Naval Mariana or Escuadrón Mariano* (the Mariana Squadron) a small fleet of three-or-four canoes with ten soldiers, most of them Filipinos, commanded by Captain Juan de la Cruz [de Santa Cruz],[85] who had at their disposal three muskets and a small cannon. The natives' ferocious resistance was correlated to the "rugged and tough" archipelago's topography, establishing a link between that "land of war" and the "frontiers people" that inhabited it.[86]

On January 27, 1670, Father Medina went to Saipan, located some three miles from Tinian, and confirmed that San Vitores's preaching had not led to profound or real conversions. A *sangley*[87] named Choco, who lived in Saipan, had spread the rumor that the holy water and oils used by the Jesuits to baptize the children and the old was meant to cause their death. Two days after arriving at the island, Father Medina was attacked in the third village that he visited by a man called Poyo, nicknamed "the killer" (he was later baptized as Luis), and his accomplice, Daon (later baptized as Vidal). At the age of thirty-three, Medina died[88] along with Hipólito de la Cruz, who was from the Visayan

---

ships approached Humåtak Bay in Guåhan, as well as the island of Zarpana or Rota (Andrés de Ledesma, S.J., *Noticia de los progresos de nuestra Santa Fe en las islas Marianas, llamadas antes de los Ladrones, y del fruto que se han hecho en ellas el padre Diego Luis de Sanvitores, y sus compañeros de la Compañía de Jesús, desde 15 de mayo de 1669 hasta 28 de abril de 1670, sacadas de las cartas que ha escrito el padre Diego Luis de Sanvitores y sus compañeros*, 1670, f. 2r).

85. Juan de Santa Cruz (or "de la Cruz") was a man of noble birth from Indang, Cavite, who was also a *panday* or master blacksmith, and is thus sometimes referred to as Juan de Santa Cruz Panday. De Santa Cruz went to the Marianas with his wife, his sister, and his baby nephew, named Pedro Juan de la Cruz (Augusto V. de Viana, "Filipino Natives in Seventeenth Century Marianas: Their Role in the Establishment of the Spanish Mission in the Islands," *Journal of the Humanities and the Social Sciences (Micronesia)* 3:1–2 (2004), p. 20).

86. Christophe Giudicelli, "Pacificación y construcción discursiva de la frontera. El poder instituyente de la guerra en los confines del Imperio (siglos XVI–XVII)," in Bernard Lavallé (ed.), *Máscaras, tretas y rodeos del discurso colonial en los Andes*, Lima, Instituto Francés de Estudios Andinos (IFEA) & Pontificia Universidad Católica del Perú, 2005, pp. 165–166; and C. Giudicelli, *Fronteras movedizas. Clasificaciones coloniales y dinámicas socioculturales en las fronteras de las Américas*, Mexico: Centro de Estudios Mexicanos y Centroamericanos, 2010.

87. The term "sangley" referred to the Chinese who settled in the Philippines, most of whom were originally traders, and their progeny. There are multiple hypotheses regarding its linguistic origin. See Henning Kloter, *The Language of the Sangleys. A Chinese Vernacular in Missionary Sources of the Seventeenth Century*. Leiden, Netherlands: Brill, 2011, p. 9.

88. ARSI, "Noticias de las islas Marianas enviadas el año de 1670," April 22, 1670, Philippinae Historiae, 1663–1734, Vol. 13, ff. 50v-52r; ARSI, "Historica narratio illorum (1668–1673), Filipinas Vol 13, fol. 95-110, transcribed in R. Lévesque, *History of Micronesia*, vol. 6, 1995, pp. 40–41; A. de Ledesma, "Noticias de los progresos de nuestra Santa Fe…," f. 10r. According to the list of martyrs, he died "in odium sacram

Islands (Cebu)—a talented harp player who tried to appease the "savages" with his music.[89] Just as Italian priest Antonio Criminal became the first Jesuit martyrized in the East Indies,[90] Father Medina became the first martyr of the Marianas. To certify his "heroic" death, the Jesuits not only recovered his body, but also took sworn statements on the glorious martyrdom of their confrere, and wrote various letters and *relaciones* to provincials and general procurators.[91]

3. Father Luis de Morales, missionary and procurator of the Philippines (1671–1689)

Father Morales was entrusted with the task of informing the authorities in Manila of these deaths. He left in July 1671 aboard the galleon *Nuestra Señora del Socorro* that had just arrived from New Spain. He was also to finish his studies in the Philippines. His superiors had asked him to learn Tagalog, and when he completed his formation, he was sent to the missions of Yndang and Maragondong, depending on the mission of Silang, near Manila.[92] On February 2, 1676, he professed the four vows. That same year, he was appointed rector of the Antipolo residence, which was some five leagues east of the capital city. He remained in this post until June 14, 1678.[93]

---

baptismati et fidei christianae praedicationis a barbari" (AHCJC, Carpeta "EI.b-9/5/2. Martirios, naufragios, &." in EI/b-9/5/1-7. "Martirios y varones ilustres." Hojas sueltas). See also F. de Florencia, *Ejemplar vida y dichosa muerte por Cristo del Fervoroso Padre Luis de Medina de la Compañía de Jesús*, ff. 38r-42r; and Francisco García, S.J., *Relación de la vida del devotíssimo hijo de María Santíssima, y dicho Mártir Padre Luis de Medina*, Cap. VI, ff. 112–113.

89. Music was an important element in Jesuit missionary work. On May 21, 1671, San Vitores wrote a letter to the procurator of the Marianas mission, Father Joseph Vidal Figueroa, asking him for musical instruments: "*todo género de instrumentos músicos, arpas, guitarras, liras, cornetas y todos los demás instrumentos que pertenecen a la música con algunos libros de punto.*" He also requested an organ and an organ player to teach the children music (A. Astrain, *Historia de la Compañía de Jesús*, Vol. VI, p. 817).

90. Javier Burrieza, "Los jesuitas: de las postrimerías a la muerte ejemplar." *Hispania Sacra*, LXI, no. 124, 2009, p. 527. On the debate that the death of Father Criminal set off in the Society, see Bernard Vincent, "Le désir du martyre: les conditions du martyre "juste" selon le gouvernement de l'ordre." In *Mélanges de l'École Fraçaise de Rome, Italie et Méditerranée*. Ecole Française de Rome – Palais Farnèse, Rome – Mefrim, Vol. 111, no. 1, 1999, pp. 324-329.

91. Juan de Santa Cruz, "Vida y martirio del padre Luis de Medina (Malaga, 25 de agosto de 1637; 29 de enero de 1670)" (AHCJC - Carpeta de mártires)–EI.b-9/6, ff. 1-13v); "Declaraciones tomadas y juradas ante el general Diego de Arévalo acerca del martirio del padre Diego Luis de Medina de la Compañía de Jesús. Islas Marianas, 27 de junio de 1671" (AHCJC, FILHIS–061. Signatura E.I, c-05 [p]. Diversos sobre islas Marianas y Carolinas / P. San Vitores. There is a copy of these statements in *Historia Missionis*, vol. III, "Copia de documentos antiguos del Archivo de la Misión de la Compañía de Jesús, no. 1," Signatura Antigua: E.I., a-13 (1593–1890), ff. 615r-622v). See also F. de Florencia, *Ejemplar vida y dichosa muerte por Cristo del Fervoroso Padre Luis de Medina de la Compañía de Jesús*, ff. 44r-46r.

92. L. de Morales, S.J., *Historia*, f. 88; L. de Morales, S.J., "The Great Chamorro War." *Pacific Profile*, 2 (3), 1964, pp. 20–21.

93. ARSI, "Primus Catalogus Personarum Anno 1716." Philipp. Catal. Trien. Philipp. 3, f. 76r; ARSI, "Catalogus Brevis Personarum Provinciae Philippinarum Anno 1681." Philipp. Catal. Trien. Philipp. 4, f. 76v.

In the meantime, the news that came from the Marianas was anything but reassuring. The robberies, rapes, and murders committed by soldiers and officers, especially those criminals exiled from the Philippines or New Spain, constituted a major obstacle for the Jesuits' evangelization efforts. Chamorros violently resisted the acculturation spearheaded by the missionaries, and there were new martyrs, including the mission's founder, Diego Luis de San Vitores in 1672; Francisco Ezquerra[94] in 1674; Pedro Diaz[95] in 1675; and Antonio Maria de San Basilio[96] and Sebastian de Monroy[97] in 1676.

In 1682, the mission's superior, Father Manuel de Solórzano Escobar, wrote to Maria de Guadalupe Lencastre, Duchess of Aveiro and Arcos (1630–1715),[98] asking for her intercession in the appointment of a good governor for the Marianas, such as loyal Don Joseph de Quiroga, for the success of the entire enterprise depended on the governor's example and authority.[99] There is no

94. Francisco Ezquerra [or Esquerra] was born in Manila on October 4, 1644, to General Don Juan de Ezquerra and Doña Lucía Sarmiento. He joined the Society of Jesus on January 17, 1661 (ARSI, "Primus Catalogus Anni Personarum Anni 1672." Philippinae Cat. Trien. 1649–1696, Tomo 2-II, f. 363r). His brother, Juan Ezquerra, was also a Jesuit, and his uncle, provincial Domingo Ezquerra (1601–1670), was the one who sent Father San Vitores to the Marianas. Upon Father Solano's death, Father Ezquerra was named superior of the mission, but on February 2, 1674, at the age of 30, he was killed by a group of Marianas' natives on a beach in Guåhan (AHCJC, Carpeta "EI.b-9/5/2. Martirios, naufragios, &c." in EI.b-9/5/1-7. "Martirios y varones ilustres." AHCJC. Loose leaves. See also P. Murillo Velarde, *Historia de la provincia de Filipinas*, pp. 334–336). See also C. R. Boxer, "Two Jesuit letters on the Mariana Mission, written to the Duchess of Aveiro (1676 and 1689)." *Philippine Studies*, 26, 1978, pp. 37–42.

95. Pedro Diaz was born in the village of Calero in Talavera de la Reina (Toledo) in 1651. He joined the Society of Jesus on April 24, 1673, in the College of Oropesa (Toledo) (AHCJC, Carpeta "EI.b-9/5/4. "Mártires de la Provincia de Filipinas de la Compañía de Jesús. Manila, 12 de mayo de 1903," contenida en EI-b-9/5/1-7. "Martirios y varones ilustres." AHCJC). See also the letter written by Father Gerardo Bouwens to General Giovanni Paolo Oliva (1676) (ARSI, Litterae Annuae Philipp. 1663–1734 (etiam de Insuli Marianis), Tomo 13, f. 129r; ARSI, "Primus Catalogus Anni Personarum Anni 1675." Philippinae Cat. Trien. 1649–1696, Tomo 2-II, f. 395r. For more on Pedro Díaz, see H. de la Costa, *The Jesuits in the Philippines*, p. 457; P. Murillo Velarde, *Historia de la provincia de Filipinas*, ff. 336r-337v.

96. Antonio María de San Basilio was born in Catania (Sicily) in 1643. He joined the Society of Jesus on January 11, 1659, and was a missionary to the Indians in the Marianas until he was killed on January 12, 1676 (ARSI, "Primus Catalogus Anni Personarum Anni 1675." Philippinae Cat. Trien. 1649–1696, Tomo 2-II, f. 392r).

97. Sebastián de Monroy was born in Seville (Andalusia) in 1649. He joined the Society of Jesus in May of 1672. Soon after, he was ministering to the Indians of the Marianas until his death in 1676 (ARSI, "Primus Catalogus Anni Personarum Anni 1675." Philippinae Cat. Trien. 1649–1696, Vol. 2-II, f. 395r). For more on Monroy's death, see the letter written by Father Gerardo Bouwens on September 6, 1676 (ARSI, Philippin. Necrologia 1605–1731, Tomo 20, ff. 316r-321v).

98. For a brief biography of the Duchess of Aveiro, see Natalia Maillard Álvarez, "María Guadalupe de Lencastre, Duquesa de Aveiro y Arcos, y su biblioteca," in Juan Luis Carriazo Rubio, José María Miura Andrades, and Ramón Ramos Alfonso (eds.), *Iglesias y conventos*, Actas de las XIV Jornadas sobre historia de Marchena (October 7–10, 2008), Marchena, 2011; Andrés Oyola Fabián & Manuel López Casquete, *Localización de las reliquias del jesuita frexnense Manuel Solórzano y Escobar (1649–1684), evangelizador de las islas Marianas*. Ponencia presentada en las XIV Jornadas de Historia, "España, el Atlántico y el Pacífico. V Centenario del descubrimiento de la Mar del Sur (1513–2013)", Llerena, 25 y 26 de octubre de 2013.

99. Manuel de Solórzano's letter to Father Francisco García, Guåhan, May 20, 1681 (R. Lévesque, *History of Micronesia*, vol. 7, 1996, pp. 445–49). See also Ward Barrett, *Mission in the Marianas. An Account of*

doubt that Father Morales, who at that time was the Vice-rector of the College of Cavite, was aware of all this.

On June 8, 1682, the *San Antonio de Padua* galleon stopped at Guåhan's port of Umatac [Humåtak]. Friar Juan Duran, auxiliary bishop of the Philippine Islands (1680) was on board, and became the first prelate to visit Guåhan.[100] There, he met with the Jesuit priests, who expressed their fears and concerns regarding the preservation of the mission, especially given Don Antonio Saravia y Villar's delicate health. Upon his arrival in Manila, Bishop Duran would have communicated these concerns to Father Morales, who had been appointed procurator of the province of the Philippines in the curias of Madrid and Rome upon the death of the provincial's first choice, Father Jeronimo de Ortega (1627–1683), rector of the College of St. Joseph in Manila.[101]

Created in 1570, the post of procurator to the Spanish court in Madrid was established to "arrange in the court the affairs that concerned principally the Spanish provinces of the Order [...]–to serve as the primary mediator between the court of Spain and the Society of Jesus."[102] The possibility of sending representatives to the court of the Austrias promoted the production of memorials and written claims, and encouraged the fusion of the expectatives of the religious orders with the interests of the urban elites. Thus, procurators were central figures in the organization of the Society whose center was in Rome and extended radially towards the provinces.[103] Procurators had to be

Father Diego Luis de Sanvitores and His Companions, 1669–1670, Minneapolis: University of Minnesota Press, 1975, p. 53).

100. Manuel de Solórzano's letter to Father Francisco García, Guåhan, May 20, 1681, cited in Marjorie G. Driver, *Cross, Sword, and Silver: The Nascent Spanish Colony in the Mariana Islands*, MARC Papers, no. 48, University of Guam, 1987, p. 21.

101. ARSI, "Primus Catalogus Anni Personarum Anni 1671." Philippinae Cat. Trien. 1649-1696, Vol 2-II, f. 353v. It is very likely that provincial Francisco Salgado (1683-1687) would have appointed Father Morales as procurator even if Ortega had not passed away on November 15, 1683, for on March 8, 1683, Ortega had been excommunicated by the Archbishop of Manila, Friar Felipe Pardo, because of his debts to the Real Audiencia (H. de la Costa, *The Jesuits in the Philippines*, p. 496; Eduardo Descalzo Yuste, "La Compañía de Jesús en Filipinas (1581–1768): Realidad y representación." PhD diss., UAB, 2015, p. 189). See the Memorial written by Father Antonio Matías Jaramillo to the King circa 1689, "Memorial al Rey Nuestro Señor por la provincia de la Compañía de Jesús de las islas Filipinas, en satisfacción de varios escritos violentos hechos con que a dicha provincia ha agraviado el reverendo arzobispo de Manila don fray Felipe Pardo de la orden de Santo Domingo" (Archivo Histórico de la Compañía de Jesús en la Provincia de Castilla (henceforth, AHCJPC), Estante 2, Caja 96. Filipinas, Legajo 1157, 4, ff. 59-62).

102. Agustín Galán García, El *"Oficio de Indias" de Sevilla y la organización económica y misional de la Compañía de Jesús (1566-1767)*, Sevilla: Fondo de Cultura Económica, 1995; Luisa Elena Alcalá, "De compras por Europa". Procuradores jesuitas y cultura material en Nueva España". *Goya. Revista de Arte*, 318, 2007, pp. 141-158; J. G. Martínez-Serna, "Procurators and the Making of the Jesuits' Atlantic Network", p. 185.

103. J. G. Martínez-Serna, "Procurators and the Making of the Jesuits' Atlantic Network", pp. 181–182. Fabian Fechner, "Las tierras incógnitas de la administración jesuita: toma de decisiones, gremios consultivos y evolución de normas," *Histórica*, XXXVIII.2 (2014), pp. 11-42.

capable of treating the most controversial and delicate affairs in the courts and curias of Europe, and that meant that only the most distinguished priests, with solid experience and education, and a thorough knowledge of their respective provinces, were appointed.[104] These influential men were authentic "cultural mediators" whose minds, as Gruzinski points out, had to encompass the immense spaces that the Spanish Catholic Crown hoped to control.[105] They were supposed to inform the members of the Court, as well as its institutions, of the state of the missions in their province.

As Joan Pau Rubiés has argued, "when the Jesuits transmitted new information about the history and ethnography of the non-European World to Europe they did so in order to obtain support for their missions."[106] Indeed, procurators had to procure the financial support necessary to sustain the missions and make them prosper. They were also charged with obtaining books, clothes, wine, and the necessary elements for the liturgical rites and the churches, such as medals, crucifixes, scapulars, rosaries, wooden sculptures, paintings on canvas, and above all, relics.[107] These cultural and political agents were also involved in ensuring the happy resolution of various local affairs, including individual petitions and legal proceedings that involved friends or proteges. And finally, they had to recruit Spanish missionaries, but especially foreigners, to inform the Father General in Rome of the Society's intercontinental affairs.[108] Faced with these difficult tasks, it is no surprise that Father Morales

---

104. Alexandre Coello and Fabian Fechner (eds.), *Political Agents and Cultural Mediators: Jesuit Procurators in a Globalizing World* (16th and 18th centuries). Leiden: Brill, 2024 (forthcoming).

105. Serge Gruzinski, "Passeurs y elites "católicas" en las Cuatro Partes del Mundo. Los inicios ibéricos de la mundialización (1580–1640), in Scarlett O'Phelan Godoy & Carmen Salazar-Soler (eds.), *Passeurs, mediadores culturales y agentes de la primera globalización en el Mundo Ibérico, siglos XVI-XIX*, Lima: Pontificia Universidad Católica de Perú & Instituto Riva Aguero & IFEA, 2005, p. 16.

106. Joan Pau Rubiés, "The Concept of Cultural Dialogue and the Jesuit Method of Accommodation: Between Idolatry and Civilization." *Archivum Historicum Societatis Iesu*, Vol. LXXIV: 147, 2005, p. 243.

107. J. G. Martínez-Serna, "Procurators and the Making of the Jesuits' Atlantic Network", pp. 194-201; Alcalá, "De compras por Europa", p. 150. Procurators took relics from the martyrs of the misions to distribute among their confreres in Europe, generating a veritable "international traffic" of new relics (L. Clossey, *Salvation and Globalisation*, pp. 188-192). For an interesting analysis of relics as sacred merchandise, see José Luis Bouza Álvarez, *Religiosidad contrarreformista y cultura simbólica del Barroco*, Madrid: CSIC, 1990, pp. 27-29; 47-56; Patrick Geary, *Furta Sacra, Thefts of Relics in the Central Middle Ages*, New Jersey: Princeton University Press, [1978] 1990; and Geary, "Mercancías sagradas: la circulación de las reliquias medievales," in Arjun Appadurai (ed.), *La vida social de las cosas. Perspectiva cultural de las mercancías*, Mexico, Grijalbo, 1986, pp. 211-239. On the importance of relics as artifacts of spatial consecration in the Catholic world, see Renato Cymbalista, "Relíquias sagradas e a construção do território cristão na idade moderna." *Anais do Museu Paulista*, Vol. 14, 2006, pp. 11-50.

108. J. G. Martínez-Serna, "Procurators and the Making of the Jesuits' Atlantic Network", pp. 207-208. The procurators of the European provinces went to Rome every three years, and those of the overseas provinces every six years (Descalzo Yuste, "La Compañía de Jesús en Filipinas", p. 162). The recruitment of foreign Jesuit missionaries, especially German, French, or Portuguese, became problematic after 1640. See Luke Clossey, *Salvation and Globalisation in the Early Jesuit Mission*, Cambridge: Cambridge University Press,

stayed in Europe until 1689[109] and postponed the writing of his *History of the Mariana Islands.*

In 1683, Father Morales left Manila for New Spain,[110] before the eruption of the Second CHamoru War, which took the lives of numerous soldiers, auxiliaries, and missionaries (among the latter, Manuel de Solórzano, Baltasar du Bois, Teófilo de Angelis, Carlos Boranga, and Agustín Strobach), and which demonstrated that the mission was far from consolidated.[111] The mission's vice-provincial, Father Gerardo Bouwens [or Bowens][112] wrote a letter to the king on May 15, 1685, notifying him of the great number of deaths caused by the CHamoru rebels.[113] He complained about Governor Don Damián de Esplana's management of the war as well as his personal conduct,[114] and he requested a larger number of missionaries to reap the great spiritual harvest that the neighboring islands promised.[115] The differences that existed between the missionaries and the governor became known at court, which made Father Morales's diplomatic object even harder.[116]

2008, pp. 26-27; 147-53.

109. A procurator appointment usually lasted from four to six years (Carlos A. Page, *Los viajes de Europa a Buenos Aires según las crónicas de los jesuitas de los siglos XVII y XVIII*, Córdoba: Báez Ediciones, 2007, p. 9; L. Clossey, *Salvation and Globalization,* p. 57).

110. The war erupted on July 24, 1684, and Father Morales does not feature in the 1684 nor 1687 *Catalogus Alphabeticus Personarum de la Provincia Mexicane* (ARSI, Mexican. Catal. Triennal (1659–1687), Vol. 5, ff. 343r-345v; ff. 394r-396v).

111. Alexandre Coello de la Rosa, "Colonialismo y santidad en las islas Marianas: los soldados de Gedeón." *Hispania,* Vol. LXX, no. 234, 2010, pp. 17–44; Alexandre Coello de la Rosa and David Atienza de Frutos (eds.), Scars of Faith. Jesuit Letters from the Mariana Islands (1668 – 1684), Chesnut Hill, Massachussets: Institute of Jesuit Sources – Boston College, 2020.

112. Gerardo Bouwens was born in Antwerp (Belgium) on September 23, 1633. He joined the Society of Jesus on September 19, 1656. He was a graduate and professed of the four vows (2/2/1675) (ARSI, "Primus Catalogus Anni Personarum Anni 1684." Philippinae Cat. Trien. 1649–1696, Vol. 2-II, f. 422v).

113. While Jesuit historian Francis Hezel regarded the intermittent outbreaks of violence known as the Spanish-CHamoru Wars (1671–72; 1684; 1690) as of secondary importance (Francis X. Hezel, *When Cultures Clash: Revisiting the 'Spanish-Chamorro Wars.* Saipan: Commonwealth of the Northern Mariana Islands [CNMI], 2015, p. 10), other scholars have gone so far as to describe these armed clashes as genuine "civil war" (B. Dixon, A. Jalandoni, and C. Craft, "The Archaeological Remains of Early Modern Spanish Colonialism on Guam and their Implications", in M. C. Berrocal y C. Tsang (eds.), *Historical Archaeology of Early Modern Colonialism in Asia-Pacific: Volume I, The Southwest Pacific and Oceanian Regions*, Gainesville: Florida UP, 2017, p. 197).

114. Born in Peru, criollo Don Damián de la Esplana (1641–1694) was a military man who fought the Mapuche Indians of Chile for 23 years. His wife was Doña Josefa de León Pinelo, a member of one of the most prominent criollo families in the Peruvian Viceroyalty (ARSI, "Relación de las islas Marianas desde el mes de junio de 1674 hasta 1675," Philipinae Historiae, 1663–1734, Vol. 13, f. 121r; M. G. Driver, *Cross, Sword, and Silver*, p. 15). Esplana was appointed governor of the Marianas in 1674 by Don Diego de Arévalo, commanding officer of the galleon *Nuestra Señora del Buen Socorro*. Esplana died of hydropsy in 1694 (A. Astrain, *Historia de la Compañía de Jesús*, Vol. VI, p. 825; 832). See also F. X. Hezel, S.J., "From Conversion to Conquest," *The Journal of Pacific History*, 17, 1982, p. 127.

115. A. Coello, "Colonialismo y santidad en las islas Marianas: los soldados de Gedeón," pp. 17–44.

116. Father Bouwens's letter was included in the documents presented by procurator Morales to the Coun-

On August 14, 1685, Father Morales reached Seville, burdened with letters, recommendations, and memorials that he had to deliver to various people, including the Duchess of Aveiro[117] and the Queen Regent, who was the main supporter of the Mariana Islands mission in the Spanish Court.[118] At that time, the ministers and members of the Council of the Indies were upset with the Jesuit procurators for bypassing the Council in their requests and going directly to the king or the court favorites (the Duke of Medinaceli from 1680–1685; the Count of Oropesa from 1685–1691 …). They knew, the councilors argued, that "whenever any of these parts go to him, he grants them what they ask."[119] As we have said before, procurators acted as the Society's ambassadors, and they had to gain the sympathy (and financial assistance) of the most important members of the courts of Rome and Madrid, which was no easy task given the intrigues and conflicts that abounded in these intricate political spaces.

One of Father Morales's objectives was to promote the beatification of Father San Vitores, who had died a martyr in Guåhan in 1671, for having a "patron saint" of their own would bring the Marianas acknowledgement as a bulwark of Catholicism in Asia. He also had to defend the causes of particular individuals—for instance, Tagalog captain Don Juan de la Cruz, Pampango *alférez* [ensign] Don Andrés de la Cruz, and Chamorro lieutenant Don Antonio de Ayihi—for whom the procurator requested a reward for their services in the conquest and colonization of the Mariana Islands.[120] Finally, Morales, like all Jesuit

cil of the Indies, asking that the Mariana Islands mission not be abandoned (AGI, Filipinas 3, Ramo 170).

117. It was no coincidence that F. García, S.J., later dedicated his hagiography on Diego Luis de San Vitores to the Duchess of Aveiro (Francisco García, S.J., *The Life and Martyrdom of Diego Luis de San Vitores of the Society of Jesus. First Apostle of the Mariana Islands and Events of These Islands from the Year Sixteen Hundred and Sixty-Eight through the Year Sixteen Hundred and Eighty-One* (trans. by Margaret M. Higgins, Felicia Plaza, M.M.B, Juan M. H. Ledesma, S.J.). Edited by James A. McDonough, S.J. Mangilao, Guam: Richard Flores Taitano & MARC & University of Guam, [1683] 2004, pp. xxii–xxiii).

118. Some of these letters and memorials have been published and transcribed in R. Lévesque, *History of Micronesia*, vol. 8, 1996, pp. 411–422; 661–668. See also *Bibliotheca Americana et Philippina*, Part III. Catálogo no. 442. Maggs Bross, 1923, pp. 198–199.

119. Colección Pastells, Fil. 2, pp. 442–443, transcribed in R. Lévesque, *History of Micronesia*, vol. 8, 1996, p. 424. There was open opposition among some members of the Society regarding what they considered the "excessive" number of procurators that the provinces sent to the peninsula. Their complaints were not only related to the expenses that this generated, but to the fact that many of them did not want to return to the Indies. This, they argued, led to murmuring in the Court, which scoffed at the fact that, "more procurators from the Society go to the Palace, than those of all religions put together." ("Algunas razones para que se vea, no ser conveniente, que las Provincias de Indias, pongan como pretenden, Procurador que venga dellas, en Madrid (sin fecha)," ARSI, Fondo Gesuitico, 721-2-1, Fasc-e, f. 32r).

120. *Memorial* del padre Morales al Consejo de Indias, August 1685 (R. Lévesque, *History of Micronesia*, vol. 8, 1996, pp. 411–17). On January 22, 1686, the Junta de Guerra [Board of War] decided to reward the three native officers (R. Lévesque, *History of Micronesia*, vol. 8, 1996, pp. 430–34). In the case of Captain Juan de Santa Cruz, Morales had asked that a pension be given to his nephews. In 1686, the Crown through the mediation of Tomás Antonio de la Cerda y Aragón, Viceroy of New Spain, Count of Paredes and Marquis of la Laguna (1680–1686), granted a medal to "Don Juan de Santa Cruz, *indio principal* of the Tagalog

procurators, had to manage numerous conflicts over jurisdiction and authority with governors and captain generals (particularly, Lima-born Don Damián de Esplana, the Marianas' governor), archbishops, and other religious orders.

Father Morales's activity was unremitting. In 1686, he went to Rome to promote the province's affairs. From there he returned to Spain, where he was requested to express his opinion on the forced evangelization of *sangleys* and their relocation in villages or towns that were adjacent to the capital.[121]

After the death of General Charles de Noyelle (December 12, 1686), Morales returned to Rome to participate in the thirteenth General Congregation of the Society of Jesus (June 21, 1687–September 7, 1687) in which Father Thyrsus González de Santalla was elected Superior General of the Society (1687–1705).[122] In the meantime, Father Morales had to defend himself and his confreres from the defamation of politico-religious rivals, such as Dominican Friar Alonso Sandin, OP (1640–1701),[123] procurator and General Diffinitor of the Philippine province, who presented "defamatory and injurious documents" at the Council of the Indies against the Society of Jesus.[124]

According to Father Morales's memorial on this matter, Manila's Dominican friars mocked the Jesuits through flyers that were handed out, "not at night, nor at resting hours, but publicly in the middle of the day, [by

---

nation, who went from Manila to the Mariana Islands with his family accompanying Father Diego Luis de San Vitores, undergoing great hardships and risking his life," and awarded "a grant to various nephews of his, named Don Ignacio Pagtacotán, Don Julián and Don Juan de la Cruz, as well as to the sergeant who also took part in that expedition, all four of them residents of the Philippines" ("... *don Juan de Santa Cruz, indio principal de la nación tagala, que con su familia pasó desde Manila a las Islas Marianas con el padre Diego Luis de San Vitores, tolerando grandes trabajos y arriesgando su vida y por haber muerto he resuelto hacer merced [entre otras] a varios sobrinos suyos, nombrados don Ignacio Pagtacotán, don Julián y don Juan de la Cruz y al sargento que también se empleó en dicha entrada todos cuatro residentes en Filipinas...*" AGN, Mexico. Instituciones Coloniales. Gobierno Virreinal. Californias–017. Vol. 26, Ex 92, f. 251v).

121. Father Morales's opinion, expressed on September 13, 1686, was that infidel *sangleys* should not be expelled from the Manila *Parian* (ghetto), for this went against the Church's duty to convert new souls to Catholicism, and also against the Royal Treasury, for it would lose the income obtained from the licenses sold to *sangleys*. He recommended that the *Parian* be kept, and that its evangelization be turned to two or more religious orders that spoke Chinese ("Memorial del padre Luis de Morales, S.J., Madrid, 13 de Septiembre de 1686," AGI, Filipinas 28, N. 131). Soon after, on May 24, 1686, there was an uprising of the *Parian sangleys* that lasted for a few days before the Spanish authorities put it down.

122. Born in 1624 and a member of the Society since 1643, Thyrsus González de Santalla was the 13th Superior General of the Society, elected on July 6, 1687, after the death of Father Charels de Noyelle in 1686. He died in 1705, and had the support of Pope Innocent XII during most of his generalship. For more on Father González's generalship, see *DHCJ*, Tomo II, pp. 1644–1650.

123. The Dominican Order is officially called the Order of Preachers (Latin: Ordo Praedicatorum), and hence the abbreviation OP used by and for its members.

124. Pedro de Espinar, S.J., *Manifiesto jurídico defensorio, en respuesta de los reparos hechos por el Padre Fray Antonio de las Huertas de el Orden de Predicadores a un memorial que en el Real Consejo de las Indias presentó el Padre Luis de Morales de la Compañía de Jesús, procurador de su religión por las provincias de Filipinas*, 1684, 17 ff. 3v. See also A. M. Jaramillo, "Memorial al Rey Nuestro Señor por la provincia de la Compañía de Jesús de las islas Filipinas" (AHCJPC, Estante 2, Caja 96. Filipinas, Legajo 1157, 4, ff. 14-15).

a] lay friar of St. Dominic, who filled a corner of the public square."[125] For their part, Dominicans accused the Jesuits of having distributed a *Relacion* that favored the Royal Audiencia of Manila against the authority of Manila's archbishop, Friar Felipe Pardo, OP (1677–1689). There was evidently a great disaffection between both orders.[126] However, what really worried the Jesuits was not the falsehood of the accusations, but that, "four of those papers, or libels" had been printed in the courts of Madrid and Rome, which suggested a defamatory intent.[127] For this reason, Father Morales asked that, "said papers be seized, and their author, even if he is a religious man, [be told] not to print them anymore, using all means necessary to secure this outcome, but not judicial means (the Society has not enough resources to try this), relying instead on your [the Council of the Indies] supreme authority and discretionary economic power.[128]

Like Morales, Father Diego Francisco Altamirano (1625–1715), General Procurator of the West Indies in Madrid and Rome (1682–1688),[129] also defended his confreres from the "calumnies and lies" that "persons with no fear of God" had spread in the Philippines, Mexico, and Spain. In a memorial that he wrote in 1685 to the king, he argued that the members of the Society and the missionaries in the Marianas, the Philippines, and other provinces were worthy of His Majesty's trust and protection.[130]

In November of 1687, Superior General Thyrsus González summarized in a letter the rumors that circulated in Rome against the Society, emphasizing the economic operations that were supposedly carried out in America, particularly in the provinces of Paraguay, Cordoba, and Buenos Aires, where yerba mate was produced and commercialized.[131] He wrote that there were accusa-

125. P. de Espinar, S.J., *Manifiesto jurídico defensorio*, f. 4r.

126. P. de Espinar, S.J., *Manifiesto jurídico defensorio*, f. 13r.

127. P. de Espinar, S.J., *Manifiesto jurídico defensorio*, f. 5r; 7v.

128. P. de Espinar, S.J., *Manifiesto jurídico defensorio*, f. 5r; 10r.

129. Father Altamirano left Buenos Aires on May 3, 1685 and arrived at Seville, his usual place of residence, at the end of the year. He took care of the preparations and offices that were necessary for Jesuit missionaries to embark for the Indies (C. A. Page, *Los viajes de Europa a Buenos Aires*, p. 12). Recently, historian J. A. Maeder argued that the third book of the *Insignes Misioneros de la Compañía de Jesús en la Provincia del Paraguay* (1687), signed by Father Francisco Jarque, was in fact written by procurator Diego Francisco Altamirano. For a brief biography of Father Altamirano, see A. Coello, "Diego Francisco Altamirano." *Diccionario Biográfico Español*, Madrid: Real Academia de la Historia, Vol. III, 2010, pp. 328–29; E. J. A. Maeder, "Estudio preliminar" a Francisco Jarque & Diego Francisco Altamirano, *Las Misiones Jesuíticas en 1687. El estado que al presente gozan las Misiones de la Compañía de Jesús en la Provincia del Paraguay, Tucumán y Río de la Plata*, Buenos Aires: Academia Nacional de la Historia & Union Académique Internationale, 2008, p. 14.

130. AGI, Filipinas 90, Ramo 3, f. 1r.

131. Juan Carlos Garavaglia, *Mercado interno y economía colonial. Tres siglos de historia de la yerba mate*, Rosario: Prohistoria Editores, [1983] 2008, pp. 147–152.

tions regarding the large-scale and profitable sale of sugar, tobacco, yerba mate, honey, and cotton. According to the papal brief of Clement IX (1667–1669), the trade of goods for the purpose of, "acquiring the necessities of life" was not banned by the Church, but trade for profit was. Indeed, the Jesuit procurators of Santa Fe and Buenos Aires acted as commercial agents of bishops and civil authorities, "in representation of the Indians" in the traffic of yerba mate, "allowing crops and other articles that the Indians might need to be stored in public deposits, rewarding the laziness and incapacity of those natives, who barely prepared for what they would need the following day."[132]

That same year, provincial Luis Pimentel (1612–1689)[133] appointed Father Jaramillo procurator for a six-year period in the XIII Provincial Congregation (Manila, 1687).[134] Like his predecessor Father Morales, Jaramillo had been a former missionary in the Marianas (from 1676 to 1680),[135] and he was considered a balanced and prudent person. In 1681, Jaramillo was in the College of Manila working as *operario* to both Spaniards and natives,[136] but he always felt attached to the Marianas mission. On August 23, 1683, he returned to Guåhan aboard a patache called *St. Francis Xavier,* in what was the first voyage across the three hundred leagues that separated the port of Cavite from Guåhan's Humåtak Bay.[137] Later, he returned to Manila, where he remained until 1684, when Father provincial Francisco Salgado (1629–1689)[138] transferred him to the Tagalog residence of San Miguel, east of the Manila *Parián.*[139] There he wrote a Memorial (1685) addressed to the King criticizing

132. A. Astrain, *Historia de la Compañía de Jesús,* Tomo VI, p. 416; Garavaglia, *Mercado interno y economía colonial,* pp. 71–72.

133. Luis Pimentel was born in Portillo, Diocese of Valladolid, on May 30, 1612. He joined the Society on March 24, 1632. He was a graduate and professed of the four vows (1/5/1650) (H. de la Costa, *Jesuits in the Philippines,* pp. 615–16).

134. ARSI, "Catalogus Brevis Personarum Provinciae Philippinarum Anno 1680." Philipp. Catal. Trien. Tomo 4, f. 83r. See also H. de la Costa, *Jesuits in the Philippines,* p. 436; R. Lévesque, *History of Micronesia,* vol. 9, 1997, pp. 181–183.

135. ARSI, "Catalogus Brevis Personarum Provinciae Philippinarum. Anno 1677, 1679 y 1680." Philipp. Catal. Trien. Tomo 4, ff. 69-72r.

136. ARSI, "Catalogus Brevis Personarum Provinciae Philippinarum. Anno 1681." Philipp. Catal. Trien. Tomo 4, f. 76r.

137. Marjorie G. Driver, *The Spanish Governors of the Mariana Islands, Notes on their activities and the Saga of the Palacio. Their Residence and the Seat of Colonial Government in Agaña,* Mangilao, Guam: Richard F. Taitano MARC, 2005, p. 8; A. Coello, "Colonialismo y santidad en las islas Marianas: los soldados de Gedeón," p. 25.

138. Father Francisco Salgado arrived in the Philippines in 1662. He was successively appointed vice-rector of the College of San Jose, rector of Silang, procurator (1675-79) and provincial of the Philippines from 1683 to 1687, when he was substituted by Luis Pimentel (P. Murillo Velarde, *Historia de la provincia de Filipinas,* f. 420r; H. de la Costa, *The Jesuits in the Philippines,* p. 601).

139. The parish of San Miguel was probably the old parish of Laguío, where the Jesuits maintained the first residence assigned to them in Manila at the end of the 16th century (ARSI, "Catalogus Brevis Persona-

{Figure 3: Diego de Silva y Velázquez, *Retrato de Felipe IV* (h. 1660). National Gallery (Londres)}

Archbishop Pardo's animosity against the Jesuits, which clearly conditioned the missionary policies to be followed.[140]

Jaramillo's devotion to the Marianas mission could only compare to that of procurator Luis de Morales. In April of 1687, the latter wrote a new *Memorial* addressed to Queen Mariana, accusing the Council of the Indies of providing insufficient support for the missions of the Pacific Islands. He wanted

rum Provinciae Philippinarum. Anno 1684 y 1686." Philipp. Catal. Trien. Vol. 4, ff. 78r-80v).

140. The parish of San Miguel was probably the old parish of Laguío, where the Jesuits maintained the first residence assigned to them in Manila at the enad of the 16th century (ARSI, "Catalogus Brevis Personarum Provinciae Philippinarum. Anno 1684 y 1686." Philipp. Catal. Trien. Vol. 4, ff. 78r-80v).

to prevent the Crown from abandoning the Marianas archipelago due to the numerous, "accidents and revolutions" that had taken place in the last few years, lamenting that, "the blood of so many martyrs" whose relics had consecrated those lands for Christendom and opened the doors to more islands, would have been shed for naught. Indeed, Morales's fear that Spain would abandon, "such conversions, which were ready to be picked, that of the thirteen reduced islands, as well as the many greatly populated isles that lie nearby, waiting to be reduced into the guild of our Holy Church."[141] Despite his excessive zeal, which could seem jarring, he obtained some support, like the promise of resettling the islands with families brought directly from New Spain, which guaranteed the monarchy's commitment to its Chamorro subjects.[142]

But there was more. On May 5, 1687, Father Morales, along with the procurator of the Indies, Father Pedro de Espinar (1630–1695),[143] presented a list with the names, surnames, and places of provenance of the forty-one individuals who were to go to the Philippine and Mariana Islands. The candidates had to be volunteers and their superiors had to approve their dispatch. All of the Jesuit provinces participated in the missionary effort, with Andalusia contributing a greater number due to its geographical location vis-à-vis the port of departure. Procurators were not always able to reach the maximum number of missionaries approved by the *Real Cédula* [Royal Decree] for each expedition. This happened to Morales and Espinar in this case, for having been allowed to embark seventy-eight missionaries, they could not recruit them all.[144] Indeed, despite the Real Cédula of December 1, 1664, in which Philip IV authorized one-fourth of the Jesuits in a missionary expedition to be recruited "outside Spain," only ten expeditions filled this quota.

During this time of scarce recruits, many Jesuit novices had written to the General of the Society asking to be sent as missionaries to the Indies. What was behind the difference in numbers between willing novices and actual candidates, despite the fact that in 1675 the Queen Regent Mariana of Austria

---

141. AGI, Ultramar 562, transcribed in R. Lévesque, *History of Micronesia*, vol. 9, 1997, pp. 29–30. The Council of the Indies' response from October 27, 1687 is transcribed in R. Lévesque, *History of Micronesia*, vol. 9, 1997, pp. 36–39.

142. "Real Decreto del 10 de abril de 1687," transcribed in R. Lévesque, *History of Micronesia*, vol. 9, 1997, pp. 34–35.

143. Little is known about Father Espinar except that in 1675, he was appointed General Procurator of the Indies by Father Provincial Javier Riquelme (1675–1678). He was assisted by Father Francisco Salgado, who years later occupied the post of provincial (1683–1687) (H. de la Costa, *Jesuits in the Philippines*, p. 436; C. Sommervogel, S.J., *Bibliothèque de la Compagnie de Jésus*, Tomo 3, p. 452; Descalzo Yuste, *La Compañía de Jesús en Filipinas*, pp. 188–189).

144. AGI, Audiencia de Filipinas 83, Ramo 10, ff. 1r–4v.

had extended boarding permission to one third of all expeditions?[145] The most evident cause lay in the reticence of the Council of the Indies to grant licenses to foreign Jesuits to go to the Indies[146] as a response to the country's vulnerable political situation.

In 1679, a weakened Spanish Crown had been forced to cede the Frenche-Comté region to France. In 1684, it lost Luxemburg through the Truce of Ratisbon, accelerating even more the decadence of the Spanish empire. Another cause seemed to be the reticence of Jesuit superiors in Spain to approve their novices or young priests' requests to be sent as missionaries to the Indies, despite the fact that in 1675 the regent Queen Mariana of Austria had extended the permit to one third of the expeditions.[147] Marie-Lucie Copete and Bernard Vincent, who have looked at Jesuit missionary activity and recruitment in Andalusia, concluded that the colleges of Seville, Granada, and Cordoba were especially active from the end of the sixteenth century to the decade of the 1630s. There is another boom of missionary activity and interest during the popular missions of Fathers Thyrsus González de Santalla and Gabriel Guillen[148] between 1669 and 1672, and then again in 1679.[149]

Concerned, Father Morales wrote to the General of the Society on September 16, 1687, a series of "postulates" in which he directly accused Spanish provincials and superiors—in particular, the provincials and rectors of the colleges of Andalusia—of blocking missionary vocations and recruits to the Indies, "with the frivolous pretext of the need for such individuals in the Spanish provinces."[150] He singled out Father Florencio de Mina, rector of the college of Cadiz, accusing him of, "dissuading and cooling" the vocation of Father Joseph de Lara. He also accused Father Joseph de Madrid, vice-provincial of Andalusia, for stopping Father Sebastian González from departing, even though, "by order of Father Vice-provincial Bartolome de Plasencia he was already approved to go

---

145. Descalzo Yuste, *La Compañía de Jesús en Filipinas*, p. 190.

146. H. de la Costa, *Jesuits in the Philippines*, p. 437; L. Clossey, *Salvation and Globalization*, pp. 149–153.

147 Descalzo Yuste, *La Compañía de Jesús en Filipinas,* p. 190.

148. Gabriel Guillen was born in 1627 in Cariñena (Zaragoza). He excelled as promoter of the popular missions in Extremadura and Andalusia (C. Sommervogel, *Bibliothèque de la Compagnie de Jesus*, T. III, 1892, pp. 1935; DHCJ, vol. II, p. 1841). See also Marie-Lucie Copete & Bernard Vincent, "Missions en Bétique. Pour une typologie des missions intérieures," in Pierre-Antoine Fabre & Bernard Vincent (coords.), *Missions religieuses modernes. "Notre lieu est le monde"* (Rome: École Française de Rome, 2007, pp. 274–75; 280; William V. Bangert, SJ, A History of the Society of Jesus. Saint Louis: The Institute of Jesuit Sources, 1972, p. 196.

149. M.-L. Copete & B. Vincent, "Missions en Bétique," pp. 274–75.

150. "Postulados que dio el padre Luis de Morales, procurador de esta provincia," September 16, 1687 (ARSI, Philipp. 12 (1660-1729), f. 106r).

to the Philippines mission."[151] There was no solution five years later. In 1691, the Society was allowed to embark sixty individuals, but the actual number of missionaries who went to New Mexico was thirty-two.[152] More than ten years later, the problem persisted, leading Father Juan Martínez de Ripalda,[153] General Procurator of the Indies, to request in 1704 that a greater number of foreign Jesuits be allowed on the missions.[154]

Moreover, those who the superiors sent were, "novices, seminarians, often times fractious or useless, to be rid of them, and thus giving God and the Holy Ministry of the Missions, their refuse, their worst men."[155] The solution, he argued, was to punish provincials for the damage that their proceedings caused, and grant special powers to procurators so that provincials and college rectors would give overseas missions the priority that they should, and so that they could not retain the best subjects who were selected to go.[156] In 1652, feeling despair beucase of the reticence of Spanish provincials to let their best subjects go to the missions, Father Miguel Solana had already suggested this solution when he asked General Goswin Nichel (1652–1664) to exempt volunteers from obeying their provincials, putting them under his jurisdiction, but the General absolutely refused.[157] Morales also recommended the public recognition of the distinguished benefactors of the missions of the Philippine and Mariana Islands, such as the Viceroy of New Spain (1680–1686) Don Tomas de la Cerda y Aragón (1638–1692), and his wife, Doña Maria Luisa Manrique de Lara y Gonzaga (1649–1729), Marquises of la Laguna and Counts de Paredes, for the

151. "Postulados que dio el padre Luis de Morales, procurador de esta provincia," September 16, 1687 (ARSI, Philipp. 12 (1660–1729), f. 106r).

152. *Bibliotheca Americana et Philippina*, Part III. Catálogo no. 442. Maggs Bross, 1923, p. 241.

153. Juan Martínez Ripalda was born in Olite (Navarra) on December 13, 1641. He joined the Society of Jesus on September 20, 1659, in Villagarcía de Campos (Valladolid), and died on December 2, 1707, in Madrid (*DHCJ*, vol. IV, p. 2526).

154. "Memorial del Padre Juan Martínez de Ripalda, procurador de las provincias de Indias al Rey, pidiendo que se permita mayor numero de jesuitas extranjeros para la misión de Pais o Palaos (1704)" (AHCJPC, Filipinas C-285, Doc. 5. ff. 2r-7v).

155. "Postulados que dio el padre Luis de Morales, procurador de esta provincia," September 16, 1687 (ARSI, Philipp. 12 (1660–1729), f. 106r). This was not, however, specific to Andalusia. Aliocha Maldavsky detected that the same issues permeated early seventeenth century expeditions organized by the procurators to the Spanish and Roman courts to the missions in the Viceroyalty of Peru, New Granada, and Paraguay (*Vocaciones inciertas. Misión y misioneros en la provincia del Perú en los siglos XVI y XVII*, manuscript, n.d., pp. 183–89).

156. This solution had already been suggested in 1652, when Father Miguel Solana, in despair at the refusal of the Spanish provincials to let his best men leave, asked General Goschwin Nickel (1652-64) to exempt the volunteers from obedience to his provincials and place them under his jurisdiction, but the General flatly refused (ARSI, Congregationum Provincialum, 73, ff. 131r, 196v, 208v y 211r, cited in Descalzo Yuste, *La Compañía de Jesús en Filipinas*, pp. 189-190).

157. Doña María Luisa Manrique de Lara y Gonzaga was cousin of another "mother of the missions", Doña María de Guadalupe, the Duchess of Aveiro. See footnote 98.

outstanding piety with which they had assisted said missions.[158]

In another memorial (August 1685), Father Morales had asked for a bell and ecclesiastical ornaments for each of the twelve churches already built in the Mariana Islands, and this request was granted through the Real Cédula dictated by Charles II on March 30, 1686.[159] That Cédula also granted San Ignacio de Agaña the title of *Ciudad*, a status that only the king could award. Along with the acknowledgement of Umatac [contemporary Humåtak] as a *Villa*, these were instances that strengthened the status of the Mariana Islands as part of the Spanish empire.[160]

Overjoyed at these successes, Society General Father Charles de Noyelle[161] (General of the Society on July 5, 1682), asked King Charles II for his support in the appointment of Father procurator Antonio Jaramillo as the first prelate of the future bishopric of Guåhan. The Jesuits argued that the mission in the Marianas was much larger than other existing dioceses, with a greater number of souls, and it was difficult to care for these souls properly without a bishop and more ministers. They also lamented that many Catholics, especially young children, died without having received the holy sacrament of Confirmation. Now that Guåhan was a city, a bishopric would mean that new priests could be ordained there to better care for the natives, strengthening and increasing Christianity in these islands.[162]

On September 8, 1685, Don Francisco Bernaldo de Quirós,[163] Spanish diplomatic agent (or *agente de preces*) together with Sancho de Losada at the court of Rome, petitioned the Pope to appoint Father Antonio Jaramillo Bishop of Guåhan, a request that he reiterated in November of 1685, and various

---

158. "Postulados que dio el padre Luis de Morales, procurador de esta provincia," September 16, 1687 (ARSI, Philipp. 12 (1660-1729), f. 107r). On the role played by the Countesses of Paredes as ladies-in-waiting and chambermaids to the queens of Spain, Frédérique Sicard, "Condesas de Paredes: señoras de su casa y camareras de la reina". *Revista de Estudios filológicos,* 26, 2014. [Available online] <http://www.um.es/tonosdigital/znum26/secciones/estudios-25-condesas_de_paredes.htm> [Consulted March 2, 2024].

159. R. Lévesque, *History of Micronesia,* vol. 8, 1996, p. 412. However, on May 23, 1687, with the fleet about to depart for New Spain, Father Morales wrote a letter lamenting that only one-of-the-twelve bells had been placed aboard (AGI, Audiencia de Filipinas 83, Ramo 11, ff. 1r-16v).

160. This petition was also included in Father Luis de Morales's *Memorial* (1685) (R. Lévesque, *History of Micronesia,* vol. 8, 1996, p. 413). See also M. G. Driver, *The Spanish Governors of the Mariana Islands,* p. 10.

161. Charles de Noyelle (1615–1686), a Belgian priest, was the 12th Superior General of the Society of Jesus, elected unanimously in 1682 to succeed Giovanni Paolo Oliva.

162. "Filipinas Marianas. Sobre la erección de una iglesia catedral y obispado a que fue presentado por su Majestad el padre Antonio Jaramillo" (ARSI, Fondo Gesuitico, 849).

163. Don Francisco Bernardo de Quirós was appointed "agent, procurator and solicitor of the interests of Castile, the Indies, and the Crusade in the Roman court" on December 1678. On September 1687, he was promoted to the Consejo de Órdenes, and Don Antonio Pérez de la Rúa substituted him in his previous role (Maximiliano Barrio Gozalo, "La agencia de preces de Roma entre los Austrias y los Borbones (1687–1730)". *Hispania,* Vol. LXXIV, nº 246, 2014, pp. 19–20).

times throughout 1686.[164] The petition was stalled until May 14, 1687, when Bernaldo de Quirós told King Charles II, "the best that you can do is to refrain from asking (…) because without [the consent of] the Congregation for the Erection of Churches and Consistorial Provisions, his Holiness was not very favorable" to it.[165] But the Jesuits insisted. On April 3, 1688, Father Thyrsus González, the new Superior General, wrote a letter in which he again asked the king to erect a cathedral in San Ignacio de Hagåtña, transforming the capital of the Marianas archipelago into a *civitas Dei*, and placing it in the mystical body of the Spanish empire.[166]

More than a simple request for the religious wellbeing of Guåhan's Christians, a bishopric in Guåhan was the Society's way of resisting the Manila archbishop Felipe Pardo's determination to impose his episcopal authority over the Jesuits and other orders that lived outside their communities. Pardo, however, was bent on applying the decrees of the Council of Trent that established that the archbishop should supervise *de vita et moribus* all of the *doctrinas*[167] in his ecclesiastical circumscription.[168] The Jesuits wanted to convince the Pope that the Mariana Islands were not really *doctrinas*, but missions, and as such, they were subject to a special juridical regime. However, the court of Madrid seemed favorable to the politics of the prelates.[169] After all, according to Manchado López, despite being two different institutions, the Pastoral Visit and the Real Patronato had a common core: safeguarding jurisdiction, "which in one case could be ecclesiastical and in the other, affect the rights acquired by the Crown in spiritual matters." That is, at the heart of this wrangling was the matter of authority, be it that of the Church, of the Crown, or of their representatives.[170]

---

164. The specific dates were: November 21, 1685; January 30, 1686; August 4, 1686; and December 12, 1686. Archivo de la Embajada de España cerca de la Santa Sede, Leg. 116, Indias. 1660–1687. "Islas Marianas: erección de obispado," ff. 280r-84r; 294r.

165. Archivo de la Embajada de España cerca de la Santa Sede, Leg. 116, Indias. 1660–1687. "Islas Marianas: erección de obispado," f. 284r.

166. AGI, Filipinas 86, Ramo 6, ff. 1r-2v. See also R. Lévesque, *History of Micronesia*, vol. 9, 1997, pp. 162–167; F. X. Hezel, *From Conquest to Colonization*, p. 19.

167. "Doctrina" was the name given to communities that were not yet parishes, consisted of converted Indians, and thus were no longer in a strictly missionary status.

168. The relations between the archbishop of Manila, Dominican Friar Felipe Pardo (1677–1689), and the Society of Jesus were never friendly. In terms of the Mariana Islands, he never visited them. See H. de la Costa, *Jesuits in the Philippines*, p. 515; and the memorial written circa 1689 by Father Antonio Matías Jaramillo ("Memorial al Rey Nuestro Señor por la provincia de la Compañía de Jesús de las islas Filipinas, en satisfacción de varios escritos violentos…") in AHCJPC, Estante 2, Caja 96. Filipinas, Legajo 1157, 4, ff. 1–150.

169. Pedro Rubio Merino, *Don Diego Camacho y Ávila, Arzobispo de Manila y de Guadalajara de México, 1695–1712*, Sevilla: Escuela de Estudios Hispanoamericanos & Ayuntamiento de Badajoz, 1958, p. 256.

170. M. M. Manchado López, *Conflictos Iglesia-Estado en el extremo oriente ibérico*, p. 21.

This episcopal measure represented, in fact, a frontal attack against the constitutions of the Society of Jesus in the Philippines. Most of the Jesuits in Manila felt proud of their independence and were unwilling to share the benefits and duties with the rest of the secular and regular clergy, whom they considered inferior in talent and religious commitment. However, there were divergent, as well as unformed, opinions regarding what the Society should do faced with this new matter of affairs.

On January 3, 1688, General Thyrsus González wrote a letter to Father Francisco Salgado, provincial of the Philippines, ordering him to investigate the convenience of abandoning, or keeping, their parishes and *doctrinas*.[171] On April 17, after listening to various expert opinions, the General wrote to procurator Father Luis de Morales, and charged him with the defense of the missions that had cost so much blood and sweat to establish, opposing all episcopal intromission. Morales's work in the court of Madrid was to convince the civil and ecclesiastical authorities that the archbishoprics' centralizing politics should be halted, lest they concentrate too much power. If he failed, the Superior General ordered him to write immediately to Salgado telling him to leave the *doctrinas* and parishes.[172]

Don Luis Francisco de la Cerda y de Aragón (1660–1710), VIII Marquise of Cogolludo, IX Duke of Medinaceli (1691) and ambassador plenipotentiary of Spain in Rome (1687–1696)[173], advised General Thyrsus González to defer his request regarding the bishopric of Guåhan until a new Pope was chosen. Pope Innocence XI, who died in June of 1689 and had been pope since 1676, did not have good relations with the Marquise, whom he considered too young and inexperienced and not modest enough in his conduct. The new pope would have to address the issue of the Guåhan bishopric, along with the request for the beatification of the bishop of Puebla de los Ángeles, Juan de Palafox y Mendoza (1600–1659), author of the renowned *Inocenciana,* a letter that contained several complaints and critiques regarding the Society of Jesus,[174] and the activation of the process of beatification of Father San Vitores, which the Jesuits had initiated in Madrid between April of 1688 and May of 1689.[175]

171. H. de la Costa, *Jesuits in the Philippines*, p. 515.

172. H. de la Costa, *Jesuits in the Philippines*, pp. 515–16.

173. Besides being ambassador to Rome, the Marquise of Cogolludo was the Viceroy of the Kingdom of Naples (1696–1702) and a member of the State Council (1699–1710).

174. Sor Cristina de la Cruz de Arteaga y Falguera, OSH, *Una Mitra sobre dos mundos. La del Venerable Don Juan de Palafox y Mendoza,* Sevilla, Artes Gráficas Salesianas, 1985, pp. 593–94.

175. The so-called "Process of Toledo" (April 1, 1688 to May 19, 1689) was presided by Cardinal Don Luis Manuel Fernández Portocarrero Bocanegra y Moscoso-Osorio (1635–1709). We suppose that Father

Papacy fell on Cardinal Pietro Vito Ottoboni (1610–91), elected on October 6, 1689, who adopted the name Alexander VIII (1689–91). Soon after his election, he activated the cause of the beatification of bishop Palafox, naming Cardinal Jerónimo Casanate as postulator (1690)[176]. He did not do the same for Father San Vitores, whose inchoate process of beatification concluded on October 6, 1985, when Pope John Paul II finally beatified him.

In early 1689, Father Morales wrote another memorial asking, again, for a license to return to the Philippines accompanied by seventy Fathers and eight coadjutor brothers, but only forty-one were allowed to embark. They all stayed in the hospice Nuestra Señora de Guadalupe in Seville until they could board.[177] With the Council's permission, the Casa de Contratación covered the expenses of provisions and gear (clothing and bedding) that the missionaries would need during the voyage. Father Morales thus returned to the Indies, leaving the new General Procurator, Father Antonio Matías Jaramillo, in the court of Madrid.[178]

4. The conflict over episcopal jurisdiction in the Philippine Islands (1689–1705)

On June 15, 1688, procurators Alejo López (1649-1693) and Antonio Jaramillo left the Philippines heading for Spain.[179] Upon his arrival in Mexico, the procurator Jaramillo learned of his appointment as provincial of the Philippines in 1687. However, he decided to continue his journey to Spain. To replce

---

Morales testified in this process. On September 26, 1689, the Cardinal sent the documents of the process to Father Antonio Zapata, S.J., who remitted them to Pope Alexander VIII. We know that these documents were introduced in the Sacred Congregation of Rites in 1695, but their present whereabouts are unknown (V. M. Diaz, *Repositioning the Missionary: the Beatification of Blessed Diego Luis de Sanvitores and Chamorro Cultural History*, Ph.D. Diss., University of California, Santa Cruz, June 1992, p. 260).

176. Archivo de la Embajada de España cerca de la Santa Sede, Legajo 79. Estado. 1689–1690. "Escrituras referentes al Conclave de 1689 y elección de Alejandro VIII," ff. 176r-89r. See also Marqués de Villa-Urrutia, *La embajada del marqués de Cogolludo a Roma y el duque de Medinaceli y la Giorgina*, Madrid: Francisco Beltrán, 1927, pp. 42–55; 80; Arteaga y Falguera, *Una Mitra sobre dos mundos*, p. 594).

177. This hospice was founded in 1686, but in 1720, it was moved to Cadiz with the procuratorial office. Like the Jesuit College of San Hermenegildo, its purpose was to house the missionaries of the Society who were waiting to embark for the Indies. As Martínez-Serna has pointed out, although the hospice received sufficient financial donations, the annual contributions of the overseas provinces paid for most of the costs related to the missionaries' upkeep (J. G. Martínez-Serna, "Procurators and the Making of the Jesuits' Atlantic Network," p. 203).

178. AGI, Filipinas 5; AGI, Filipinas 86, f. 1r.

179. A. M Jaramillo, SJ, *Memorial al Rey Nuestro Señor con varios reparos, sobre otro, que fray Raimundo Verart, del Sagrado Orden de Santo Domingo, y como poder aviente del R. Arzobispo de Manila, presentó a su Majestad*, Madrid: Imprenta de Antonio Román, 1691. Biblioteca de la Universidad de Sevilla, Fondo Antiguo, A 109/094(10) f. 17

him, Father Jose Sanchez took charge of the province.[180]

Just before leaving Manila, the procurators learned of the uprising that had taken place in the Marianas. In a letter dated on December 28, 1689, written from the galleon Santa Maria de Guadalupe, Father Jaramillo notified the Duchess of Aveiro y Arcos of the revolt that had taken place on the island of Guam.[181] On May 28, 1687, the soldiers of the presidio, led by a soldier named Manuel Salgado, had rebelled against Sergeant Don Joseph de Quiroga, putting him in shackles for three months.[182] The criticisms against the governor of the Philippines for his inaction were part of a set of grievances that the procurator Jaramillo intended to discuss in the courts of Madrid and Rome.

Procurators Jaramillo and López arrived at the end of 1689, soon after Father Morales had departed, with the purpose of debating in the Council of the Indies the Society's abandonment of their Indian parishes in the Philippine archipelago.[183] Father Jaramillo wrote a *memorial* in which he gave nine reasons, "to leave and place those parishes in His Majesty's hands," including the hostility of the Dominican order towards the Jesuits, especially given the great power that Dominican bishops enjoyed in the Philippines.[184] Archbishop Felipe Pardo, described as having Jansenist sympathies, was accused of lending credence to secret accusations leveled by the natives against a group of Jesuit priests. These religious men were seized, and the parish of Mariquina or Jesús de la Peña (1686), and later that of Cainta (1688), were taken from them.[185]

However, as Horacio de la Costa first argued, the two procurators had divergent opinions, revealing that the monolithic character of the Society has been greatly exaggerated.[186] When General Thyrsus González de Santalla asked Father López, who had gone to Rome to meet with him, to write a report

---

180. Descalzo Yuste, *La Compañía de Jesús en Filipinas*, p. 189.

181. C. R. Boxer, "Three Unpublished Jesuit Letters on the Philippine and Mariana Missions, 1681-1689". *Philippiniana Studies,* 10:3 (1962), pp. 440-442.

182. Alexandre Coello de la Rosa, *Jesuits at the Margins. Missions and Missionaries in the Marianas (1668-1769).* London & New York: Routledge, 2016, p. 100.

183. A. M Jaramillo, *Memorial al Rey Nuestro Señor con varios reparos...*, ff. 26-27.

184. A. M. Jaramillo, "Memorial al Rey Nuestro Señor por la provincia de la Compañía de Jesús de las islas Filipinas, en satisfacción de varios escritos violentos..." (AHCJC, FILEXP 01, E.I – d. 01, ff. 145-50. This *Memorial* was printed in Spain ca. 1689. There is a printed copy in AHCJPC, Estante 2, Caja 96. Filipinas, Legajo 1157, 4, ff. 1-150).

185. The Society had administered Mariquina, or Jesús de la Peña, since 1633 (A. M. Jaramillo, "Memorial al Rey Nuestro Señor por la provincia de la Compañía de Jesús de las islas Filipinas, f. 135. See also A. M. Jaramillo, *Memorial al Rey Nuestro Señor con varios reparos...*, f. 22). In 1694, King Charles II demanded that the parishes of Cainta and Mariquina be returned to the Society through the Real Cédula of March 31, 1694 (H. de la Costa, *Jesuits in the Philippines*, pp. 515; 524).

186. H. de la Costa, *Jesuits in the Philippines*, p. 517.

on how the matter stood (1690), López also warned against the dangers of administering the Indian parishes in the present circumstances, but he also acknowledged the material and spiritual rewards that *doctrinas* and parishes provided. His conclusion was that if they had to leave the *parishes*, they could and should retain the *missions* under their full authority, and the *doctrinas* in the Visayas administered by the Jesuits fell under the latter category.[187]

While López went to Rome, Father Jaramillo went to Madrid, where he met with members of the Spanish court and Jesuit superiors, and tried to influence the Council. He also wrote to the Superior General (1689) manifesting his opinions and doubts regarding the matter of the *doctrinas*. He finally met with the Society's General in Rome on April 13, 1690, and gave him another memorial in which he refuted Father López's arguments. According to Jaramillo, the religious orders were administering parishes in the Philippines not because they had chosen to do so, but because there was a lack of secular clergy. Illustrious missionaries, such as Father Francis Xavier, had never acted as parish priests.[188] Neither had Father Diego Luis de San Vitores, the first apostle of the Marianas, who promoted mobile missions to the lands of the infidels, and demanded that curates or doctrines be abandoned, for they did not correspond with the rules and statutes established by General Claudio Acquaviva (1543–1615, in office 1581-1615).[189]

In July of 1662, San Vitores had been destined to the doctrine of Taytay, some 25 kilometers east of Manila, where he learned the Tagalog language.

---

187. "Memorial del padre Alejo López, sustituto de procurador para la Compañía, a nuestro General Tirso González sobre algunas cosas que juzga conviene se manden ejecutar en la provincia de Filipinas" (1690) (ARSI, Philip. 12 (1660–1729), ff. 132r-138v.); "Papel que dio a N. P. General el padre Alejo López, sustituto de Procurador, que conviene retener las doctrinas de Bisayas y Tagalos, y que si algunas se ha de hacer dejación, se dejen las de Tagalos (1690)" (ARSI, Filipinas 12 (1660–1729), f. 138r-148v). See also H. de la Costa, *Jesuits in the Philippines*, pp. 517–21.

188. In 1556, when Francis Xavier had already died, John III, King of Portugal, ordered the Viceroy of Goa to collect testimonies for the opening of an informative process. The death of the king (1557) caused a series of delays, often attributable to the skepticism of the Jesuits themselves about the alleged miracles. It was not until the appointment of Cardinal Pamphili as promoter of his cause that his beatification began, first (Paul V, In sede principis, 1619) and later his canonization (Gregory XV, 1622) (Jean-Robert Armogathe, "La fábrica de los santos. Spanish causes and Roman processes from Urban VIII to Benedict XIV (17th-18th centuries)", in Marc Vitse (eds.), *Homage to Henri Guerreiro. La hagiografía entre historia y literatura en la España de la Edad Media y del Siglo de Oro*. Madrid: Universidad de Navarra & Vervuert & Iberoamericana, 2005, p. 155

189. Born in Septembrer 14, 1543, and a member of the Society since 1567, Claudio Acquaviva was the fifth Superior General of the Society from 1581 to 1615. He had determined that once missions turned into parishes, they could be ministered for 20 years, after which they should be turned over to their dioceses's Archbishop, and the Jesuit priests were to found new ones or revitalize old ones (Francisco de Borja Medina Rojas, "Métodos misionales de la Compañía de Jesús en América hispana y Filipinas." *Mar Océana. Revista del Humanismo Español e Iberoamericano*, 4, 1999, p. 189).

From there, he had informed General Goswin Nickel[190] on the state of the Philippine province, carrying out incursions "to the mountains of the island of Mindoro and to the mountains called San Pablo in the same island of Manila."[191] In his view, superiors were too busy administering the doctrinas, and thus neglected the non-evangelized, such as the natives of Burney (Borneo) as well as those in the archipelago known as the Ladrones Islands.[192] This obviously went against the primary objective of a missionary, which was to convert infidels to the true faith.

The dangers of ministering parishes were well known: priests relaxed their discipline, and this was followed by scandals regarding affairs of chastity, poverty, trade for profit, etc.[193] Moreover, Father Jaramillo pointed out that missionaries who were supposed to be evangelizing infidels had nothing to do with priests who lived "tied to their parishes," where:

> ..., things are settled in the manner of the Indians, as is the case in Tagalos and the Visayas, where there is barely any hope of acquiring more progress beyond what is now enjoyed, for the dangers of dying and the reasons for infidelity are remote, and in the most faraway places innumerable days go by without the people asking for a single ministry, and the minister is thus forced to appeal for them himself to keep from falling in idleness and its evils.[194]

Like Fathers San Vitores and Morales, procurator Jaramillo was a clear defender of the missionary model in Asia, and therefore, in the Philippines. However, this did not mean that he thought the Jesuits were incapable or unfit for the administration of a parish. On the contrary, he argued that the Jesuits were better *doctrineros* for the Indians than secular priests or friars. For one

---

190. Born in 1582, Goswin Nickel was the tenth Superior General of the Society, from 1652 to 1664.

191. ARSI, Filipinas 12 (1660–1729), f. 169r.

192. F. de Borja Medina, "Métodos misionales de la Compañía de Jesús," p. 189.

193. Father Jaramillo pointed out the dangers of doing business with the large amount of wax that was collected in the parish churches of the Visayas, where "the *alcaldes* become rich, and if the ministers or priests want to join the trade, they also acquire much profit." Although he recognized that by keeping these doctrines they were exposed to accusations of responding to "matters of greed," he argued that "the reasons given, that matter so much to religion, are what should guide us, and not what people may say, even if it concerned the King himself" (ARSI, Filipinas 12 (1660–1729), f. 181r). Dominicans, particularly Friar Alonso de las Huertas, had accused the Jesuits of amassing wealth in stores of cotton and other products of their parishes or *reducciones* [reductions]. In their defense, the Jesuits argued that gold and silver were lacking in the Philippines, and for this reason, the natives paid their tributes in kind (Pedro de Espinar, S.J., *Manifiesto jurídico defensorio*, f. 8v).

194. ARSI, Filipinas 12 (1660–1729), f. 172v.

thing, Jesuits were chaster and better prepared. Their studies certainly did not prepare them for becoming parish priests to the Indians, but if they were obligated by circumstances, Jaramillo argued, they had no problem in administering them for the good of the souls.[195] Jesuits were also more self-sacrificing and long-suffering than friars, who were so, "destitute of the fear of God and of the pangs of conscience, that they only ministered well when they are close to Manila, and not in the Visayas."[196] Finally, the Society destined more ministers to these doctrines than the rest of the secular orders, even though the Bishop of Cebu, Friar Diego de Aguilar (1680–1692), supported his confreres.[197]

On the other hand, the parishes of Tagalos and the Visayas operated as missionary schools, in the image of those of Juli[198] and Santiago del Cercado[199] in Peru, where the Jesuits learned the native languages and customs. But, the procurator argued, if forced to, the Society could do without these curates, just as it had done without those of "Suaraga [Panay] in the Visayas; Cainta; Mariquina; a village beyond Antipolo; Siao; Terrenate; and Zamboanga," without having their missionary spirit and work reduced in the least.[200] Contrary to Father López, who defended the Visayas as platforms from which to revive "old Christianities," such as those in Zamboanga, Terrenate and Siao, and launch new ones, Jaramillo argued:

> I do not see those new expeditions, nor any succor being offered to those supposed Christianities, and I do not think that any will be undertaken, if the king does not devote his money and his weapons to them.[201]

---

195. In 1700, Jaramillo would repeat this argument in "Memorial presentado por los procuradores Mimbela, San Agustín y Jaramillo al rey Carlos II (1700)," ff. 4r-4v. DIGIBUG: Repositorio Institucional de la Universidad de Granada, Fondo Antiguo, Siglo XVII. [Available online] <http://hdl.handle.net/10481/5187> [Consulted March 2, 2024].

196. ARSI, Filipinas 12 (1660–1729), f. 177v. Thyrsus Gonzalez used this argument regarding the worthiness and probity of Jesuit parishioners in a memorial he sent to the King in the early eighteenth century ("Memorial al Rey, del P. Tirso González, Prepósito General de la Compañía de Jesús sobre el paso de Misioneros a Indias," RAH, Colección Jesuitas, Tomo 9/3714, ff. 26-27).

197. Indeed, San Vitores considered that so many Jesuit priests were destined to parishes, that the Society's missionary capacity was reduced (F. de Borja Medina, "Métodos misionales de la Compañía de Jesús," pp. 189–90).

198. For a study of the parish of Juli, in Chucuito (Puno, Peru), see Norman Meiklejohn, "Una experiencia de evangelización en los Andes: los jesuitas de Juli (Perú): Siglos XVII-XVIII." *Cuadernos para la historia de la evangelización en América Latina*, no. 1, 1986, pp. 109–85; and N. Meiklejohn, *La Iglesia y los lupaqas durante la Colonia*, Cuzco, Peru: Colegio de Estudios Regionales Andinos "Bartolomé de Las Casas," 1988.

199. For a study of the parish of Santiago del Cercado (Lima, Peru), see Alexandre Coello de la Rosa, *Espacios de exclusión, espacios de poder: el Cercado en Lima colonial (1568–1606)*, Lima: Pontificia Universidad Católica del Perú & Instituto de Estudios Peruanos, 2006.

200. ARSI, Filipinas 12 (1660–1729), f. 182r. These parishes were placed under the control of the Augustinian Recollects.

201. ARSI, Filipinas 12 (1660–1729), f. 178r.

In sum, he argued for the conservation of all *doctrinas*, including those in the Marianas, where the revolt of the presidium soldiers (from May to August of 1688) threatened the mission's stability.[202] According to the report written by Father López to General González, Jaramillo's opinion in this regard was emphatic:

> ... the Marianas ... cannot [be abandoned], because they are in extreme need, and the Faith is not yet fully established in them, and nobody else wants these missions and the immense work that they entail; and beacause we are extremely necessary in said Mariana[s'] missions, [and] it seems that God concurs for He prevents us from falling.[203]

If they were to keep only some of them, he preferred those of Tagalos to those in the Visayas, for the latter were too far from the capital.[204] Father López disagreed in this regard, for he insisted that keeping the *doctrinas* in the Visayas favored the reestablishment of the presidium and mission of Zamboanga in the island of Mindanao, which had been abandoned in 1663 by orders of the governor, who had recalled the soldiers to defend the capital from a hypothetical Chinese invasion.[205] Indeed, San Vitores also believed that the Jesuits should direct their efforts to the conversion of the "moors" of Mindanao.[206] Moreover, López argued that unlike the doctrina of Tagalos, where the natives lived in fixed settlements in the outskirts of Manila and could therefore be considered ordinary parishes, the natives of the Visayas lived in small villages across various islands. López stated that this kind of arrangement was better suited for the Jesuits than the one demanded by parishes (*stabilitas loci*), and his conclusion was that if they had to leave the doctrinas, they should nonetheless retain those

---

202. For more on this revolt, see the letter sent by Father Diego de Zarzosa to Father procurator Antonio Jaramillo, Hagåtña, May 22, 1689 (R. Lévesque, *History of Micronesia*, vol. 9, 1997, p. 298).

203. ARSI, Filipinas 12 (1660-1729), f. 178v.

204. H. de la Costa, *Jesuits in the Philippines*, p. 540.

205. Indeed, in 1690, Father López asked the Jesuit provincials in the Philippines to visit the Visaya doctrines and make sure that the Indigenous people were properly indoctrinated, and the churches had the necessary implements to celebrate mass and the sacraments. "Memorial del padre Alejo López, sustituto de procurador para la Compañía, a nuestro General Tyrso González sobre algunas cosas que juzga conviene se manden ejecutar en la provincia de Filipinas" (1690), in ARSI, Filipinas 12 (1660-1729), f. 134v.

206. San Vitores had strongly censured the abandonment of the fort of *Nuestra Señora del Pilar* in Zamboanga, considering that this left the Christian converts exposed to the pressure and violence of the moors, and thus in danger of apostasy (F. de Borja Medina, "Métodos misionales de la Compañía de Jesús," p. 190). The Jesuits did not return to Zamboanga until 1718 (Miguel Saderra Massó, *Misiones jesuíticas de Filipinas (1581–1768 y 1859–1924)*, Manila: IF, 1924, p. 35; H. de la Costa, *Jesuits in the Philippines*, pp. 540–41).

in the Visayas, under their full authority.[207]

Established in the Viceroyalty of New Spain (1690–1697), Father Morales adamantly defended the conservation of the Jesuits' Philippine missions, including the Marianas. He adopted the theses set forth by procurator Father Pedro de Espinar (1675–1681),[208] who for his part had accepted Morales's arguments for the conservation of the Marianas archipelago in the dispute in 1684 against the Dominican friars Cristóbal de Pedroche and Antonio de las Huertas, who defended Archbishop Pardo's theses.[209] It was procurator Espinar who gave Morales the instructions that General Thyrsus González had issued on June 11, 1689, in Rome, in which he asked him to explore the views held by the religious orders in the Philippines regarding the abandonment of their parishes. Procurator Morales received these instructions on July 16, 1690, and he himself wrote that it would be unwise to abandon the doctrinas in the Philippines and the Marianas to the friars. He also rejected submitting them to the ordinary jurisdiction of the bishops, arguing that the Council of the Indies should, on the contrary, acknowledge the independence of the Society of Jesus given the "special reasons [of] our institution and forms of government."[210]

Not much is known about Father Morales's activities during his Mexican stay.[211] However, there is evidence that he continued looking out for the

207. "Memorial del padre Alejo López, sustituto de procurador para la Compañía, a nuestro General Tirso González sobre algunas cosas que juzga conviene se manden ejecutar en la provincia de Filipinas" (1690) (ARSI, Philip. 12 (1660–1729), ff. 132r-138v.); "Papel que dio a N. P. General el padre Alejo López, sustituto de Procurador, que conviene retener las doctrinas de Bisayas y Tagalos, y que si algunas se ha de hacer dejación, se dejen las de Tagalos (1690)" (ARSI, Filipinas 12 (1660–1729), f. 138r-148v). See also H. de la Costa, *Jesuits in the Philippines*, pp. 517–21.

208. H. de la Costa, *Jesuits in the Philippines*, pp. 435–36.

209. Antonio de las Huertas, OP [Order of Preachers: Dominicans], *Reparos al Memorial que estampado ha publicado el padre Luis de Morales, de la Compañía de Jesús, que sobre su contenido hace fray Antonio de las Huertas, del orden de los predicadores*, Biblioteca de la Universidad de Sevilla (ca. 1680), Fondo Antiguo, A 096/082(05), ff. 1r-7r; P. de Espinar, S.J., *Manifiesto jurídico defensorio*, 17 ff. Copies of these documents are available at the University of Santo Tomas Archives (Manila, Philippines), Books, Lib. 15.4 y 15.5. "Notas o reparos del P. Fr. Antonio de las Huertas a un Memorial del P. Luis Morales, SJ", ff. 167-172v; "Respuestas a los Reparos del P. Huertas por el P. Pedro de Espinar, SJ", ff. 173-189v).

210. "Instrucciones del General Tirso González a los procuradores Andrés de Espinar y Luis de Morales, con fecha en Roma, 11 de junio de 1689; Mexico, 16 de julio de 1690" (ARSI, Filipinas 12 (1660–1729), f. 249r). In 1691, Jaramillo wrote new letters and memorials in which he repeated his accusations against Archbishop Felipe Pardo, who had passed away on December 31, 1689 ("Señor. Antonio Xaramillo, de la Compañia de Jesus, Procurador General por la provincia de Filipinas; dize, que las continuadas molestias, que su religion ha padecido, y padece en aquella islas, assi en el comun, como en muchos de sus individuos...," Universidad de Sevilla, Fondo Antiguo, Signatura: A 109/094(08); A. M. Jaramillo, *Memorial al Rey Nuestro Señor con varios reparos*, ff. 1r-28r.). On this particular issue, see Alexandre Coello de la Rosa, "Pasquines, libelos y corrupción: los conflictos jurisdiccionales entre el arzobispo de Manila, fray Felipe Pardo y la Compañía de Jesús (1677-89)," Colonial Latin American Historical Review (CLAHR), vol. 1, n° 2 (2013), pp. 113-45.

211. His name appears in the Society's Catalogue of the Mexican Province of April 23, 1690, as a member of the college (residence) of San Ildefonso (founded in 1574) ARSI, Mexico 6, "Catalogus Alphabeticus

Marianas' mission.[212] Moreover, he followed the apostolic progresses in the archipelago closely, gathering information and concluding the *Historia de las Islas Marianas*.[213] In 1690, he wrote the king asking him to send the stipend that corresponded to the archipelago's *situado* of that year, which ascended to 5,425 pesos, as well as the 5,000 pesos that corresponded to the two boys and girls seminaries that operated in Guåhan, reminding His Majesty of the great necessities that afflicted His subjects in the archipelago.[214] It should not be surprising, then, that on April 18, 1691, he wrote a letter to the Duchess of Aveiro, one of the main benefactors of the Marianas mission, congratulating her for the birth of her second grandchild, and using the occasion to tell her of the deplorable state of the Philippines, where violence and greed ran rampant, and the number of missionaries was insufficient.[215] Father Morales evidently hoped that the Duchess would again send a monetary stipend to help missionaries with their work.

In 1693, Father Morales was transferred to the San Andres house of probation, which served as a residence, juniorate, an exercises retreat, and a space for procurators for the missions of the North and the Philippines.[216] In early March of 1697, before leaving Acapulco for Manila, Father Morales had finished the *Relación* that described the tragic events of the Second Chamorro War.[217] He had enough time to complete the *History* translated in this text,

Personarum Provincia Mexicana. 23 de abril de 1690," f. 35r. The students of the Colegio Máximo de San Pablo in Mexico City were lodged in the residence of San Ildefonso (Elsa Cecilia Frost, "Los colegios jesuitas," in Antonio Rubial García, ed., *Historia de la vida cotidiana en México, tomo II: La ciudad barroca*, Mexico: Fondo de Cultura Económica, 2005, p. 311).

212. For instance, in 1693, Father Morales challenged General Joseph Madrazo, future interim Governor of the Marianas (1696–1699), for wanting to load the *situado* and the cargo destined to the Marianas in the hold, instead of on deck, claiming that this could compromise their proper unloading in the islands (R. Lévesque, *History of Micronesia*, vol. 10, 1997, pp. 582–583).

213. ARSI, "Catalogus Brevis Personarum Provinciae Philippinarum. Anno 1690, 1694 y 1695." Philipp. Catal. Trien. Tomo 4, ff. 89r; 91r; 93v.

214. Archivo General de la Nación (henceforth, AGN), Mexico, Instituciones Coloniales, Indiferente Virreinal, Sección Jesuitas, Caja 4866, Exp. 027, ff. 1-2. In 1692, Father Morales wrote another memorial to the King, denouncing that, "what had been decreed had not been executed, so that those who were supposed to execute it betrayed their duty, causing great harm to the missionaries and soldiers who should be treated with the utmost compassion and care" (AGN, Mexico, Instituciones Coloniales, Gobierno Virreinal, Sección Californias–017, Vol. 26, Expediente 37, ff. 206r-207v). On January 21, 1694, Morales again asked that the situados to the Mariana Islands of 1693 and 1694 (63.600 total pesos) be liberated. (AGN, Mexico, Instituciones Coloniales, Indiferente Virreinal, Sección Filipinas, Caja 790, Exp. 021, f. 1v-6v).

215. *Bibliotheca Americana et Philippina*, Part III. Catálogo no. 442. Maggs Bross, 1923, p. 241.

216. ARSI, Fondo Gesuitico 630-c, "Catalogus Provinciae Mexicane Societatis Iesu. 2 de mayo de 1693." f. 88r; ARSI, Mexico 6, "Catalogus Provinciae Mexicana Societatis Iesu. 30 de mayo de 1696," f. 93r; 118r; ARSI, Mexico 8, "Catalogus Provinciae Mexicana Societatis Iesu. 2 de mayo de 1693," f. 326r-328r. See also Gerard Decorme, S.J., *La obra de los jesuitas mexicanos durante la época colonial, 1572–1767*, Tomo I. Fundaciones y obras, México, Antigua Librería Robredo de José Porrúa e Hijos, 1941, p. 92.

217. In 1692, Domingo Abella confirmed the opinion of bibliographers' José Toribio Medina, Wenceslao E. Retana, and Pedro Vindel regarding the authorship of the *Relación de 1684*, which was not signed nor addressed to anybody. He argued that it had been written by Father Morales between 1689 and 1690

which began with the arrival of Father San Vitores in 1668, to the last events that happened in 1695. It is very likely that he was asked to write it by the General of the Society before leaving Europe for New Spain.[218] Once in Mexico, he found the last reports of the mission, which were incorporated into the final draft. He then must have sent the original to Father Juan de Palacios, provincial of New Spain, who remitted it to Father General González.[219]

On June 23 of 1697, Father Morales was in the island of Guåhan, where he met with the mission's superior and told him that the book had been completed.[220] Before then, on May 5, 1697, the Dominican, Franciscan, and Recollect provincials, along with the Vicar Provincial of the Calced Augustinians, the Vice-Provincial of the Jesuits, and the Vicar General of the Order of San Juan de Dios, met in the Augustinian convent of San Pablo in Manila to secretly sign the agreement or *Concordia de las Religiones*, an 18-clause document that was supposed to strengthen the ties between the different religious corporations, safeguarding their common interests vis-à-vis external challenges. They undertook that when a papal brief or a royal decree affected the various orders in its execution, and they would have to meet and decide the course to follow. As for episcopal visitations, whether total or partial, they agreed to oppose it. In short, the orders agreed to settle their differences through arbitration.[221]

On September 15, 1697, the new Archbishop of Manila, Don Diego Camacho y Ávila (1697–1705),[222] took possession of his episcopal see, which

(D. Abella (ed.), *Vignettes of Philippines-Marianas Colonial History,* Pamphlet no. 1, Manila: International Association of Historians of Asia, 1962, pp. 41–42).

218. According to Murillo Velarde, "There [in New Spain], [Morales] happily dispatched the affairs of this Province, and of the Marianas' mission" (transcribed in R. Lévesque, *History of Micronesia,* vol. 12, 1998, p. 47).

219. Since *Historia de las islas Marianas* was not going to be published in New Spain, we suppose that he did not need a licence from the Viceroy Gaspar de la Cerda y Mendoza, Count of Galve, nor from the Holy Inquisition, to write it. Therefore, these institutions had probably no awareness of this text's existence, and therefore did not censor it, like they did with Father Francisco de Florencia's *Historia de la Provincia de la Compañía de Jesús de Nueva España* (1694).

220. R. Lévesque, *History of Micronesia,* vol. 10, 1997, pp. 133–135.

221. P. Rubio Merino, *Don Diego Camacho y Ávila,* pp. 114–15; H. de la Costa, *The Jesuits in the Philippines,* p. 524; Marta María Manchado López, "La "Concordia de las religiones" y su significado para la historia de la iglesia en Filipinas," in Florentino Rodao (coord.), *España y el Pacífico,* Madrid: Agencia Española de Cooperación Internacional (AECI) & Asociación Española de Estudios del Pacífico (AEEP), 1989, pp. 65–79; Descalzo Yuste, *La Compañía de Jesús en las Filipinas (1581–1768),* pp. 225–226.

222. P. Rubio Merino, *Don Diego Camacho y Ávila,* p. 105. Don Diego Camacho y Ávila (1652–1712) had been canon of the Cathedral of Badajoz before this appointment (P. Rubio Merino, *Don Diego Camacho y Ávila,* pp. 78–80). He was consecrated as Archbishop of Manila in the Cathedral of Puebla de los Ángeles (Mexico) on August 19, 1696, but he did not leave for Manila until March 25, 1697. He remained in this office until 1705 (having been relieved in 1704, and substituted by Francisco de la Cuesta, OSH) and was sent to New Spain again as Bishop of Guadalajara (Rubio Merino, *Don Diego Camacho y Ávila,* pp. 103–05; M. M. Manchado López, "La Concordia de las religiones," pp. 65–66).

was vacant since 1689. The galleon that carried him stopped in Guåhan on June 23, before reaching the port of Cavite. The Archbishop visited the island then; and again on June 25, where he saw for himself that a great number of soldiers were dedicated to private agricultural activities whose benefits accrued not to the community, but to the personal gain of General Joseph de Madrazo, interim governor of the archipelago since 1696.[223] Meanwhile, there was still a large population in the northern isles that had not been reduced to the main islands of Guåhan, Rota, and Saipan.

Upon Camacho's arrival in Manila, his impression was not good either. He was a strict and righteous man, zealous of his authority, whose first priority was promoting morality.[224] With this in mind, on December 16, he embarked upon an intense reorganization of his dioceses, initiating a pastoral visit *in officio oficiando,* that is, in virtue of his episcopal authority, in five churches administered by regular clergy to know in situ the ministers who watched over the flock of God. He visited Toledo first (on December 17, 1697), which was administered by the Augustinian friars, who, according to Rubio Merino and Manchado López had so resisted his authority, that the visit failed. He then went to Binondo unannounced (on December 25, 1697), taking the Dominican friars by surprise, with the purpose of naming an interim parish priest, causing such a serious conflict that Governor Fausto Cruzat y Góngora (1690–1701)[225] had to intervene.

Like the superiors of the other orders, Father Morales, who was by then provincial of the Philippines and (in 1703) rector of the College of Manila,[226] reacted against the violence exercised by the prelate, who, "violently broke the doors of those two churches, while soldiers and secular ministers sieged them, carrying shackles as if these places were prisons that housed delinquents or

223. "Diego Camacho y Ávila, arzobispo de Manila, escribe a S.M. de su llegada a las islas Marianas y estado de estas," *San Joseph* Galleon, July 10, 1697 (AGI, Filipinas 15, 1r-3v); "Diego Camacho y Ávila, arzobispo de Manila, da cuenta a S.M. de su visita a las islas Marianas y de la necesidad de más soldados y misioneros allí," Manila, January 18, 1698 (AGI, Filipinas 17, Ramo 4, ff. 1r-2v). General Joseph de Madrazo was governor until 1699.

224. Mª Lourdes Diaz-Trechuelo López-Spínola, "Filipinas en el siglo de la Ilustración," in Leoncio Cabrero (coord.), *Historia General de Filipinas,* Madrid, Ediciones de Cultura Hispánica, 2000, p. 272.

225. P. Rubio Merino, *Don Diego Camacho y Ávila,* pp. 144–51; M. M. Manchado López, "La Concordia de las religiones," p. 74; and M. M. Manchado López, *Conflictos Iglesia-Estado en el extremo oriente ibérico: Filipinas (1767–1787),* Murcia: Universidad de Murcia, 1994, p. 25; Descalzo Yuste, *La Compañía de Jesús en las Filipinas (1581–1768),* pp. 226–227.

226. He features as Rector on February 4, 1703. ARSI, Philipp. Catal. Trien. 3, "Primus Catalogus Personarum Anno 1707," f. 24v. The *Catalogus Brevis Perssonarum Provinciae Philippinarum Societatis Iesu,* of October 9, 1699, is signed by Father Ludovicus Morales (ARSI, "Catalogus Brevis Personarum Provinciae Philippinarum Anno 1699." Philipp. Catal. Trien. Tomo 4, ff. 98r-100v).

{Figure 4: *Insula Palaos Seu Nova Philippina*. Probst, Palau (1748). Courtesy of the Micronesian Area Research Center (MARC)} criminals."[227] Right after, he removed the Society's priests from their parishes, informing Governor Cruzat.[228] Manchado López points out that the religious orders' resistance revolved around two strategies. First, they threatened to abandon their parishes and assume a conventual life. This was what procurators López and Jaramillo did at the Council of the Indies when they were defending the Society from the attacks of Dominican procurator friar Raimundo Verart, Archbishop Pardo, and the "Dominican clan."[229] Such a course of action would have dramatic consequences, for, "in this land where most towns are only of Indians, and the only Spaniard in them is the religious minister of the provinces and the divine cult and vassalage, the towns would be so dead without their religious ministers, as a body without a soul for its vital actions."[230] The second was writing letters and memorials to the king and the pope manifesting their disagreement with the archbishop's policies, which would reduce the privileges and exemptions of the regular clergy and augment episcopal power by establishing control over the entire jurisdiction.[231]

The regular friars' resistance to being visited by the archbishop was the norm in the Indies, contravening the dispositions of Trent and the papal bulls that followed. Episcopal visits were meant to make sure that priests did not commit abuses or mistakes in their administration of the parish. But the problem was that unlike secular priests whose superior was the bishop of their jurisdictions, regular priests were part of distinct religious institutions with their own hierarchies and superiors.[232] Since there were not enough secular priests to take over the parishes and minister the sacraments and holy offices, members of the religious orders, as an act of "charity", had temporarily assumed the duties

---

227. "Memorial presentado por los procuradores Mimbela, San Agustín y Jaramillo al rey Carlos II (1700)," f. 8r.

228. Between 1691 and 1697, the governor carried out the first *visitas de la tierra*. Soon after his arrival in Manila (1691), he asked the oldest *oidor* of the Audiencia and the interim governor, Don Alonso de Abella Fuertes to inspect the provinces of northern Luzon (Cagayan, Panganisan, Ilocos, and Pampanga), watching out especially for irregularities related to tribute collection (Nicholas P. Cushner, S.J., *Spain in the Philippines. From Conquest to Revolution*, Quezon City, Philippines - Vermont - Tokyo: Ateneo de Manila University & Charles E. Tuttle, 1971, p. 108; Antonio Molina Memije, *América en Filipinas*, Madrid, Fundación Mapfre América, 1992, pp. 131–32).

229. A. M. Jaramillo, *Memorial al Rey Nuestro Señor con varios reparos...*, ff. 26-27.

230. "Memorial presentado por los procuradores Mimbela, San Agustín y Jaramillo al rey Carlos II (1700)," f. 1r.

231. M. M. Manchado López, "La Concordia de las religiones," p. 68. Martínez-Serna argues that "many bishops, conscious of canonical prerogatives and potential tithes–from which the Company was also exempt– often sided with the secular state against the Jesuits. Feuds between bishops and the Jesuits were resolved on the part of the Society by its procurators, who acted in the name of their provincial" (J. G. Martínez-Serna, "Procurators and the Making of the Jesuits' Atlantic Network", p. 189).

232. P. Rubio Merino, *Don Diego Camacho y Avila*, pp. 227–28.

of the secular clergy. Theirs, they argued, was a special situation borne out of their generosity and therefore not subject to ordinary jurisdictional norms.[233] Be that as it may, Archbishop Camacho indeed did not have enough parish priests to replace recalcitrant friars,[234] and he had no option but to suspend the ecclesiastical visits and return the parishes to the regular clergy to continue administering "temporarily".[235]

In early 1698, hoping to divide the signatories of the *Concordia*, the Archbishop offered the Jesuits half of the parishes administered by the other religious orders. Secular priests would administer the other half. Since the Jesuits refused, he wrote to the king on January 19, 1698, complaining about the uncomfortable situation he found himself in. He renounced the archbishopric and, "asked to be allowed to come and die, retired in a cell."[236] However, as the religious orders' superiors suspected, he had other plans in mind.[237] Not only did he not renounce his miter, he in fact battled the orders again, this time in the juridical field. In 1699, the Ecclesiastical Chapter of Manila wrote a letter to the king on June 27, 1699, informing him of the prelate's difficulties in imposing his episcopal authority over the orders. The writers underlined, "how absolute regulars are in these parts, for it seems that they do not even acknowledge Your Majesty and His Royal Laws, nor the first seat with its decrees, wanting to be so independent that what they want, they execute, even if it is in detriment of Your Majesty's sovereignty."[238]

Informed of the content of this letter, the order's superiors asked its procurators in the courts of Madrid and Rome to finally solve the matter of the parishes. In early 1700, the Council of the Indies was studying the docu-

---

233. M. M. Manchado López, "La Concordia de las religiones," pp. 68–69.

234. As Bangert pointed out, Archbishop Camacho had only 53 secular priests, with whom he could not counterbalance the power of the regular priests (Bangert, *A History of the Society of Jesus*, pp. 196, 356).

235. This interim character was recognized by the King through the Real Cédula of December 30, 1696 that was sent to Philippine Governor Don Fausto Cruzat y Góngora. An unintended consequence of this disposition was that young men in the archipelago were discouraged from joining the orders. Moreover, when Charles II asked Governor Cruzat in 1697 about the erection of a seminary in Manila, ordered by a 1677 Real Cédula, he was told that this seminary had never been constructed, precisely because the Catholic orders opposed it (P. Rubio Merino, *Don Diego Camacho y Avila*, p. 231; Lucio Gutiérrez, *Historia de la iglesia en Filipinas*, Madrid: Fundación Mapfre, 1992, pp. 207–08).

236. "Memorial presentado por los procuradores Mimbela, San Agustín y Jaramillo al rey Carlos II (1700)," f. 2r. See also P. Rubio Merino, *Don Diego Camacho y Avila*, p. 242; M. M. Manchado López, "La Concordia de las religiones," pp. 73–74.

237. "Memorial presentado por los procuradores Mimbela, San Agustín y Jaramillo al rey Carlos II (1700)," f. 2r.

238. AGI, Filipinas 294, "Carta del Cabildo Eclesiástico sobre visitas de las doctrinas por el arzobispo (Camacho)," Manila, 27 de junio de 1699. N. 8, ff. 1r-2v).

mentation submitted by the Archbishop and the procurators.[239] Dominican procurator Friar Jaime de Mimbela; the Augustinian Recollects' procurator, Friar Juan Antonio de San Agustín; and Jesuit procurator Father Antonio Matías Jaramillo met in Madrid and wrote a memorial on February 13 of that year that appealed directly to the king.[240] This was probably Jaramillo's last act as procurator to the court, for on May 10 of 1701, he was back in Manila with Father Morales, awaiting the conflict's resolution.[241]

While these matters were debated in Madrid and Rome, Jesuits in the Philippines continued looking for new islands to evangelize and spill their own blood for Christ. Marianas Governor Joseph de Madrazo organized a fleet of 112 ships commanded by Captain Sebastian Luis Ramón to reduce the natives of Gani—the name given to the northern islands of the Marianas archipelago— to the three main islands of Rota, Saipan, and Guåhan. This operation took eight months.[242] Once they had submitted to the *Pax Hispanica*, the Marianas' natives were under the definitive control of thirteen Jesuit priests, four coadjutor brothers, and an oblate, whose authority was secured by a relatively large number of soldiers and officers. This was precisely what Archbishop Camacho had recommended when he had visited the islands in 1697.[243] Wars, epidemics (1671; 1684; 1695), migration, and natural disasters (typhoons of 1671 and 1693) had truly decimated the Mariana Islands' population. But Fathers Luis de Morales and Charles Le Gobien both chose to downplay this fact, concluding their *History of the Mariana Islands* (Paris, 1700) with the hope that the news that came from Palau would promote even more missions across the Pacific:

> A new field for the preaching of the Gospel has been opened. More than thirty islands have been disocered to the south. Nobody knows their extension or size, but they are sure that there is an infinite number of people who do not know Jesus

239. P. Rubio Merino, *Don Diego Camacho y Ávila*, pp. 242–44.

240. "Memorial presentado por los procuradores Mimbela, San Agustín y Jaramillo al rey Carlos II (13/2/1700)," ff. 1r-10v.

241. ARSI, "Secundus Catalogus Personarum Provincie Insularum Philippinarum Anno 1701." Philipp. Catal. Trien. Tomo 3, f. 13-14r.

242. ARSI, "Relación de la última reducción…," Philipp. Suppl. 1584–1750, Tomo 14, f. 91v. See also A. Astrain, *Historia de la Compañía de Jesús*, Tomo VI, 1920, p. 834; Vicente Muñoz Barrreda, *La Micronesia española o los archipiélagos de Marianas, Palaos y Carolinas*, Manila: Tipografía "Amigos del País," 1894, p. 70; M. G. Driver, *Cross, Sword, and Silver*, pp. 44–45; D. A. Farrell, *History of the Northern Mariana Islands*, p. 176.

243. "Diego Camacho y Ávila, arzobispo de Manila, da cuenta a S.M. de su visita a las islas Marianas (23-25 de junio de 1697) y de la necesidad de más soldados y misioneros allí," Manila, January 18, 1698 (AGI, Filipinas 17, Ramo 4, f. 2r).

Christ and who have never heard of him. You can read how this discovery was made in the letter added at the end of this History. All that is needed are saints and fervent missionaries to go in the name of Jesus Christ to take possession of these infidel lands and bring the light of the Gospel to these people buried for so many centuries in the deepest darkness of paganism. *Messis quidem multa operari* (folio 394) *vero pauci. Rogate ergo dominum messisut mittst operarios in messem suam.* [The harvest is abundant, but there are few workers; pray, then, so that the owner of this harvest sends workers to his field].[244]

And so it was. In December of 1696, a group of twenty-nine starving men, women, and children from the Caroline Islands arrived in two small vessels at Guiguan, "which is close to the promontory or Punta formed by the island of Samar y Babao [Eastern Samar], place of the old Society's missions."[245] There, Father Paul Klein (or Pablo Clain)[246] met with them and, determined to initiate these islands' spiritual conquest, he wrote to the Society General Thyrsus González on June 10, 1697, relaying the information that was then known of these islands and stressing the necessity of sending missionaries to them. For his part, the General informed Pope Innocent XII, and the French and Spanish kings, Louis XIV and Philip V, hoping to gain their support in the enterprise.[247]

---

244. C. Le Gobien, *Histoire*, book X, pp 393–394. My translation.

245. Andrés Serrano, S.J., *Noticia de un gran número de islas de gentiles, por nombre País o Palaos, muy pobladas de gente de muy dócil natural, y sin especie, que se haya conocido de idolatría, ni mezcla de la infame secta de Mahoma...* (RAH, Fondo Jesuitas, Tomo 149, 9/3722, Documento 22, f. 1r-4v). Most of Serrano's information is based on a letter written by Father Klein on June 10, 1697 to Society General Thyrsus Gonzalez, which included a map (transcribed in R. Lévesque, *History of Micronesia*, vol. 10, pp. 95–105).

246. Paul Klein was born in Bohemia (Czech Republic) on January 25, 1652. He joined the Society of Jesus on September 16, 1669, and professed the four vows on February 2, 1685. Father Klein died in Manila on August 30, 1717 (ARSI, Philipp. 3, "Primus Catalogus Anni Personarum Anni 1716." Philippinae Cat. Trien. 1701–1755 1768, Tomo 3, f. 77r; "ARSI, Philipp. 3, "Suplementum Catalogorum Provincia Philippinarum Societatis Iesu a die 17/7/1713 usque ad diem 21/07/1719," f. 113r). On September 8, 1675, while he was in the College of Prague, he wrote a letter to the General of the Society expressing his desire to become a missionary in the East Indies (ARSI, Index Eorum Quorum Epistolae Indipetae, "Extra "F. G.," Adservantur, Vol. 756, no. 119). He renewed this request to the General on January 12, 1678 (ARSI, Index Eorum Quorum Epistolae Indipetae, "Extra "F. G.," Adservantur, Vol. 756, no. 125). He was provincial of the Philippines from 1708 to 1710 (R. Lévesque, *History of Micronesia*, Vol. 10, p. 576). In 1708, he wrote a pharmaceutical text entitled *Remedios fáciles para diferentes infermedades por el P. Pablo Clain de la Compañía de Jesús para el alivio, y Socorro de los PP. Ministros Evangelicos de las Doctrinas de los Naturales* [*Easy Remedies for Different Illnesses by Father Paul Klein, S.J., to Assist Ministers Evangelising the Natives*] that circulated widely even before its publication in 1712 (Arnel E. Joven, "Colonial Adaptations in Tropical Asia: Spanish Medicine in the Philippines in the Seventeenth and Eighteenth Centuries," *Asian Cultural Studies* vol. 38 (2012), pp. 177–78 [Available online] <https://icu.repo.nii.ac.jp/records/2911> [Consulted March 2, 2024].

247. Pablo Klein's letter to General Thyrsus Gonzalez regarding the Carolinas; Manila, June 10, 1697 (transcribed in R. Lévesque, *History of Micronesia*, vol. 10, pp. 95–100). See also P. Murillo Velarde, S.J., *Historia de la provincia de Filipinas de la Compañía de Jesús*, ff. 375v-377r; A. Astrain, *Historia de la Compañía*

These diplomatic activities were successful. Provincial Morales authorized the mission, and in September of 1697, Governor Don Fausto Cruzat sent a galleon under the command of Dutch Jesuit Jacobus Xavier (or Jaime Javier)[248] which left Manila headed towards Samar, where it should have met up with a ship commanded by Father Francisco Prado. [249] This expedition failed, for the first vessel shipwrecked in the province of Leyte, and more than twenty natives died.[250] However, General Tirso González was determined to establish new missions in the Golden East.[251] On December 26, 1699, he wrote a letter to Father Antonio Tuccio[252] asking him about the islands of Palau, and encouraging him to, "find out the disposition in them regarding the introduction of the Holy Gospel."[253] Father Andrés Serrano[254] was appointed General Procurator to the court of Madrid, and he presented a memorial in 1706, asking for "a good number of missionaries who can cultivate that great field of the Lord which He has placed before us, to gather a most abundant harvest. "[255] The Council of the Indies demanded that the Jesuit missionaries sent to these islands commit themselves exclusively to missionary activities. Thus, they were forbidden from

---

de Jesús, Tomo VII, pp. 764–65; H. de la Costa, *The Jesuits in the Philippines*, pp. 549–550.

248. Jacobus Xavier was born on June 13, 1665. He joined the Society of Jesus in January 1668, becoming a temporal coadjutor in 1699. Brother Xaiver died in Manila on November 27, 1729 (ARSI, Philipp. 3, "Primus Catalogus Personarum Ano 1701," f. 8v; "Vertius Catalogus Rerum Provinciae Insularum Philipinarum Anno 1731," f. 192r). See also H. De la Costa, *Jesuits in the Philippines*, p. 618.

249. Francisco Prado was born in Badajoz (Extremadura) on October 4, 1650. He joined the Society of Jesus on August 5, 1668, professing the four vows in August 1687. Father Prado died in Manila on October 15, 1709 (ARSI, "Primus Catalogus Personarum Ano 1701." Philipp. 3, f. 3v; "Tertius Catalogus Rerum Provincia Insularum Philippinarum. Anno 1713," f. 73).

250. L. Ibáñez y García, *Historia de las islas Marianas*, pp. 162–63; L. Gutiérrez, *Historia de la iglesia en Filipinas*, pp. 262–63; Patricio Hidalgo Nuchera, (eds.), *Redescubrimiento de las islas Palaos*, Madrid: Miraguano Ediciones & Ediciones Polifemo, 1993, p. 10.

251. A. Coello, *Jesuits at the Margins*, chapter 4, pp. 177-221.

252. Antonio Tuccio was born in Messina (Sicily) on April 16, 1641. He joined the Society of Jesus on May 18, 1658. He left for the Philippines in 1672, and was appointed rector of the colleges of Cavite and Manila. On August 15, 1677, he professed the four vows. He was provincial of the Philippines in two different periods (from 1696 to 1699; and again from 1707 to 1708). According to Horacio de la Costa, he died in 1708 (H. de la Costa, *The Jesuits in the Philippines*, pp. 435; 553–554; 580). C. Sommervogel says he died on February 4, 1716 (*Bibliothèque de la Compagnie de Jésus*, Tomo 8, p. 265).

253. Andrés Serrano, S.J., *Noticia de un gran número de islas de gentiles*, f. 3v). See also *Breve Noticia del nuevo descubrimiento de las islas Pais, o Palaos, entre las Filipinas y Marianas, y del ardiente y fervoroso celo con que les promueven la santidad de N.M.S.P Clemente papa XI por sus breves apostólicos; el cristianísimo rey de Francia Luís XIII, el grande por su real carta; y nuestro piadosísimo y católico monarca Felipe V por su decreto y reales cédulas en Consejo de Indias* (circa 1705–07) (Biblioteca Nacional (henceforth, BNM), Sala Cervantes, Libros Filipinos de don Antonio Griaño, R/33089, 37 f. s).

254. Andrés Serrano was born in Murcia on October 9, 1655. He joined the Society of Jesus on June 13, 1670. He was a graduate and professed of the four vows (2/2/1689) (ARSI, "Primus Catalogus Anni Personarum Anni 1701." Philippinae Cat. Trien. 1701–1755 1768, Tomo 3, f. 4r).

255. "Memorial del padre Andrés Serrano al Felipe V (1706)" (AHCJPC, Filipinas C-285, Doc. 2, f. 1r; BNM, Sala Cervantes, VE/1465/3).

living in colleges and "being preachers, teachers or superiors," lest they lose the mission.[256] General Thyrsus González opposed these restrictions, arguing that they were contrary to the Society's institution and rules.[257] Father Juan Martinez de Ripalda, procurator of the province of New Granada and Quito supported his confrere, Father Serrano, insisting on the need to send missionaries regardless of their national origin to the Palau Islands for their natives "die in their infidelity because the Holy Gospel has not been preached to them."[258]

Until then, and throughout, provincial Morales sent periodical reports on the state of abandonment in which the Marianas found themselves to procurator Jaramillo. On June 9, 1700, Father Morales wrote a letter informing Father Jaramillo of the appointment of Don Francisco de Medrano y Asiaín as captain general and interim governor of the Marianas (from 1700 to 1704).[259] In Morales's own words:

> ...this man has a good age, prudence, experience, zeal, disinterestedness, and supports us [Jesuits]; and above all, he is a good Christian, and since those islands need a man like him, as I already wrote to you, and to Father Quirós, and I repeat it here, procure for him the property of government of the Marianas from His Majesty.[260]

However, to Father Morales's dismay, Governor Medrano's plans for the Marianas mission were quite different from his own. In 1701, Governor Medrano wrote to the king, suggesting the transfer of 2,600 Marianas' natives

---

256. "Memorial al Rey, del P. Tirso González, Prepósito General de la Compañía de Jesús sobre el paso de Misioneros a Indias" (RAH, Colección Jesuitas, Tomo 9/3714, ff. 26-27).

257. Ibidem.

258. "Memorial del padre Juan Martínez de Ripalda, procurador de las provincias de Indias al Rey, pidiendo que se permita mayor número de jesuitas extranjeros para la misión de Pais o Palaos (1704)" (AHCJPC, Filipinas C-285, Doc. 5, f. 7v). Father Martinez had argued the same thing regarding the Spanish American missions, asking in a memorial of July 5, 1702, that eight German Jesuits be sent to them without reticence, for the Jesuit superiors and rectors there were all Spanish. *DHCJ*, vol. IV, p. 2526. See also L. Clossey, *Salvation and Globalization*, p. 152.

259. This same letter also informed procurator Jaramillo that in 1698, the Bishop of Troy and Crown auditor, Don Juan de Sierra y Osorio had perished in the galleon that was taking him from Manila to Acapulco. Father Morales also said that Governor Cruzat was not only uninterested in the Marianas' mission, but that in 1698, he had decided not to send any pataches with provisions to the islands. Besides the two annual voyages from the Philippines to the American continent, there were important trade and travel routes between the Philippine Islands, as well as between Cavite and Guåhan. A patache with provisions was usually sent from the Philippines to the Marianas in April or May, although in 1699, the Governor sent the patache in July, which was storm season, and much was lost at sea (RAH, Fondo Cortes 567, 9/2669 46, ff. 2r-2v. See also Emma Helen Blair and James A. Robertson, *The Philippine Islands, 1493–1898*, Vol. 42, Mandaluyong, Rizal: Cacho Hermanos, 1973, pp. 207–08). At other times, as we can learn from Father Morales's letter, Philippine governors suspended the patache altogether.

260. RAH, Fondo Cortes, 9/2669/46, f. 1r.

to one of the Philippine Islands, arguing that their small number did not justify the maintenance of the presidium nor that of the Jesuit schools and houses. He also argued that the transfer would not be difficult, for in the last four years, the natives of Gani had been reduced without incident.[261] Most intellectuals and political experts did not share Medrano's opinion. Don Francisco de Seijas y Lobera (1650–1705), for instance, argued that if the Monarch optimized commerce between the Philippines and New Spain, repopulation would soon follow.

For Father Lorenzo Bustillo, commissary of the Holy Office, the difficulties associated with the Marianas mission were not an economic matter, but a pastoral one[262]—indeed, in the monarchical absolutism of the period, prosperity, politics, and the Christian faith were inseparable.[263] Sharing this concern, the Jesuits were not interested in abandoning the Marianas, but in using them to launch expeditions that would result in the foundation of new missions. Indeed, numerous priests, brothers, and auxiliaries had met violent deaths in the Islands, or had been transferred elsewhere (like Father Morales himself), but fewer and fewer missionaries were sent to replace them.[264] Procurator Andrés Serrano complained that it had been fourteen years since the last Jesuit missionaries had come to the Philippines, and that meant that many natives died without the consolation of having received the holy sacraments. In his view, King Philip V could not betray the obligations of evangelization contracted by his predecessors through the Spanish Patronato Regio, abandoning the natives of the Marianas and the surrounding islands to their fate. His duty as a temporal sovereign and spiritual vicar was to contribute to the salvation of souls and to prevent them from dying as infidels.[265]

The king did not disdain these concerns. Moved by these and similar exhortations, on December 13, 1701, he wrote to Father Juan de Palacios,[266]

---

261. ARSI, Litterae Annuae Philippinae, Tomo 13 1663–1734 (etiam de Insuli Marianis), ff. 326-332v.

262. AGN, Inquisición 543, Exp. 49, f. 435.

263. Pablo Fernández Albadalejo, *Fragmentos de monarquía*, Madrid: Alianza Universidad, 1992, pp. 60–85; P. Fernández Albadalejo, *Materia de España. Cultura política e identidad en la España moderna*, Madrid: Marcial Pons, 2007, p. 98.

264. Alexandre Coello de la Rosa, "Colonialismo y santidad: la sangre de los mártires (1668–1676)." *Hispania Sacra*, Vol. LXIII, no. 128, 2011, pp. 707–745.

265. Lorenzo Bustillo's letter to General Thyrsus Gonzalez, April 14, 1702 (ARSI, Philipp. 13. Litterae Annuae Philipp. 1663–1734 (etiam de Insuli Marianis), ff. 326r-326v; transcribed in R. Lévesque, *History of Micronesia*, vol. 10, 1997, pp. 337–339). See also "Reparos sobre el arbitrio y lo imposible de su ejecución," April 10, 1702 (ARSI, Philipp. 13. Litterae Annuae Philipp. 1663–1734 (etiam de Insuli Marianis), ff. 327r-332r; transcribed in R. Lévesque, *History of Micronesia*, vol. 10, 1997, pp. 341–349). The monarch's moral obligation regarding the salvation of infidels is expounded in the 1704 memorial written by procurator Juan Martínez de Ripalda (AHCJPC, Filipinas C-285, Doc. 5, f. 7r).

266. Juan de Palacios was born in Alfaro (La Rioja, Spain). He was admitted to the Society of Jesus in 1658. Eventually, he was sent to New Spain, where he served as rector of the College of Guadalajara. In

Jesuit provincial of New Spain, ordering him to send twelve priests and three coadjutor brothers to the Philippines, which included the Marianas.[267]

Meanwhile, on May 5, 1700, the Council of the Indies met to give their final decision regarding the jurisdictional conflict between the Archbishopric of Manila and the regular orders, after having received the expert opinion of the Indies' *fiscal* on February 20 of that same year. They concluded that Archbishop Diego Camacho had acted within the law, as dictated by canon rights. A dispatch signed by the king was immediately sent to the archbishop, dated May 20, 1700, informing him that he could retake his visits and supervise parish priests (including those from the religious orders).[268]

Another such dispatch was sent to the president of the Audiencia of Manila and the Philippine governor, approving the assistance that they had lent the prelate. Finally, each of the Superior Generals of the five religious orders in the jurisdiction of Manila received letters in which they were openly censured for the incidents that had taken place in the parishes of Tondo and Binondo. The king not only praised the archbishop's work, he urged the orders to submit themselves to the episcopal jurisdiction, reminding them that as protector of the Council of Trent, he could (and would) impose his authority over them. On the other hand, the Council asked its ambassador in Rome, Don Juan Francisco Pacheco Téllez-Girón, IV Duke of Uceda (1649–1718), to procure a swift solution to the jurisdictional conflict from the new pope, Clement XI.[269] They also awaited a sanction to the religious orders' superiors for showing that their ministries had responded to, "ex voto iustitiae" and not, "ex voto caritatis."[270]

On January 19, 1705, the Trent Council Congregation, constituted by independent cardinals, arrived at its own verdict. On January 30, Pope Clement XI issued papal brief, "Ad futuram rei memoriam," which finally put an end to the conflict over jurisdiction launched by the Archbishop of Manila.[271] This was

1698, he was appointed provincial. Father Palacios died in Lima on January 15, 1715 (C. Sommervogel, S.J., *Bibliothèque de la Compagnie de Jésus*, Vol. 6, p. 98).

267. King Philip V's letter to the Jesuit provincial of Mexico; Barcelona, December 13, 1701 (Lucas Alamán Papers, 1598–1853, Benson Latin American Collection, University of Texas at Austin, Doc. 21). The provincial, however, could not execute the Monarch's orders ("Informe del provincial de la Compañía de Jesús en la Nueva España al rey sobre la imposibilidad de dar entero cumplimiento al mandato relativo al envío de doce sacerdotes y tres coadjutores a las Filipinas," Mexico, April 7, 1703 (Lucas Alamán Papers, 1598–1853, Benson Latin American Collection, University of Texas at Austin, Doc. 21).

268. P. Rubio Merino, *Don Diego Camacho y Ávila*, pp. 250–52.

269. Born Giovanni Francesco Albani in 1649, Pope Clement XI succeeded Pope Innocent XIII on November 23, 1700, to his death in 1721.

270. P. Rubio Merino, *Don Diego Camacho y Ávila*, pp. 249–50.

271. P. Rubio Merino, *Don Diego Camacho y Ávila*, pp. 277–78; H. de la Costa, *Jesuits in the Philippines*, p. 527.

sent along with a Real Cédula signed by King Philip V, to the new governor of the Philippines, Don Domingo de Zubálburu (1701–09), so that he could see to, "its complete compliance."[272]

As Horacio de la Costa has said, the final decision was not what the orders' superiors had expected. The papal commission certified the prelates' right to visit the more than 700 parishes in their dioceses, acknowledging their ecclesiastical jurisdiction in all matters related to the care of the souls and administration of the sacraments, and limiting this jurisdiction in affairs of correcting parish priests, such as censures or transfers, which needed the consent of the orders' superiors.[273] Ecclesiastical causes could not be appealed to Rome, and the Jesuits, along with the other orders, had to accept the verdict. In exchange for their immunity, the orders' priests had to commit themselves to the parishes at present under their command, administering the sacraments and carrying out all of the duties normally undertaken by the secular clergy.

Despite disagreeing with it, considering it illegitimate for bishops to intervene in the relationship between regular priests and their orders' superiors, General Thyrsus González accepted the verdict. If until then, the orders had considered abandoning the parishes, they were no longer capable of doing so. Parishes in the Philippines were not simple missions, but actual parishes. They not only received, albeit irregularly, a stipend from the king—a congrua portio—they were also under episcopal jurisdiction.

After the papal commission passed down its verdict, Archbishop Camacho was promoted to the Diocese of Guadalajara in New Spain. He could only go there two years after his 1704 appointment (news and people travelled slowly between Madrid, the Philippines and New Spain), and he was substituted in Manila by Hieronymite Friar Francisco de la Cuesta.[274] Father Jaramillo remained in Spain, retiring to the town of Ocaña, in Toledo, where he passed away on December 30, 1707.[275]

---

272. P. Rubio Merino, *Don Diego Camacho y Ávila*, p. 279; M. M. Manchado López, "La Concordia de las religiones," pp. 74–75; M. M. Manchado López, *Conflictos Iglesia-Estado en el extremo oriente ibérico*, pp. 26–27.

273. H. de la Costa, *Jesuits in the Philippines*, pp. 527–29.

274. Camacho served as Archbishop of Guadalajara from 1707 until his death in 1712, despite being named in 1704. Francisco de la Cuesta would remain as Archbishop of Manila from 1707 until he was transferred to New Spain as bishop of Michoacán, in 1723 (P. Rubio Merino, *Don Diego Camacho y Ávila*, p. 366).

275. H. de la Costa, *Jesuits in the Philippines*, p. 612.

5. Reasons behind the publication of the *History of the Mariana Islands*

As it becomes clear to those who read it, the *History of the Mariana Islands* is an apologia destined to secure a favorable opinion towards the Jesuits' missions in the Pacific that would consolidate the Crown's economic and military support. Since his arrival in these islands, Father Luis de Morales undertook a comprehensive investigation of the Marianas archipelago, paying attention to its geography, botany, and climate (natural history), as well as to the customs and beliefs of its inhabitants (a type of thick description), looking for the hidden presence of God in those faraway lands. After he left, he continued his investigation through his confreres' relaciones, letters and Cartas Anuas, especially related to their evangelizing labors (moral history), as well as other texts or manuscripts that featured the islands. Only through this kind of work could the hard work and heroic deeds of the Society of Jesus in the Far East be written and made known. The order, after all, had numerous enemies and even more numerous critics across Europe.

One of the most common criticisms levelled against the Jesuits was that they preferred to attend to the wealthy or prosperous among the infidels—such as the societies of Japan, Siam, or China. Charles Le Gobien's book of a Jesuit mission in the, "poor and abandoned" islands of the Marianas revealed that these criticisms were undeserved. The Marianas were located far from the centers of power of the Spanish Empire (Manila and Mexico) as well as those of the other European powers in the Pacific, and they were also poor. There was no gold or silver to attract settlers or provide the missions with economic self-sufficiency. For the natives, the most valuable objects were conches, which they, "valued more than they value pearls and precious stones in Europe."[276] Moreover, given its orography, there were little probabilities that it could develop a profitable, large-scale agricultural production. The Marianas' natives, who would come to be called Chamorros, were neither cannibals nor the founders of a great civilization like that of the Incas or Aztecs. According to Morales's observations, they were simple fisherfolk and horticulturalists, who had a very basic system of trade across the archipelago. Their perceived savagery and barbarian character was attributed to their resistance to civilization and their physical strength and violent "nature." The fact that the Society consecrated some of its best men to these islands and their inhabitants would demonstrate, "that

---

276. L. de Morales, *Historia*, ff. 33–34.

Jesuits were not only martyrs for the faith, but that they had a truly universal apostolic vocation."[277]

Throughout the eighteenth century, Jesuits had attempted to prove that savages could be peacefully civilized. In places where nature was blessed with a mild climate, natives could perfect themselves following the steps to civilization.[278] This was supposed to be the case in the Marianas. Although they were in the tropics, their skies were serene and beautiful, "the air was pure, and heat was never excessive." [279] A temperate climate, regarded as conducive to human progress, was already present; therefore, the missionaries, as agents of Christianity, would also act as agents of civilization who would transform the savages into loyal subjects of the Crown.[280] This transformation could be achieved with mechanisms of assimilation, and without the unfortunate violence that had been applied in the early conquest of America.

However, this kind of civilizing process did not take place. While San Vitores, like a new "Francis Xavier",[281] dreamed of peacefully liberating the Chamorros from their alleged ignorance and guiding them towards civilization, both the Chamorros, on the one hand, and the Spanish soldiers and officers on the other, had different plans. First, the soldiers and officers (governor, captain general) forced the natives to pay a type of tribute-in-kind-or-in-labor, a tribute that the Chamorro were legally exempted from. The exploitative working conditions imposed on a population already threatened by new diseases, warfare, and displacement–and the alterations that this forced labor regime produced in the Chamorro socioeconomic fabric–increased food scarcity and fed into the revolts and social disruption.

In a spiral of increasing violence and exploitation, the natives' resistance led to the construction of a presidium (1683), which meant that a greater number of soldiers was sent to the islands. Most of these men were either forcibly recruited (often levied en masse) in New Spain and Manila, or they were condemned delinquents—men found guilty of bigamy, gambling, and other such

---

277. See the Prologue by J. P. Rubiés in this book.

278. M. Duchet, *Antropología e historia en el Siglo de las Luces*, pp. 182–196.

279. L. de Morales, *Historia*, f. 25.

280. Various Enlightenment intellectuals, such as Georges-Louis Leclerc, Count of Buffon (*Histoire naturelle, générale et particulière*, 1749–1788) and François Marie Arouet, alias Voltaire (*Essai sur les moeurs et l'esprit des nations*, 1756), believed that the Jesuits were the most appropriate agents for the "civilization" of the Indians (M. Duchet, *Antropología e historia en el Siglo de las Luces*, pp. 183–184).

281. Ulrike Strasser, "Copies with Souls: the Late Seventeenth-Century Marianas Martyrs, Francis Xavier, and the Question of Clerical Reproduction," *Journal of Jesuit Studies,* 2:4 (2015), pp. 558-585.

petty crimes or misdemeanors—and some, but very few, were volunteers.[282]
Being Christian, and soldiers, many of these men looked upon the Chamorro
as inferior creatures, and as such, they behaved atrociously, taking Chamorro
women as unwilling wives, lovers, or domestic workers, and enslaving or forcing Chamorro men to work for them.

Thus, not long after Father San Vitores's arrival, the relations between
the missionaries, the native population, and the soldiers, were characterized by
violence. The first insurrection, the so-called First Great War of Guam, broke
out on September 11, 1671, a few weeks before Father Morales left for Manila
to inform the authorities, precisely, of the events that had led to the martyrdom
of Fathers Medina, San Vitores, Monroy, and San Basilio, and their auxiliaries.
In 1684, another insurrection (the Second Great War), led to new loss of life.
The following wars (1690; 1694–1699) contributed to such a brutal decline
in the native population (between 5,000 and 3,000 souls), that the future of
the mission was compromised. As if feeling that the hell described by the missionaries was already upon them, hundreds of men and women committed
self-immolation, using their bodies to record the heteronomy of defeat and its
bitter consequences.[283]

Before delving into these tragic developments, Father Morales's *History*
dedicates the second book to the Chamorros as he came to know them before
violence and war disrupted the mission. He describes some of their customs in
great detail and represents them as naturally simple people—in the philosophical, stoic tradition—who lived in a world of absolute intellectual and moral
freedom.[284] In his own words, "never a people lived in a more complete liberty
and a more absolute independence. Each is the master of his actions from the
moment they are old enough to reason."[285] Here is a nostalgic image of the
"good savage" that argued for the viability of non-Christian societies; that is,
their peoples' right to live in accordance to their own social mores, for, despite
their imperfections and ignorance, the evident "natural goodness" reflected the

282. María Fernanda García de los Arcos, *Forzados y reclutas: los criollos novohispanos en Asia (1756–1808)*, Mexico: Potrerillos Editores, 1996.

283. L. de Morales, *Historia*, f. 39. Suicide was an accepted custom in precolonial Chamorro society. See David Atienza, "Suicide in Guam: an anthropological view," paper presented at the Roundtable Meeting on Suicide in Guam, Blessed Diego Luis de San Vitores Catholic Theological Institute for Oceania, Yona, Guam, 30 Nov. 2010.

284. On the Stoic element, which provided the basis of the understanding of human society, see Anthony Pagden, *The Enlightenment and Why it Still Matters*, Oxford: Oxford University Press, 2013, pp. 53–78.

285. L. de Morales, *Historia*, f. 31.

presence of the Creator.[286]

The notion that there were "innate ideas" implanted by God that allowed human beings to comprehend the world and organize it according to moral principles or virtues had Aristotelic and scholastic roots.[287] Given the intrinsic unity of the human genre, the Chamorro distinguished good from evil, virtue from vice, natural from anti-natural, etc., through a natural sociability, or *primae praeceptae* of natural law. If everything comes from God, as Thomas Aquinas argued, then there could be no doubt that the Chamorro were an integral part of humanity.[288] Despite their "barbarity", they were social beings who shared with the civilized members of their species a moral sense based on affective recognition, including pain, pity, compassion, and love. They were, therefore, perfectly capable of understanding the Gospel, and thus, the missionaries' intervention was justified.[289]

Reason thus defined corresponded to a natural human capacity that, exercised with proper guidance, was the defining characteristic of civilization; and it contrasted with a notion of human history as a field determined by divine and demonic interventions. In several suggestive works, Joan Pau Rubiés and José Antonio Cervera Jiménez (2023) have underlined the capacity of Jesuit and Dominican missionaries such as Alessandro Valignano (1539-1606), Mateo Ricci (1552-1610), and Juan Cobo (1547-92) to act as cultural mediators between modern Europe and the oriental civilizations of China and Japan, respectively.[290] Rejecting José de Acosta's theses regarding the demonic nature of native religions as idolatries, Jesuits who were true strategists of cultural

---

286. In the early eighteenth century, some Jesuits, such as Father Luis de Morales, adopted principles of stoicism, such as a universal concept of humanity, that gave them a different perspective from which to criticize the corruption of "civilized" man as a deviation from the first Christian communities. But we would be wrong to identify these principles as precursors of Rousseunian thought. After all, Rousseau's savage was an abstraction, situated in a time that preceded civil society (M. Duchet, *Antropología e historia en el Siglo de las Luces*, pp. 188–189; 289–292; Joan-Pau Rubiés, "Ethnography, philosophy and the rise of natural man, 1500–1750," in *Europe's New Worlds. Travel Writing and the Origins of the Enlightenment* (forthcoming).

287. Anthony Pagden, *The Fall of Natural Man. The American Indian and the Origins of Comparative Ethnology* (Cambridge: Cambridge University Press, 1982), pp. 94–97; A. Pagden, *La ilustración y sus enemigos. Dos ensayos sobre los orígenes de la modernidad*, Barcelona: Península / HCS, 2002. See also M. Duchet, *Antropología e historia en el Siglo de las Luces*, p. 12.

288. A. Pagden, *The Fall of Natural Man*, p. 96.

289. These arguments were in line with those of the Saxon historian and jurist Samuel Pudendorf, *De iure naturae et gentium* (1672), whose defense of the concept of universal humanity led him to question the separation between the "state of nature" and "civil society" posited by Grotius, Hobbes, and John Locke (A. Pagden, *La ilustración y sus enemigos*, pp. 58–62; Pagden, *The Enlightenment and Why it Still Matters*, pp. 56–63; J. P. Rubiés, "Ethnography, philosophy and the rise of natural man, 1500–1750," pp. 116–18).

290. José Antonio Cervera Jiménez, "Los misioneros católicos como mediadores científicos y filosóficos entre Europa y China: los casos de Juan Cobo y Matteo Ricci." *Hispania. Revista Española de Historia*, vol. 83, nº 274, 2023, pp. 1-21.

adaptation (Valignano, Ricci) regarded the non-idolatrous ethical system of Confucianism as compatible with the moral and religious doctrines of Christianity, which was in a "superior state". According to Rubiés, the identification of Confucian moral philosophy with ancient stoicism was the theological key that allowed for the accommodation of oriental civilizations in Christian humanist thought.[291]

In *The Enlightenment and Its Enemies*, Anthony Pagden analyzed the debate between "epicureans" (Hugo Grotius, Rene Descartes, Claude-Adrien Helvétius, Thomas Hobbes, and John Locke) and "stoics" (the Count of Shaftesbury, Samuel von Pufendorf, Denis Diderot, and Immanuel Kant) as the dominant issue in late eighteenth century political theory. For the first, human beings were motivated by moral selfishness. As purely material beings, in Helvetius's words, social living forced men to permanent confrontation. The latter, especially Shaftesbury and Diderot, argued that sociability and kindness were the same thing.[292] They praised the simplicity and austerity of "natural man", and rejected the epicureans' materialist vision of humanity.

The *History of the Mariana Islands* shows that the Jesuits in the late seventeenth century Mariana Islands attributed their right (and duty) to evangelize the natives less to Thomist social and moral values (such as those defended by Acosta),[293] than to the universal theories of neo-stoicism espoused by Ricci and Valignano. These were based on a universal, natural law that allowed for communication and understanding between human societies, and the ulterior conversion of all infidels to Christianity.[294] In the sixth book of his *History*, Marianas' natives are represented as far from that Hobbesian image of "wolves" inclined to hurt those of their own species. Although Father Morales admits

---

291. J. P. Rubiés, "The Concept of Cultural Dialogue and the Jesuit Method of Accommodation," pp. 242; 257; J. P. Rubiés, "The Concept of Gentile Civilization in Missionary Discourse and its European Reception," in Charlotte de Castelnau-L'Estoile, Marie-Lucie Copete, Aliocha Maldavsky, Ines G. Županov [eds.], *Missions d'Évangélisation et Circulation des Savoirs*, p. 328.

292. A. Pagden, *La ilustración y sus enemigos*, p. 61; Pagden, *The Enlightenment and Why it Still Matters*, pp. 53–78; M. Duchet, *Antropología e historia en el Siglo de las Luces*, pp. 326–411.

293. While Fernando Cervantes (*The Devil in the New World. The Impact of Diabolism in New Spain*. New Haven & London: Yale University Press, 1994) highlighted the role of Father Acosta's nominalism, I believe that his Thomism was much more significant. The fundamental Thomist principles held by Acosta were, first, the unity of the human genre; second, that all human beings were capable of overcoming the limitations of their physical habitat and understand and follow the teachings of the Gospel; and third, that objective reason and experience were key in learning historical facts. Johan Leuridan Huys (*José de Acosta y el orígen de la idea de misión, Perú, siglo XVI*. Centro de Estudios Regionales Andinos "Bartolomé de Las Casas" & Universidad de San Martín de Porres, Facultad de Ciencias de la Comunicación, Turismo y de Psicología, 1997, pp. 91–96) and Claudio Burgaleta (*José de Acosta, S.J. (1540–1600). His Life and Thought*. Chicago: Loyola Press, 1999, pp. 87–96; 121–126) have also defended Acosta's Thomism regarding *De Procuranda Indorum Salute* (1577).

294. L. Clossey, *Salvation and Globalisation*, p. 253.

that they were easily irritated and were quick to resort to arms, "they make peace just as quickly and their wars never last very long (...) It would seem that they fear getting hurt or bloodying the battlefield, for two-or-three men dead or seriously wounded are enough to determine who is victorious. They fear bloodshed so much that they run off and disperse whenever they see it. " [295]

The introduction of foreign elements, such as property, money, and the Christian religion, altered the liberty, simplicity, and independence from outside influence that had until then characterized Chamorro relations. Their desire to preserve these elements in their culture made them mistrust Governor Don Damián de Esplana (1674–1676), who, after negotiating the peace of 1674, told one of the main leaders or *magaˈlåhi* of San Ignacio de Hagåtña that in order to be happy, the natives had to befriend the Spanish.[296] Father Morales, moreover, was convinced that the moral corruption of his compatriots would destroy both felicity and friendship, because most of the governors and captain-generals that ruled the archipelago used (and abused) the islands' inhabitants to acquire personal fortunes and the social prestige that went along with the position. Their cruel mistreatment of the natives, whom they regarded as beasts of burden, produced such consternation and discouragement in Morales, that they determined the structure and coherence of his *History*.

At various points, the *History* adopts a critical discourse towards the notion of "civilization" by equating the barbaric impulses of the "bad Indians" to those of the "bad Spaniards." The efficient collaboration between the sword and the cross, between the strict Christian piety predicated by the Society and the strict military rigor that (should have) characterized the Spanish forces in the archipelago, encouraged the constitution of mission communities, but collaboration was not always effective. In fact, it sometimes seemed irreconcilable.[297]

Following Russian critic Mikhail Bakhtin's analysis on the dialogical dimension of texts (*Problems of Dostoevsky's Poetics*, 1963), some *magaˈlåhi*, such as [Diego de] Aguarín [Aguaˈlin] and Hurao, became moral subjects imbued with the power of the word.[298] Since the so-called discovery of the New

295. L. de Morales, *Historia*, ff. 31-32

296. L. de Morales, *Historia*, ff. 125–126.

297. For an example of the tensions between military and religious power in New Spain, see Bernd Hausberger, "La conquista jesuita del noroeste novohispano." *Memoria Americana–Cuadernos de Etnohistoria* (Buenos Aires), 12, 2004, pp. 131–168.

298. Bakhtin, c. in C. Ginzburg, "Alien Voices...," p. 77. For an application of Bakhtin's dialogical and intertextual networks in an analysis of Gonzalo Fernández de Oviedo y Valdés' *Historia General y Natural de las Indias* (1535), see the pioneering article by Kathleen A. Myers, "History, Truth and Dialogue: Fernández de Oviedo's Historia general y natural de las Indias (Bk XXXIII, Ch LIV)." *Hispania*, 73, 1990, pp. 616-625.

World, Spanish chroniclers and historians, such as Pedro Mártir de Anglería (1457–1526), Francisco López de Gomara (1511–1566), Gonzalo Fernández de Oviedo (1478–1557), and Bartolome de Las Casas (1484–1566), as well as radical skeptics like Michel de Montaigne (1532–1592) and Jean de Léry (1536–1613), used the *good savage* as a rhetorical device through which to critique the behavior of the Spanish conquerors and colonists.[299]

In a similar vein, Father Morales appears sympathetic towards the tragic loss experienced by the Chamorro through the words that he attributes to their leaders. Without justifying their barbaric use of violence, Morales understood native resistance as a call of attention to the Crown and the Council of the Indies towards the failure of the Jesuit missionaries' attempt at imposing a peaceful project of evangelization as a first step in the full incorporation of infidel nations into the Catholic monarchy.

Other official or propagandistic texts, such as hagiographies, charted the islands as a "savage and ferocious" periphery in which the martyrs of the faith—San Vitores, Medina, Solórzano, Monroy, etc.—had given their lives in, and for, the service of God.[300] Their object was not to establish a "historical truth," but instead to educate, entertain, and most especially, edify, the readers, praising the heroic actions of the missionaries (a "rhetorical truth").[301] Although the reports and official *relaciones* described the Chamorro as "barbaric, fierce, Jesuit-killers" who destroyed churches and villages, they were now presented as peaceful and almost noble personages.[302] By manipulating their voices, Morales/Le Gobien (as well as García and Aranda, as Joan Pau Rubiés points out) located themselves in a dialogical dimension that allowed them to express

---

299. These last two writers saw "savages" in the flesh. Montaigne encountered some Tupi in the port of Rouen in 1569; and Léry lived among the Tupinamba after abandoning, with the rest of the Protestant settlers, the Brazilian colony of France Antarctique. See Michel de Certeau, *La escritura de la historia*. México: Universidad Iberoamericana, [1975] 2006. 2ª ed., pp. 203–233; Frank Lestringant, *Le cannibale. Grandeur et décadence*, Paris: Perrin, 1999; J. Rabasa, *Inventing A-m-e-r-i-c-a*, pp. 178–179; C. Ginzburg, "Alien Voices…," pp. 79–81; Bernat Castany Prado, "Francisco López de Gómara y Jean de Léry: escepticismo moderado y escepticismo radical en las crónicas de Indias," en Guillermo Serés & Mercedes Serna Arnaiz (coord.), *Los límites del océano. Estudios filológicos de crónica y épica en el Nuevo Mundo*, Bellaterra: Centro para la Edición de los Clásicos Españoles & Universitat Autònoma de Barcelona, 2009, pp. 18–22. In *Supplement au voyage de Bougainville* (1796), French philosopher Denis Diderot (1713–1784), praises the lives of the savages only to critique the hypocritical customs of European societies. In *Esquisse d'un tableau historique des progres de l'esprit humain* (Paris, 1794), Nicolas de Condorcet (1743–1794), thought that education could redeem savages from their ignorance (C. Ginzburg, "Alien Voices…," pp. 73–76).

300. A. Coello, "Colonialismo y santidad: la sangre de los mártires," pp. 707–745.

301. N. Durán, *Retórica de la santidad*, pp. 69–83.

302. V. M. Diaz had already identified the discourse of the "noble savage" in Le Gobien's *Histoire* in "Pious Sites…," pp. 319–320; Vicente Miguel Diaz, *Repositioning the Missionary: Rewriting the Histories of Colonialism, Native Catholicism, and Indigeneity in Guam*. Honolulu, University of Hawai'i Press, 2010, pp. 143–44. See also C. Ginzburg, "Alien Voices…," p. 80.

{Figure 5: Adario solicitando la expulsión de los franceses (François-René Chateaubriand, *Oeuvres illustrées*, Paris: Librairie Centrale des Publications Illustrées, 1858)}

their own perspectives.[303] In 1670, a Chamorro leader named Hurao delivered an eloquent speech in front of two thousand warriors, in which he voiced his doubts regarding the superiority of European culture, while defending the ancestral customs of his people.[304] Thus, he exclaimed, with certain sarcasm:

> The Europeans would have done better to remain in their own land. We had no need of them to be happy. Satisfied with what our country provided us, we lived without desiring anything else. The knowledge that they have given us has only served to increase our necessities and excite our ambition. They think our nakedness is wrong, but if clothing were necessary nature itself would have provided us with it. Why should we burden ourselves with clothes, a superfluous thing that limits the freedom of our arms and our legs, because they allege that we must cover our bodies? They treat us as a coarse people and regard us as savages. Must we believe them? Do we not see that under the pretext of instructing and civilizing us, what they do is corrupt us? That they have made us lose the primitive simplicity with which we lived, taking from us our liberty, which is dearer than life itself? They want to convince us that they bring us happiness, and many among us have been blind enough to believe them. Yet they would believe them no longer if they realized that, all our miseries and maladies appeared with the arrival of these foreigners, who came to disturb our peace. Did we know any of the insects that cruelly pursue us, before their arrival on these islands? Did we know mice, flies, mosquitoes, and all these other pests whose only purpose is to torment us? Those are the gifts that they have brought us in the floating houses in which they sail! Did we know rheumatism and colds? What diseases we had, we also had the means to cure them, but they have given us their ills without teaching us their remedies. Were iron and the other bagatelles that they have brought, which have no intrinsic value, were they worth seeing ourselves surrounded by so many calamities?

With its rhetoric virtuosity, the previous speech, which today is regarded as an icon of Chamorro identity in Guåhan,[305] suggests that some Jesuit mis-

303. See the Prologue by J. P. Rubiés in this book.

304. As Rubiés points out, the hagiographies of Francisco García (San Vitores) and Gabriel de Aranda (Monroy) attribute this speech not to Hurao, but to Agualin, leader of an earlier revolt (1670), demonstrating the free use that Jesuits made of their sources (Ibidem).

305. Jonathan Blas Diaz has recently vindicated Hurao's figure, arguing, "the remarkable speech of Hurao

sionaries, like Father Morales, developed a critical discourse regarding the abuses of colonization. The veiled critique of the Spaniards' behavior, ably put in the mouths of Chamorro leaders, laid out the complexities of missionary work, presenting the missionary as a figure with a "double consciousness". Hurao (and/or Agua'lin) are presented as leaders of a disinherited culture, exclaiming that

> They reproach us our poverty, our ignorance, and our lack of industry, but if we are as poor as they say we are, what is it that they seek among us? Believe me, if they had no need of us, they would not expose themselves to so many perils or work so hard to settle in our midst. Why do they insist on subjecting us to their laws and customs, forcing us to abandon the precious gift of liberty that our parents bequeathed us? In sum, why give us sorrow in exchange of a good that we can only hope to enjoy after our deaths? They regard our histories as fables and fiction. Do we not have the same right to regard as equally fictitious what they teach us and preach as incontestable truth? They abuse our good faith and our simplicity. All of their arts are directed towards fooling us, and all of their science tends only to make us wretched. If we have ever been as ignorant and blind as they would have us believe, it was when we overlooked their pernicious desires until it was too late, and when we allowed them to settle among us. Let us not lose courage in the face of our many misfortunes; they are only a handful of men and we can be rid of them easily. Even though we do not have those murderous weapons that spread terror and death all over, we can finish them off because we greatly outnumber them. We are stronger than we think, and we can soon free ourselves of these foreigners, and regain our primitive liberty.[306]

Historical allegories constituted the best way to bypass censure and transmit the truth to the King and the Council of the Indies. Thus, mingling the voices of Hurao/Agualin//Morales/Le Gobien is a discursive strategy that serves as, "a tragic counterpoint to the triumph of civilization, produced by civilized writers who were capable of feeling nostalgia (of a stoic character) for a golden age of simplicity and innocence that was irrevocably lost. "[307]

provoked a revolution amongst the *Chamoru* to win back their homeland, to win back their freedom from the *estrangherus* (...) The spirit of Hurao continued to infuse *Chamoru* thought in the islands. For some, Hurao's message is poignant today as we continue to fight for justice in our homelands" (Jonathan Blas Diaz, *Towards a Theology of the Chamoru. Struggle and Liberation in Oceania*, Manila: Claretian Publications, 2010, p. 66).

306. L. de Morales, *Historia*, ff. 80–81.

307. See the Prologue by J. P. Rubiés in this book. See also J. P. Rubiés, "Ethnography, philosophy and

His ideas on the primeval liberty and simplicity of the Chamorro came from Michel de Montaigne (1533–1592), an author whose work was in the Index of Forbidden Books since 1676, which questioned the supposed "blessings of civilization."[308] In Ginzburg's opinion, "[Le Gobien] turned the arguments of Montaigne into a harangue, and the unnamed 'noble Savage' into Hurao, the nobleman from the Mariana Islands, full of hatred for European civilization."[309] Ginzburg argues that many French Jesuits turned to Montaigne's skepticism to refute the Jansenist licentiousness and insistence in the depravation of humanity since Adam's fall.[310]

Morales did not deny that the CHamoru were dominated by passions; that they were vengeful[311] (p. 33) and proud[312] (p. 36), but they were not more so than many Spanish governors and sergeant majors, who placed their personal interest above that of the mission, betraying the Christian principles that were supposed to govern the colonization of the archipelago. This kind of exercise in reason expanded the mental boundaries of Christian humanism, revealing that the knowledge of other societies, with ethical principles as sound, or sounder, than those of the Europeans, could contribute to the moral enhancement of the latter. Were Morales/Le Gobien wrong in denouncing that the CHamoru were but lambs who had been exposed to the "cruelty of merciless wolves and lions"? Was not Christian civilization responsible for having planted the seeds of corruption, greed, and other evils in the Western and Eastern Indies?

In a later article, Anthony Pagden analyzed diverse texts of the eighteenth century in which there was an inversion of certain stereotypes associated with "savages" that turned the latter into "noble" critics of colonialism.[313]

---

the rise of natural man, 1500–1750," pp. 97–127.

308. C. Ginzburg, "Alien Voices…," pp. 79–82.

309. C. Ginzburg, "Alien Voices…," p. 80.

310. John McManners, *Church and Society in Eighteenth-Century France. Volume 2. The Religion of the People and the Politics of Religion*. Oxford: Clarendon Press, 1998, pp. 521–22; Bangert, *A History of the Society of Jesus*, pp. 204-217. For more on the "radical skepticism" of certain French intellectuals, such as Michel de Montaigne and Jean de Léry (1536–1613), see Castany Prado, "Francisco López de Gómara y Jean de Léry…," pp. 9–24.

311. Luis de Morales, *History of the Mariana Islands*, p. 33 [manuscript page].

312. Luis de Morales, *History of the Mariana Islands*, p. 36 [manuscript page].

313. Anthony Pagden, "The Savage Critic: Some European Images of the Primitive." *The Yearbook of English Studies*, "Colonial and Imperial Themes." Modern Humanities Research Association, Special Issue, Vol. 13, 1983, pp. 32–45. David Allen Harvey has pointed out the importance of the historical and cultural context that surrounded the construction of the myth of the "noble savage." The new elite of state employees, magistrates, and traders arose during the reign of Louis XIV questioned the privileges of the French aristocracy, arguing that various other social groups shared, and perhaps surpassed, the moral qualities of the nobility (D. A. Harvey, "*The Noble Savage and the Savage Noble: Philosophy and Ethnography in the Voyages of the Baron de Lahontan.*" *French Colonial History*, 11, 2010, pp. 161–191).

Besides Montesquieu's *Lettres persanes* (1721), Pagden looked at the Baron de Lahontan's *Dialogues curieux entre l'auteur et un sauvage du bon sens qui a voyagé* (1703). In what the sixteenth century rhetorists would have called an antitheta, this French author contrasted his own views, as representative of the imperialist ideology, with those of a native Huron from Canada named Adario, who defended the moral values of his people.[314] As Rubiés reminds us, Lahontan's intention was not a mere critique of imperialism. Like Morales/Le Gobien, he did not idealize the savages to censure European colonists' cruel behavior, nor to question the hegemonic values of a supposedly corrupt Europe, as Jean Jacques Rousseau argued in his *Discours sur l'origine et les fondements de l'inégalité parmi les hommes* (1755). They criticized the fundamental institutions of civilization, such as laws and private property.[315]

In the early eighteenth century, Europe had developed a universalist conscience of itself as harbinger of the unquestionable moral and intellectual superiority that accrued naturally with progress and civilization. For Chamorro leaders, there was obviously nothing natural about European laws and customs, but there was also little of moral superiority and civility in the despotic character of their new rulers. Indeed, the "barbarity of the civilized", in the words of the Count of Buffon, showed that humanity had "degenerated" into an inferior state of moral development.[316] In the fifth book of his *History*, Morales acknowledges this in a rebuke to the soldiers pronounced by the Jesuit missionaries, whose misconduct was compromising the mission:

> You have come here [...] with the desire to serve God and assist in the conversion of these peoples, and so far we have looked upon you as the defenders of this new Christianity. But your conduct and your attitude is contrary to this! Are the barbarians not right when they say that you only tend to their destruction? How can they trust us if you vex them with your untoward behaviors? What is most deplorable is that you are offending God by hamper-

---

314. A. Pagden, *European Encounters with the New World. From Renaissance to Romanticism*, New Haven & London: Yale University Press, 1993, pp. 120–126. See also M. Duchet, *Antropología e historia en el Siglo de las Luces*, pp. 30; 90–91; D. A. Harvey, *"The Noble Savage and the Savage Noble," pp. 163–164*.

315. J. P. Rubiés, "Travel Writing and Ethnography," in Rubiés, *Travellers and Cosmographers. Studies in the History of Early Modern Travel and Ethnology*, London: Ashgate, 2007, pp. 32–35. See also the Prologue by J. P. Rubiés in this book; and J. P. Rubiés, "Ethnography, philosophy and the rise of natural man, 1500–1750," p. 108.

316. French naturalist, cosmologist, and encyclopedist Georges Louis Leclerc, Count of Buffon (1707–1788) argued that in their brutal treatment of the savages, the civilized renounced their higher rank and regressed to a "natural" state of barbarity and savagery. He was not the only Enlightenment naturalist to hold similar views. See M. Duchet, *Antropología e historia en el Siglo de las Luces*, p. 241.

ing our missionary visits with your disorder. Where would we be
if the Lord, who is our strength and our support, abandoned us?
For, what can twenty or thirty men do against more than thirty
thousand barbarians bent on destroying us?[317]

At this point, there should be no doubt that the missionaries in the Marianas clashed with the Augustinian principles of Christianity, adopting instead
more flexible notions regarding the liberty and equality of the human genre
that were at the basis of Enlightened thought.[318] In his *History*, Father Morales
projected a moral or didactic dimension—an *exemplum*—which condemned
his compatriots misdeeds through different Chamorro leaders' voices.[319] As
procurator of the Society, his intention was to appeal to the monarch as ultimate
administrator of justice—Cicero's *moderador republicae*—in the world of men.
The king should promulgate laws based on the Christian principles of justice
and equity; and to preserve legality, he had to be accessible to all his subjects,
regardless of their condition or distance. Regarded as a just and powerful father
who looked after his subjects' ultimate well-being, was the king not obligated
to make sure that those who had recently converted to Christianity and had
been incorporated into his realm be protected from both predatory officers and
soldiers, as well as violent infidels who would destroy them?

So, then, why was the *History* not published in Spain? Was it because
of the tense relations Father Morales had with the Council of the Indies? Or
those he had with important Jesuit rectors in Andalusia? Or was it perhaps that
his superiors considered his labor as a historian incompatible with the tasks of
procurator and provincial to which he was supposed to be dedicated?[320] The
death of Popes Innocent XI (1689) and Alexander VIII (1691) did not help the
project; neither did the palace intrigues around the succession of King Charles
II, the last of the Spanish Hapsburgs.[321]

In fact, the dynastic change that took place in Spain, along with the

---

317. L. de Morales, *Historia*, ff. 103–104.

318. J. P. Rubiés, "Concept of Cultural Dialogue," p. 244; J. P. Rubiés, "Ethnography, philosophy and
the rise of natural man, 1500–1750," pp. 97–127.

319. Peter Burke, "Del Renacimiento a la Ilustración", in Jaume Aurell, Catalina Balmaceda, Peter Burke
and Felipe Soza, *Comprender el pasado. Una historia de la escritura y el pensamiento histórico*, Madrid: Akal,
2013, p. 172.

320. In 1699, Father Morales succeeded Father Antonino Tuccio as provincial of the Philippines (R.
Lévesque, *History of Micronesia*, vol. 10, 1997, p. 137).

321. After King Charles II died in 1700, Louis XIV sent his grandson to Spain to be crowned King
Philip V (Henry Kamen, *La España de Carlos II*, Madrid: Biblioteca Historia de España, [1980] 2005, pp.
494–506).

decisive support lent by Louis XIV to the Jesuit missions in Asia, led to a change in the Society's strategy to obtain political support for their missions. Let us not forget that the Church had a universal jurisdiction over all Christian lands; a jurisdiction that subjected political powers to spiritual or religious powers. However, from 1690 onwards a period of intense nationalistic rivalry within the Society of Jesus emerged, thus fracturing the diversity or international character of the Jesuits sent overseas.[322] Nonetheless, it was not surprising that, under the aforementioned circumstances, the General of the Society would bet on the French monarch, who not only favored the publication of the History of the Mariana Islands in Paris in 1700, but also incorporated it into the national patrimony.[323] On October 18, 1685, the Sun King promulgated the edict of Fontainebleau,[324] which forbade Protestantism in his domains.[325] As national monarchies became stronger, a "nuptial" commitment took shape between the monarch (*rex*) and his domains (*regio, regnum*) through the "confessionalization" of the absolutist state.[326] The king's religion became obligatory for all his subjects (*religio regis, religio regiones seu regni*), as he took it upon himself to act against those who threatened national religious unity.[327] The Sun King did not admit dissent. French Protestants were subjected to a campaign of harassment and forced conversion. Those who did not convert to Catholicism left the country or were persecuted, and a diaspora of 200,000 French Huguenots settled in the Netherlands and other northern European countries, including many French intellectuals and artists.[328] One of them, Pierre Bayle (1647–1706), continued

322. Frederik Vermote, "Travellers Lost and Redirected: Jesuit Networks and the Limits of European Exploration in Asia". *Itinerario*, Vol. 41, n1 3, 2017, p. 501

323. For an analysis of how "official history" was used by the absolutist regime of Louis XIV as propaganda, see C. Grell, "Les historiographes en France XVIe – XVIIIe siècles," in Chantal Grell (dir.), Les Historiographes en Europe de la fin du Moyen Âge à la Revolution, pp. 127-156.

324. This edict revoked the edict of Nantes, proclaimed by Henri IV on April 13, 1598, which granted certain liberty to Protestants, protecting them from religious persecution. Henri IV had in fact renounced his Protestant faith five years earlier (supposedly saying "Paris is well worth a Mass" in the process).

325. W. J. Stankiewicz, *Politics and Religion in Seventeenth-Century France*, Berkeley & Los Angeles: University of California Press, 1960, pp. 52–91.

326. For an understanding of the concept of "confessionalization," see the volume edited by Paolo Prodi, *Disciplina dell'anima, disciplina del corpo e disciplina della società tra medioevo ed età moderna*, Bologna: Società editrice il Mulino, 1994, especially Pierangelo Schiera' "Disciplina, Stato moderno, disciplinamento: considerazioni a cavallo fra la sociologia del potere a la storia constituzionale" (pp. 21–46) and Wolfgang Reinhard's "Disciplinamento sociale, confessionalizzazione, modernizzazione. Un discorso storiografico," pp. 101–123.

327. Josep-Ignasi Saranyana, "Por qué el Renacimiento y el Barroco persiguieron las heterodoxias," in Ricardo Izquierdo Benito and Fernando Martínez Gil (coord.), *Religión y heterodoxias en el mundo hispánico. Siglos XIV-XVIII*, Madrid: Ediciones Sílex, pp. 17–18.

328. This measure sank Francia in an unprecedented political and economic crisis (Warren C. Scoville, *The Persecution of Huguenots and French Economic Development, 1680–1720*, Berkeley & Los Angeles: California University Press, 1960; J. McManners, *Church and Society in Eighteenth-Century France*, pp. 565–88).

contributing to the corpus of French letters during his exile in Rotterdam.[329]

In 1687, two years after his arrival in that city, Bayle began his celebrated *Dictionarie historique et critique* (2 vols. 1695–1696; 2 vols. 1704), which was a reworking of the *Grand Dictionnaire Historique* (1674) written by Father Louis Moréri (1643–1680), S.J. Like a stoic philosopher, "whom no passion can disturb," a historian, according to Bayle, must forget his political and religious sympathies and stick to fact, "sacrificing his indignant anger at affronts to it; as well as the memory of favors owed, and even the love he may feel for his country (*patria*).[330] Ironically, this French historian is known for having justified his plagiarism of foreign texts to "increase" the national patrimony.[331]

Županov reminds us that although (or perhaps because) they had a corporative scriptorial culture, Jesuits were among the most cited and plagiarized writers of the seventeenth century.[332] But plagiarism was not only carried out by those outside the Society. As already stated, Le Gobien's *Histoire des Isles Marianes* is based not only on the writings of Luis de Morales, but also on those of the first Jesuits who worked in the mission (Solórzano, Bouwens, Cuculino), and who, as eyewitnesses, had legitimacy and authority. Le Gobien gathered these testimonies left by these "Sons of St. Ignatius" and redacted a coherent narrative.[333] Paradoxically, it was the French priest, who had never been to the East Indies, who became the *author* of the first *History of the Mariana Islands*,

329. The literature on the figure of this French Huguenot ("the philosopher of Rotterdam") and the controversy over tolerance is too broad to cite fully, but some of the most interesting works include Julián Arroyo Pomeda, *Bayle o la ilustración anticipada*, Madrid: Ediciones Pedagógicas, 1995; Arroyo Pomada, *Pierre Bayle (1647–1706)*, Madrid: Ediciones del Orto, 1997, pp. 34–41; Fernando Bahr, "John Locke y Pierre Bayle: sobre la libertad de conciencia." *Tópicos*. Universidad Católica de Santa Fe, 12, 2005, pp. 43–67; and W. J. Stankiewicz, *Politics and Religion in Seventeenth-Century France*, pp. 206–242.

330. According to P. Bayle, "a Historian as such, like Melchisedec, is without father, without mother, and without pedigree. If he be asked, 'Of what country are you?' He ought to answer, 'I am neither a Frenchman, nor a German, nor an Englishman, nor a Spaniard, etc. I am an inhabitant of the world, neither engaged in the service of the Emperor, nor of the King of France, but only in that truth. She is my sole Queen, and I have taken no oath of allegiance to any besides her. I am her sworn Knight, and the collar of the order which I wear is the same ornament worn by the chief justice, and the high priest of the Egyptians.' [...] So far as he yields to the love of his country, so far does he deviate from the rules of his history, and by how much he showeth himself a good subject, by so much he becomes a bad Historian" (Pierre Bayle, 1734 (2nd ed.), *The Dictionary Historical and Critical of Mr Peter Bayle*, trans. P. Desmaizeaux, London: Knapton et. al., vol. V, p. 531). See also Pagden, *The Enlightenment and Why it Still Matters*, pp. 100–104.

331. Charles Nodier, *Questions de Littérature Légale. Du plagiat, de la supposition d'auteurs, des supercheries qui ont rapport aux livres*. Paris : Barba, 1821. See also Kevin Perromat Augustin, *El plagio en las literaturas hispánicas: Historia, Teoría y Práctica*, Ph.D. Diss., Université Paris-Sorbonne, Ecole Doctoral IV, Centre de Recherches Interdisciplinaires sur les Mondes Ibériques Contemporains, 2010, p. 531.

332. I. G. Županov, *Disputed Mission*, p. 103.

333. Charles Le Gobien's editing style and policies have not been sufficiently studied, but we know that he took some liberties. On February 2, 1705, the *Journal des savants* published an article that was critical towards his *Lettres édifiantes*, complaining that he "enhanced" the style of his confreres' letters so much that he compromised their veracity (I. G. Županov, *Disputed Mission*, p. 13).

while Morales, as writer, was forgotten by posterity.[334]

The definitions that we have today for the notions of authorship, plagiarism, and citation of historical texts are very different from that of the Jesuits at the end of the seventeenth century. It was not important to be the first to write a particular idea or text, but to be supported in the writing process by the authorities—it was not until after the eighteenth century that plagiarism was considered morally reproachable. We cannot know why the copy of the manuscript found at the Provincial Jesuit Archive of Catalonia is incomplete: is it because the original manuscript was unfinished, or because it was missing the last books? What is most probable is that Luis de Morales finished the *History* in New Spain, right before embarking again for the Philippines in 1697. He would have therefore included information gathered there on the last governments of Don Joseph de Quiroga y Losada (1680; 1688–1690; 1694–1696) and Don Damián de Esplana (1683–1687; 1690–1694), which correspond to the last three books. While its author went to the Philippines, the text was sent to Europe, and General Thyrsus González, for the political and strategic reasons stated above, decided that it would serve its purpose more fittingly if it were published in France.

In any case, although we cannot rule out that this unfinished *History* is, as Jon Pau Rubiés argues, a Spanish translation of Le Gobien's *Histoire des Isles Marianes*, it is clear that the French Jesuit used materials written in Spanish (and Latin) by the Jesuit missionaries who labored in the islands, especially Father Luis de Morales, many of which had already been used (and published) by hagiographers Francisco Garcia and Francisco de Florencia. And after the expulsion of the French Huguenots, the publication and distribution of works that reinforced the state religion—Catholicism—became a national priority in France, part of its "official history." And the contribution of French Jesuits to the diffusion of French literary culture was extraordinary, especially in their schools and educational institutions, where French authors such as Corneille were studied enthusiastically with the purpose of reinforcing a French identity in tune with the absolutist state.[335] From his exile, Bayle rejected the use of religion for political purposes, arguing, "monarchs cannot make their current religion a political law."[336] But it was not his opinion that mattered: the publication of

334. Michel Foucault, "What's an Author?," in Paul Rabinow (ed.), *The Foucault Reader*, New York: Pantheon Books, p. 108.

335. J. McManners, *Church and Society in Eighteenth-Century France;* Marc Fumaroli, "Between the Rigorist Hammer and the Deist Anvil: The Fate of the Jesuits in Eighteenth Century France," in John W. O'Malley, et. al., *The Jesuits II. Cultures, Sciences, and the Arts, 1540-1773,* Toronto & Buffalo & London: University of Toronto Press, 2007, pp. 684-85.

336. W. J. Stankiewicz, *Politics and Religion in Seventeenth-Century France*, p. 219.

Le Gobien's book was authorized by Father Jean Dez, provincial of France, on September 18, 1699 in Vannes, and it became the first—and only—official history of the Mariana Islands for hundreds of years.

6. Concluding remarks

The *History of the Mariana Islands* was an apologetic work published in a context in which the jurisdictional conflicts between the religious orders and the secular prelates in the Philippines had led to a political crisis in the Spanish empire. The Society of Jesus was in the crossfire between the mendicant orders who resented what they saw as its privileges; and the secular prelates who wanted to submit the Jesuit missionaries to the episcopal jurisdiction. The book was therefore part of the Society's efforts to vindicate the importance of its missionaries in the Asiatic territories of the Spanish empire.

We do not know when he started to write it, but it is very likely that Father Morales did so during his return trip to Spain at the end of 1689, and that he finished it in Mexico towards 1697, before departing again towards the Philippine islands—a voyage that stopped over in Guåhan. His work had a panegyric character in regards to the archipelago where he had labored as a missionary along with many other holy men who were martyrized, including Father San Vitores and others who had founded the mission in 1668.

On the contrary, the Jesuit priest Charles Le Gobien was located in a global context marked by the Chinese and Malabar Rites controversy as well as by the strong opposition of the Society, and especially its General, Thyrsus González Santalla, to the beatification of the venerable bishop of Puebla, Don Juan de Palafox.[337] As editor of the *Lettres édifiantes et curieuses*, his approach was that of a publicist who had the Society's prestige in France and Europe as his topmost concern. Unlike Father Morales, for whom the Marianas were an eminently Spanish "problem," Le Gobien's perspective was broader, and he located the Islands, and the missions in the Pacific more generally in the context

---

337. After the Relator of the cause of beatification for the Venerable Juan de Palafox died in 1699 (Cardinal Jeronimo Casanate, OP, 1620–1699), General Thyrsus Gonzalez wrote several letters to the King and the Count of Altamira, Spanish ambassador in Rome, asking them to intervene so that the new Relator, celebrated lawyer Bernardino Peregrini, was suspended (Arteaga y Falguera, *Una Mitra sobre dos mundos*, p. 596). See "Oposizión Hecha al Progreso en las Causas, procesos de la Beatificación y Canonización del V.S. de Dios el Ilustrísimo y Reverendísimo Señor D. Juan de Palafox y Mendoza, Obispo que fue de la Puebla de los Angeles en la Nueva España, y despues de Osma en los Reynos de Castilla. Y Satisfacción de ella" (Rome, July 26, 1698) (AHCJPC, Estante 2. Caja 92. El obispo Palafox (IV). Leg. 1193, 39, ff. 1r-14v). A few years earlier, procurator Antonio Jaramillo had written an *Apologia* (Valencia, 1695) against Palafox that he had signed with the pseudonym "Don Matías Marín, theology professor" (Arteaga y Falguera, *Una Mitra sobre dos mundos*, pp. 594–95).

of the Society's evangelical expansion across Asia and Oceania.[338]

Le Gobien's point of view is evidenced in some of the paratexts that accompanied his *Histoire*. In a letter addressed to the Bishop of Ypres, Monsignor Martin de Ratabon (1654–1728),[339] Father Le Gobien praised his confreres for having propagated the Christian faith in such faraway places as China, Japan, Tonkin, Cochinchina, and Canada. They had done all this, he said, moved solely by the desire to spread and defend the Catholic religion. Defending the Society against the accusations that circulated against it in Rome in which the order was presented as economically profiting from its missions in Paraguay, Cordoba and Buenos Aires, Le Gobien wrote:

> The history that I present to Your Eminence can serve more than any other, to show the purity, the disinterestedness of the zeal of the missionaries employed by the church to convert the infidels. The peoples who inhabit the Mariana Islands have no treasure, no precious stones to offer them; the only thing to be found among them is that evangelical pearl, the value of which only those who appreciate the gifts of God can know. Nothing could have attracted the Workers of the Gospel towards that barbaric nation, perhaps the least civilized that exists in all of the orient, but the desire to suffer for Jesus Christ and gain souls for him.[340]

Since Bishop Ratabon had been one of the great defenders of Catholicism before the protestant heresy in France, he could partake in the defense of the Jesuit missions in Asia. And indeed he did. The Jesuits did not abandon the Marianas until they were expelled from the Spanish empire.[341] They were succeeded by Augustinian Recollect friars, but dwindling resources, the islands' isolation, and complications during the Spanish War of Independence led the friars to abandon the mission in 1814, replaced by two Philippine-born secular

338. As Vermote pointed out, "the late seventeenth century ushered in an era of competing jurisdictions and loyalties to state networks within the Society of Jesus that further fragmented the overall structure of an organization struggling to link Europe and Asia" (Vermote, "Travellers Lost and Redirected…", p. 501).

339. It is worth mentioning that Cornelius Jansen, a bitter enemy of the Jesuits, was Bishop of Ypres (Belgium) from 1636 until his death in 1638.

340. Charles Le Gobien's letter to Monsignor Ratabon, Bishop of Ypres, in C. Le Gobien, *Histoire*, pp. iii–xiii. Martin de Ratabon was Bishop of Ypres from 1693 to 1713, when he resigned and was appointed Bishop of Viviers (France) until his death in 1728.

341. The Pragmatic Sanction of 1767 expelled the Jesuits from the peninsula on April 2, 1767; the order reached New Spain in June of that year, and the Philippines in May of 1768. Despite having been notified of their expulsion nearly a year before, the Jesuits remained in the Marianas until the first Augustinian Recollects arrived to succeed them on November 2, 1769.

priests.[342] The Recollects returned in 1820.[343]

Today, Catholicism is one of the hallmarks of identity in Guåhan, so much so that according to University of Guam President Dr. Robert A. Underwood, Catholicism and Chamorro identity are, "two parts of the same whole."[344] Perhaps paradoxically, another important symbol of identity is constituted by the omnipresent seventeenth century leaders that resisted colonization, such as the "noble Hurao". Thus, Perez Viernes staunchly defended *maga'låhi* Hurao's famous speech, which was delivered during the first CHamoru rebellion in 1671, as an example of indigenous resistance and agency. When delivering his speech, Hurao, transmuted into a national archetype, who not only "inspired the masses", Perez contends, but also "contributed to the making of his people's history"[345]. As he often reminds us, Hurao will be a source of inspiration for future generations, thereby turning him into an icon of CHamoru identity. One thing is certain: the words he uttered are lost in translation, so that it is impossible to fully know what he once said. Nonetheless, it is evident that what was recorded by Jesuit misisonaries in Guam had nothing to do with Hurao's own wording because that "way of uttering" was a "rhetoric truth".

In any case, the truth is that the speech was recorded in writing by Father Luis de Morales before it could be turned into a major symbol and reference of Chamorro identity. Today, it is printed on the walls of the Supreme Court of Guam and in the panels at the entrance of the Guam Museum. On March 9, 1993, Underwood, who had then been recently elected Guam's Delegate to the U.S. Congress, read it to the U.S. House of Representatives,[346] and it became known and used by nationalist groups but also by US civil society, turning

342. In the context of the Napoleonic invasion of Spain and the revolutionary junta processes in the American colonies, the Marianas had stopped receiving the situado from Mexico in 1813, for there was no longer a Manila-Acapulco galleon ("Documentos relativos a los curatos de islas Marianas y sucesos en ellos en el año 1829: siendo gobernador el señor don José Medinilla y Pineda con otros sobre el pago de estipendios a los ministros de dichas islas y un estado de las mismas." Archivo Provincial de los Agustinos Recoletos (Marcilla, Navarra), Legajo 74, no. 2).

343. See Marjorie G. Driver, *The Augustinian Recollect Friars in the Mariana Islands, 1769–1908,* Mangilao, Guam: Richard Flores Taitano & MARC & University of Guam, 2000.

344. Robert A. Underwood, "The Practice of Identity for the Chamorro People. The Challenge of Hispanicization," in Miguel Luque Talaván, Juan José Pacheco Onrubia, and Fernando Palanco (eds.), 1898, *España y el Pacífico: interpretación del pasado, realidad del presente*, Valladolid: Congreso Internacional de la Asociación Española de Estudios del Pacífico, 1999, pp. 527–534.

345. Perez Viernes, J. (2016). Hurao Revisited: Hypocrisy and Double Standards in Contemporary Histories and Historiographies of Guam. 3rd Marianas History Conference. Mangilao, Guam: *Guampedia*, [Available online] <https://www.guampedia.com/mhc3/>, p. 126. [Consulted March 2, 2024].

346. Robert A. Underwood, Congressional Record–House, Tuesday, March 9, 1993, 103rd Congress, 1st Session, 139 Cong. Rec. H 1075, cited in *Hale'-ta Hinasso': Tinige'Put Chamorro–Insights: The Chamorro Identity*. Agaña: Political Status Education Coordinating Commission, 1997, pp. 13–14.

Hurao into the leader of postcolonial subalternity.[347]

Although *maga'låhi* Hurao really existed, it was the Jesuits who transformed him into a rhetorical figure, using him to reflect on the abuses that underlay the civilizing mission. In the Jesuit representation, Hurao faces a terrible existential drama: the impossibility of recovering his "natural" world, the one before the conquest, and the impossibility of a peaceful coexistence with the Spanish that is based on respect for Chamorro culture. And if there was indeed an eyewitness to those sorrowful events, that was Father Luis de Morales, who was in the islands as a missionary, and was later procurator and provincial of the Philippines. In 1707, he was relieved from his appointment of provincial, and he died on June 14, 1716, in Manila.[348]

In conclusion, the *History of the Mariana Islands* is an invaluable source for those interested in the conquest and evangelization of the Marianas archipelago, no matter who the actual author might be. The text has served for many years as a unique and relevant snapshot of this critical time in the history of the Marianas, as well as in the history of the Jesuit mission in the Micronesian islands and the roles that particular missionaries such as Father Luis de Morales and Diego Luis de San Vitores played.

---

347. For an analysis of "subalternity" (or subalterity) as a category that refers to the "local voices" silenced by the hegemonic powers, see J. Jorge Klor de Alva, "Colonialism and Post Colonialism as (Latin) American Mirages". *Colonial Latin American Review*, Vol. 1, nos.1-2, 1992, pp. 3–23; Ranajit Guha, *Las voces de la historia y otros estudios subalternos*, Barcelona: Crítica, 2002.

348. ARSI, "Suplementum Catalogorum Provincia Philippinarum Societatis Iesu a die 17 Jul. 1713 usque ad diem 21 Jul. 1719". Philipp. Catal. Trien. 3, f. 133r. See also P. Murillo Velarde, *Historia de las Filipinas*, Tomo IV, Cap. 28, transcribed in R. Lévesque, *History of Micronesia*, vol. 12, 1998, pp. 46–51; vol. 7, 1996, p. 663.

Editor's Note

This is an English version of the first Spanish edition of *History of the Mariana Islands* published in Madrid in 2013 by Ediciones Polifemo. I want to thank my Spanish editor, Ramón Alba, for providing me with the facilities to do it. For the present English edition, I took the opportunity to correct some orthographical errors in the transcription of the manuscript as well as to add up-to-date historiographical information on Micronesia.

The page numbers of the original text appear in brackets. The punctuation and grammar of the manuscript have been modernized to make it more legible for the contemporary reader, except for the spelling of placenames, which have been kept as they appeared in the manuscript, with the Chamorro spellings bracketed next to them. Abbreviations have been substituted by whole words. Important dates or facts that help contextualize certain historical characters have been bracketed next to them. Needless to say, footnotes that contextualize, analyze, or provide relevant or necessary information have also been added. To differentiate them from the footnotes that were part of Luis de Morales's manuscript and/or Le Gobien's *Histoire*, the latter are clearly identified as such. Finally, there is a brief chronology of the governors of the Mariana Islands that correspond to the period addressed in the text (1668–1700); a list of the Viceroys of New Spain during this same period (1664–1701); a list of Governors and Captain-generals of the Philippine Islands (1668–1701); and an onomastic index.

In 1964, Jesuit historian Paul V. Daly translated Charles Le Gobien's *Histoire des Isles Marianes* in Hagåtña, Guam, and published it as a series in Guam's Catholic weekly journal, *Umatuna Si Yu'us* (Hagåtña, July 26–December 20, 1964). He wanted to make the first history of the Mariana Islands available to the population of Guam—of course, he had no way of knowing that the first *History* of the islands was, in strictu sensu, another text, written in Spanish by Father Luis de Morales, S.J. However, Daly's translation, which was never published in book form, was not considered appropriate for this English edition of the *History of the Mariana Islands*: it was not a translation of Father Morales's manuscript, but of Le Gobien's French version of that manuscript (among other texts). The present English translation was done by Yesenia Pumarada Cruz, but Daly's version was an invaluable reference.

History of the Mariana Islands

Luis de Morales and Charles Le Gobien of the Society of Jesus

(Note: this copy has many transcription errors, but they are easily discovered)[349]

Foreword

(page 1) Although the Spanish discovered the Mariana Islands in their first voyages to the Orient at the beginning of the previous century, thirty years went by before anything was known regarding the character and customs of the peoples who inhabited them, and that [became known] thanks to the missionaries who took the faith to them in these latter times. It is with the *relaciones* written by these apostolic men, most of whom have had the joy of giving up their lives for Jesus Christ, that I have written the History that I now present to the public. I have put no words of my own in this text; only what I have found in their letters and *relaciones*, which have been sent to me from Rome, Spain, and the Low Countries.[350]

Since until the present, the situation, number, and even the very name of these islands was unknown, separated as they are from the rest of the world by the vast seas that surround them, I have thought it necessary to publish exact maps to make them known. These maps were drawn on the field by Father Alonso López himself, a Spanish Jesuit who traversed these islands at various times and who worked for many years in the conversion of their people.

To complete this knowledge, I will add a brief memoir that another missionary who came from Rome to look after the mission's affairs left with one of his friends. It contains the new place names given by the Spaniards.

---

349. AHCJC, FIL HIS–061, E.I, c-05/2/0, 149 folios. There is another copy in the same archive (AHCJC, FIL HIS–061, E.I, c-05/3/0). This note is at the top of the manuscript.

350. This refers to the letters and relaciones included in the *Cartas Anuas* or Annual Letters written by the mission superiors of the Mariana Islands' Vice-province.

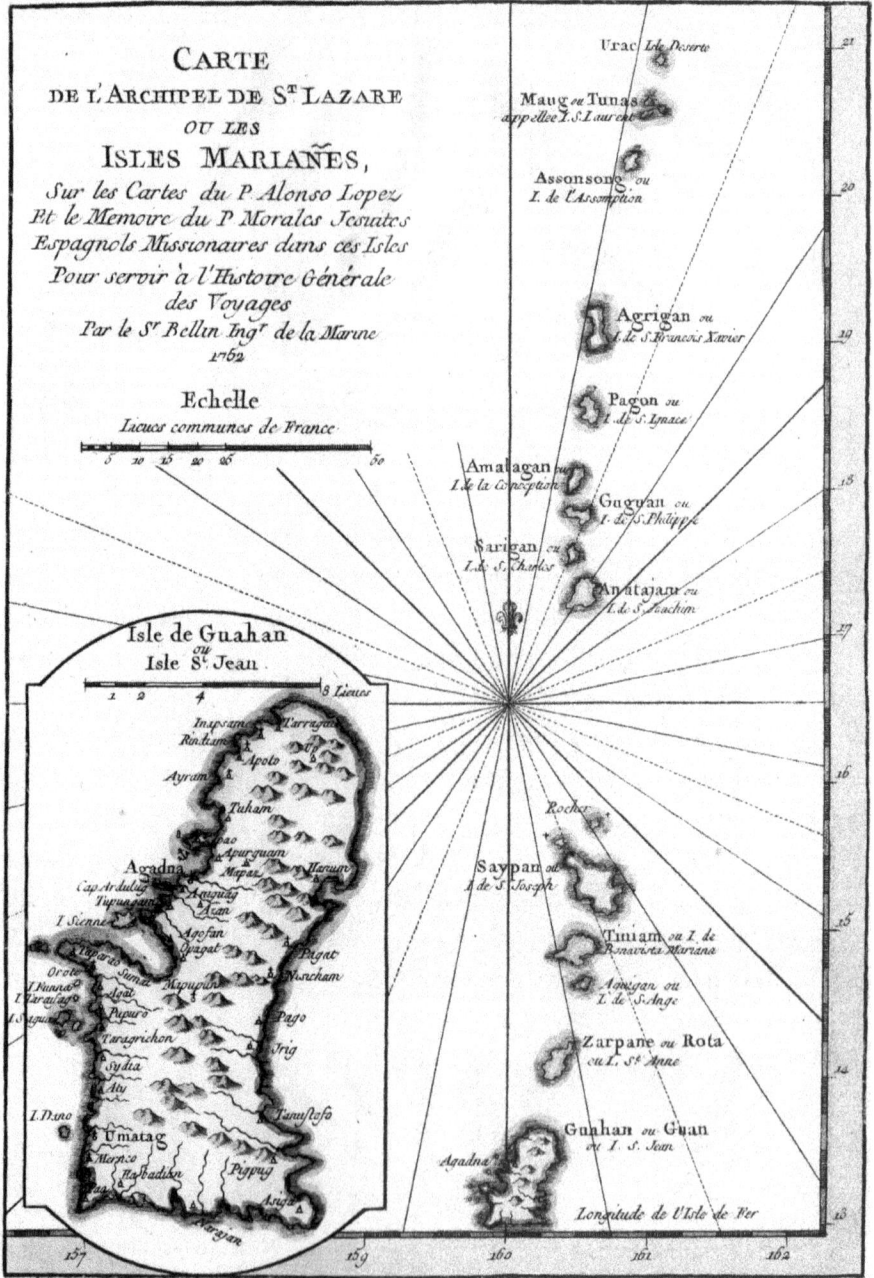

CARTE
DE L'ARCHIPEL DE S.^T LAZARE
OU LES
ISLES MARIAÑES,
Sur les Cartes du P. Alonso Lopez
Et le Memoire du P. Morales Jesuites
Espagnols Missionaires dans ces Isles
Pour servir à l'Histoire Générale
des Voyages
Par le Sr Bellin Ing^r de la Marine
1762

Echelle
Lieues communes de France

Urac *Isle Deserte*

Maug *ou* Tunas
appellée I. S. Laurent

Assonson *ou*
I. de l'Assomption

Agrigan *ou*
I. de S. François Xavier

Pagon *ou*
I. de S. Ignace

Amalagan *ou*
I. de la Conception

Guguan *ou*
I. de S. Philippe

Sarigan *ou*
I. de S. Charles

Anatajan *ou*
I. de S. Joachim

Isle de Guahan
ou
Isle St Jean

Rocher

Saypan *ou*
I. de S. Joseph

Tinian *ou I. de*
Bonavista Mariana

Aguigan *ou*
I. de S. Ange

Zarpane *ou* Rota
ou I. St Anne

Guahan *ou* Guan
ou I. S. Jean

Agadna

Longitude de l'Isle de Fer

N.° 2. Tome X. in 4.° Page 364.

Tome 4. in 8.° Page 359.

{Figure 6: Map of the Ladrones or Mariana Islands, based on the map by Father Alonso López and the Memoir by Father Luis de Morales. Extracted from *Le Petit Atlas Maritime Recuil de Cartes et de Plans des Quatre Parties du Monde*. Tome III (1764), no. 69, courtesy of the Micronesian Area Research Center (MARC)}

Account of the islands' situation and extension by Father Luis de Morales, S.J.

(page 2) Guahan or Guam [Guåhan], the largest and southernmost of these islands, has a circumference of forty leagues.[351] The Spanish call it San Juan [Bautista] Island. It is located at thirteen degrees 25 minutes northern latitude (and seven leagues from the Island of Zarpana).

Zarpana or Rota [Luta], which the Spanish call Santa Ana Island, has a circumference of 15 leagues. It is located at 14 degrees and 13 leagues from the island of Aguiguan.

Aguiguan [Aguijan] or Santo Ángel is 13 leagues in circumference. It is located at 14 degrees 43 minutes and one league from the island of Tinian.

Tinian, or the Buena Vista Mariana Island, has a circumference of 15 leagues. It is at 14 degrees 50 minutes and 3 leagues from the island of Saipan.

Saipan, or San José Island, has a 25-league circumference. It is at 15 degrees 20 minutes and 35 leagues from the island of Anatajan.

Anatajan [Anatahan] or San Joaquín Island, has a 10-league circumference. It is located at 17 degrees 20 minutes and 3 leagues from the island of Sarigan.

Sarigan, or San Carlos Island, has a circumference of 4 leagues and is at 17 degrees 35 minutes and 6 leagues from the island of Guaguan.

Guaguan or San Felipe Island [Guguan] has a 3-league circumference and is located at 17 degrees 10 minutes and 10 leagues from the island of Pagón.

Pagón [Pagan] or San Ignacio Island, has a circumference of 14 leagues, and is located at 19 degrees and 10 leagues from the island of Agrigan.

Agrigan [Agrihan] or the Island of San Francisco Javier, has a 16-league circumference. It is located at 20 degrees 4 minutes and 20 leagues from the island of Asonsong.

(page 3) Asonsong or Asunción Island [Åsunsion], has a circumference

---

351. The island of Guåhan (or Guam) is the main and most meridional of the isles and islands that comprise the Marianas archipelago, a set of fifteen volcanic and coral islands that extend from north to south, forming a wide arc of more than 800 kilometers in the western Pacific, between the Tropic of Cancer and the Equator (Fernando Ciaramitaro, "Política y religión: martirio jesuita y simbolización monárquica de las Marianas," *Convergencia. Revista de Ciencias Sociales,* 78 (2018), p. 198).

of 6-leagues. It is located at 20 degrees and 5 leagues from Mang.

Mang or Tunas [Maug] is composed of three rocks, each with a circumference of 3 leagues. The Spanish call them the Isles of San Lorenzo. They are at 20 degrees 35 minutes and 5 leagues from Urach [Uracas, also called San Agustin Volcano],[352] the last and northernmost of these islands.

---

352. Today Maug, or Mang or Tunas (note by L. de Morales).

History of the Mariana Islands Recently Converted to Christianity, and of the Glorious Death of the First Missionaries Who Preached the Faith in Them

Luis de Morales, S.J. & Charles Le Gobien, S.J.

BOOK ONE

The islands whose history I am about to write are situated in the Far East, in that great extension of sea that lies between Japan, the Philippine Islands, and the Kingdom of Mexico that the Spanish call New Spain. Magellan, renowned for his voyages and for the strait that bears his name, discovered them at the beginning of the past century [1521] during his famous expedition around the world.[353] Nevertheless, since his intentions were to make new discoveries, he did not sojourn in them, but contented himself with naming this multitude of islands the Archipelago of St. Lazarus[354] and continuing on to the Philippines, where he had the misfortune (page 4) of dying.

Upon their return to Spain, his companions told such magnificent descriptions of the countries that they had discovered, that the excited Spaniards decided to join the East with the West by conquering all of these islands that were hitherto unheard of in Europe.[355]

It was a tremendous project, but nothing intimidated Emperor Charles V, who, seeing himself the master of almost all of Europe and a great part of America, wished to extend his dominion to the far corners of the earth, and subject the whole world to his empire. Moreover, although he sent General Ruy López de Villalobos [1500–1544] to the Philippines to verify the news given by Magellan's companions, the emperor could not fulfill his desire.[356] His son,

353. Portuguese Ferdinand Magellan departed Seville for this expedition on August 10, 1519. He happily crossed the strait that bears his name and entered the Pacific Ocean on October 28, 1520. He died on April 27, 1521 in the Philippine Island of Mactan, fighting against one of the kings of the country. His companions returned to Seville on September 8, 1522. Pigafetta, *Relación* (note by L. de Morales/C. Le Gobien, *Histoire,* book 1, p. 2). For more on Magellan's arrival at the Mariana Islands, see Robert F. Rogers and Dirk Anthony Ballendorf, "La llegada de Magallanes a las islas Marianas." *Revista Española del Pacífico,* 2, 1992.

354. Magellan named this multitude of islands the Archipelago of St. Lazarus because he discovered them on the day when the story on St. Lazarus's resurrection was read in Mass (Francisco Colin, S.J., *Historia de las Filipinas,* T. I, C. I) (note by C. Le Gobien, *Histoire, book* I, p. 2).

355. Knight Antonio Pigafetta was one of the companions of Magellan who completed the trip around the world. His *Relazione del primo viaggio intorno al mondo* (1524), dedicated to the Grand Master of Rhodes, Philippe de Villiers de L'Isle-Adam, was published in Venice in 1536 and was incorporated in Ramusio's Tome 2 (note by L. de Morales). The text clearly embodies the scientific curiosity of the Renaissance, stressing the geography of the lands as well as the customs and even languages of their inhabitants, all with great doses of imagination (J. P. Rubiés, "The Early Spanish Contribution to the Ethnology of Asia," pp. 418–419). Chilean historian José Toribio Medina wrote the first Spanish translation (1888) and included it in the *Colección de documentos inéditos para la historia de Chile.* The last known translation, written by Leoncio Cabrero Fernández was published as *Antonio Pigafetta, Primer viaje alrededor del mundo* (Madrid: Dastin Historia, 2002).

356. See José María Ortuño Sánchez Pedreño, "La expedición de Ruy López de Villalobos a las islas del Mar del Sur y de Poniente. Estudio histórico-jurídico." *Anales del Derecho,* 23, 2005, pp. 249–292.

Philip II, who succeeded him on the throne and wanted to carry out his father's unfinished plan, ordered the Viceroy of New Spain, Luis de Velasco y Ruiz de Alarcón [1511–1564], to equip a fleet and conquer the islands that Magellan and Villalobos had discovered in the Eastern seas, thus opening the way to greater conquests.

This important mission was given to Admiral Don Miguel López de Legazpi [1502-72][357]. He left Mexico in the beginning of 1563 during monsoon season, that is, when the winds blow constantly from East to West with notable strength. The first islands that he found on his way were the Marianas, which Magellan had unfairly named (page 5) the Ladrones Islands, because their inhabitants, who had never seen iron, took some pieces of this metal, as well as various instruments also made of iron, but which had little value.[358] The great number of small boats with billowing sails that surrounded the Spanish ships caused them to be christened the Islands of the Lateen Sails [Islas de las Velas Latinas]. But this name was lost and today we know them by a different one, for it was only after the now deceased Queen of Spain, Mariana of Austria,[359] sent missionaries there to preach the Gospel, that they were called Mariana Islands.

Legazpi disembarked and, cross in hand, took possession of them in the name of the king his lord, but seeing that they were lacking in all comforts, he did not remain there long. Instead, he contented himself with befriending their inhabitants and promising them preachers who would instruct them in the true religion, and continued his voyage to the islands of Luzon and Mindanao, and which the Spanish called the Philippines in honor of Philip II.[360]

---

357. The Miguel López de Legazpi-Andrés de Urdaneta expedition took formal possession of the island of Guåhan on January 22, 1565 (V. Muñoz Barrreda, *La Micronesia española*, p. 40; A. Molina, *América en Filipinas*, p. 28; Amancio Landín Carrasco, "Descubrimientos españoles en la Micronesia," in Javier Galván Guijo (ed.), *Islas del Pacífico: el legado español*, Madrid: Ministerio de Educación y Cultura, 1998, pp. 17–25; Omaira Brunal-Perry, "Las islas Marianas enclave estratégico en el comercio entre México y Filipinas," in Leoncio Cabrero (ed.), *España y el Pacífico. Legazpi*. Vol. I, Madrid: Sociedad Estatal de Conmemoraciones Culturales, 2004, pp. 543–46).

358. Quimby points out that for the Chamorro, iron represented "a continuum of encounters—interactions through time that were informed by previous meetings and outcomes—which generated a mediating contact culture, creating trade relationships and political entanglements which facilitated subsequent colonial intrusion" (Frank Quimby, "The Hierro Commerce: Culture Contact, Appropriation and Colonial Entanglement in the Marianas, 1521–1668." *The Journal of Pacific History*. Vol. 46, no. 1, 2011, p. 2).

359. She died in Madrid on the night of May 16 or 17 of 1696 (note by C. Le Gobien, *Histoire, book* I, p. 4).

360. General Ruy López de Villalobos had first named some of these islands the Philippines in 1543, but for political reasons the Spanish preferred to call them Islas del Poniente (Sunset Islands), due to the division between the East and West Indies that Pope Alexander VI had formalized for Portugal and Spain through the renowned demarcation line. The Spanish did not officially christen them the Philippine Islands until Admiral Legazpi conquered them in 1564 (note by L. de Morales/C. Le Gobien, *Histoire, book* I, p. 5).

It took Legazpi very little time to win these islands. The Spaniards founded important establishments and committed themselves to converting their peoples, sending a great number of missionaries every year from New Spain to cultivate this land of infidels. For more than a century, Luzon, Mindoro, Mindanao, and other neighboring islands served these apostolic men as grounds where they could exercise their zeal (page 6), reaping extraordinary fruit and winning almost all of these peoples for Jesus Christ.

Only the Marianas were abandoned, even though their inhabitants were of good character and seemed willing to embrace the Gospel. The missionaries frequently asked the governors of the Philippines to send preachers to instruct those lands' natives in the truths of our Holy Faith, but since these men were more interested in their commercial affairs than in extending the kingdom of Jesus Christ, they put off executing this necessary task. Nothing more was said regarding this enterprise and the missionaries exercised their zeal in the conversion or instruction of the peoples in the Philippines, until God sent a new apostle to make his name known to the inhabitants of the Archipelago of St. Lazarus.

The man chosen by God was Father Diego Luis de San Vitores of the Society of Jesus, more distinguished by his virtue than by his birth, for[361] he belonged to one of the most illustrious families of Burgos, the capital of Old Castile.[362] God, by whose (page 7) will he was destined to apostolic duties, endowed him with extraordinary graces, called him to the Society in a miraculous way,[363] and elevated him before long to a most exalted sanctity.

Although he had a great talent for the sciences, and even though he taught Philosophy in one of Spain's most famous universities,[364] missionary work appealed to him. Since he was young, God had instilled in him the desire to devote himself to it. He joined the Society with this aspiration, and this

361. It should say "even though." See the Prologue by J. P. Rubiés in this book.

362. San Vitores has been an illustrious family name in Spain for more than 250 years. It derived from the Castle of San Vitores in the mountains of Burgos [Sierra of Tuio]. His father, Don Jerónimo de San Vitores de la Portilla, a knight of the Order of St. Juan, held distinguished and trusted positions in the Spanish court, and died while he was a member of the Consejo Real de Hacienda [Royal Treasury Council]. His mother, Doña Francisca Alonso Maluenda was from an even more illustrious family, for she descended from Don Juan Fernando Alonso Antolínez, a relative of the Cid, so famous in Spanish history. She had first been married to Don Juan de Quintabal, and gave birth to Don Juan de Quintabal, knight of Malta, who after having been in Malta joined the Society of Jesus and was destined to the missions in Japan. He embarked on a ship in Lisbon, and fell ill shortly thereafter while assisting passengers who were struck with a contagious disease, dying on board in 1637 (note by L. de Morales/C. Le Gobien, *Histoire, book* I, p. 7).

363. See the letter that he wrote to his General asking him for permission to consecrate himself to the mission in the Indies at the end of this *History* (note by L. de Morales/C. Le Gobien, *Histoire,* book I, p. 7). This letter was an appendix in the end of Le Gobien's book, but it was not found in Luis de Morales's manuscript.

364. In Alcalá in 1655 (note by C. Le Gobien, *Histoire, book* I, p. 7).

ardor, far from cooling down, in time grew more and more, if that was possible. This zeal led him to go on missions to the small country towns, exercising this saintly ministry and taking advantage of the company of Father Jerónimo López, another Jesuit.[365] This missionary, famous across all of Spain, had worked for nearly forty years converting souls with a success that was unheard of since St. Vicente Ferrer [1350–1419]. Admiring the talents that God had granted Father San Vitores for the duties of the apostolic life, he applied himself to cultivating them. He thus encouraged Father San Vitores's association with Father Thyrsus González de Santalla, who was so celebrated in the Society for his zeal, his science and his virtue, that he was elevated to the position of General of the Society,[366] an office that he still holds with great dignity.

Father López had set his eyes on Father San Vitores as his successor in the missions, but God, who had other plans for his servant, called him to convert the infidels. Father San Vitores had told his superiors, frequently beseeching them to allow him the fulfillment of his desire. But since the Society's superiors are accustomed to test the vocation of the apostolic men that they send to the missions for a long time, they did not agree to Father San Vitores' desires, and this afflicted and saddened him so, that he lovingly complained to God our Lord in the fervor of his prayers (page 8), saying:

> *Is it possible, Lord, that they will always oppose our plans, and that the ardent desires that you have given me during my infancy will remain sterile and fruitless? Speak [to them] yourself, Lord, [for] they no longer listen to me, and make your will known to those upon whom I depend, so that, abetting your designs they will allow me to*

365. This famous missionary's biography has been published in Spanish by Father Martín de la Naja (note by C. Le Gobien, *Histoire, book* I, p. 8). Valencia-born Jesuit Jerónimo López (1589–1658) popularized the Society's missionary apostolate through a nocturnal procession in the most populous towns and cities—an ascetic practice better known as the "act of contrition"—to get close to parishioners who did not go to Church (Paolo Broggio, *Evangelizzare il mondo. Le missioni della Compagnia di Gesù tra Europa e America (secoli XVI-XVII)*, Roma: Carocci Editore, 2004; Broggio, "L'Acto de Contrición entre Europe et nouveaux mondes. Diego Luis de Sanvitores et la circulation des stratégies d'évangelisation de la Compagnie de Jésus au XVIIe siècle," in Pierre-Antoine Fabre and Bernard Vincent (eds.), *Missions religieuses modernes. "Notre lieu est le monde."* Rome: École Française de Rome, 2007, pp. 229–59). During his stay in the mission of Taytay (1663–1668), Father San Vitores translated Father López's "act of contrition" to Tagalog (H. de la Costa, S.J., *The Jesuits in the Philippines...*, p. 471). As Broggio notes, "Sanvitores ne se limita pas a transplanter en Extrême-Orient des cérémonies déjà en usage dans la mère patrie: il conçut des techniques nouvelles qui tiraient profit de l'expérience acquise dans les missions intérieures" (Broggio, "L'acto de contrición...", p. 251). See also F. García, S.J., *The Life and Martyrdom of the Venerable Father Diego Luis de San Vitores*, book III, chapter VIII, p. 203. A recent biography can be found in *DHCJ*, vol. III, p. 2415.

366. He was elected General in Rome on July 6, 1687, after having founded missions in Spain and held the Chair of Theology in the University of Salamanca for many years (note by C. Le Gobien, *Histoire, book* I, p. 9).

carry your name to the furthest reaches of the world.

God heard his servant's pleas. Father San Vitores fell ill, and in a few days, his malady reduced him to the end. He waited only for death and, completely resigned to the will of God, prepared for it in the most complete peace and tranquility, when he received a letter from the Cardinal of Toledo[367] who, unaware of the state of San Vitores's health, invited him to a mission and exhorted him to devote himself absolutely to this holy office. Upon hearing this letter read to him, Father San Vitores awakened from a profound lethargy, and recalled from the bottom of his heart all of his thoughts on the missions. He manifested to the school's rector how he had always held the desire of working in the conversion of infidels and spilling his blood for Jesus Christ. He asked permission (*permissu superiorum*) to take a vow to consecrate himself to that saintly ministry under the protection of St. Francis Xavier and the martyr government of Father Marcel Francis Mastrilli,[368] who having gone to Japan for a few years had the blessed joy of giving his life in the defense of the faith [after having suffered for days the terrible torment of the water[369] and the trial of the pit, or *fossa*[370]].[371]

367. His name was Baltasar de Moscoso, Cardinal of Sandoval, Archbishop of Toledo, and Great Chancillor of Castile. He died on September 17, 1665. He was the son of Don López Moscoso Osorio VI, count of Altamira, grandee of Spain; and Doña Leonor de Sandoval y Rojas (note by L. de Morales/C. Le Gobien, *Histoire, book* I, p. 10).

368. Father Mastrilli (1603–1637) was educated in the College of Naples, where he finished his novitiate and studied the Classics (1620–1621), philosophy (1624–1627) and theology (1628–1632) until he embarked on a ship in Lisbon in 1635 to India with 32 Jesuits. On the way to Macao, he stopped in Manila and Mindanao (1636) where he worked as a missionary and a harbinger of peace. He was in Japan in 1637, when persecution against Christians intensified. According to Le Gobien, "they cut his head on October 17, 1637 in Nagasaki. [San Vitores's] recovery, which came to him in Naples on January 4, 1653 through the intercession of St. Francis Xavier, is one of the most brilliant miracles of this century. Jesuit Father Juan Eusebio Nieremberg has written his *Vida* in Spanish" (note by L. de Morales/C. Le Gobien, *Histoire,* book I, p. 11). For more on Mastrilli, see *Varones Ilustres de la Compañía de Jesús,* Bilbao: Administración del "Mensajero del Corazón de Jesús," vol. I, 1887, pp. 480–646; *DHCJ,* vol. III, pp. 2566–67.

369. The victim is made to lie on a ladder, and water is forced down his throat through a funnel. Then a board is placed on the stomach, and it is weighted down with such heavy objects, that water and blood stream out of every place with incredible violence and pain (note by C. Le Gobien, *Histoire,* book I, p. 11).

370. The torment of the pit is horrible. A great hole is dug and filled with filth and venomous creatures, like snakes, frogs, etc. The patient is suspended upside down inside the hole, where he is left to die (note by C. Le Gobien, *Histoire,* book I, p. 11). These two notes and the bracketed text that they refer to were written by Le Gobien, not by L. de Morales.

371. In the biography of Nicholas Keian Fukunaga, the first victim of this new method of torture, can be found the following description, which according to Cieslik, corresponds to the 1633 Jesuit annual letter from the Philippines: "They dug a pit some feet deep, and above it they erected a frame from which the body was hung up by the feet. To prevent the blood flowing into the head and causing death too quickly, they tied the body tightly with ropes and cords. The hands were tied behind the back, and the prisoner was lowered into the pit down to his belt or navel or even down to his knees and legs. The pit was then closed by two boards which were cut in such wise that they surrounded the body in the middle and let no light enter. In this fashion they kept the man hanging upside down, without food, poised between life and death and in doubt about the final outcome, until the slowly rising blood pressure brought about complete exhaustion, or

He had barely pronounced this vow when he found his health notably improved and, within a few days, he was completely cured (page 9). Everyone thanked God for such a speedy cure and such an evident vocation. No longer doubting that God called him to the conversion of infidels, his superiors granted him what he had long been requesting and destined him to the mission of the Philippines, where there was much work and suffering to endure. The new missionary thought only of carrying out the plan that God had for him. What finally confirmed his resolution was that he was cured in a similar way from two other illnesses. He went to Cadiz where he boarded the fleet that was leaving for New Spain on May 15 of 1660. The voyage was brief and pleasant. He arrived in Mexico at the end of the month of July, and stayed in the capital of that kingdom for some time before departing for the Philippines. Since his zeal burned him with the desire to win souls for Jesus Christ, he began to carry out his apostolic mission in New World's most beautiful and populous city. With his preaching, he secured extraordinary conversions, and the success with which God crowned his efforts was so magnificent, that many tried to keep him from leaving for the Philippines when the time to embark came.

The Viceroy,[372] who had frequently gone to his sermons and been greatly moved, believed that he was serving religion and the state by keeping a man so full of zeal and permeated by God. He communicated his designs to San Vitores and did everything in his power to persuade him to work reforming the customs in the city of Mexico, which was mired in luxury and debauchery, than to go to the Philippines to preach the faith to a small number of barbarians, for this was preferable for religion's good and the well-being of souls. He argued that New Spain was a vast field where he could exercise his zeal, and that he had an abundance of sinners and infidels to convert without having to go find them somewhere else. The Viceroy's words left no impression in the new apostle's spirits. God was calling him to go elsewhere, and his only thought was to go to the Philippines, so that nothing could deter him.

He embarked [on the *San Damián* galleon] in the port of Acapulco[373]

---

else hunger entirely sapped his physical strength. Or until, worn out by the torment, loneliness and solitude, he finally succumbed to this deadly torture and renounced his faith while there was still life left in him. Nicholas hung three days in the pit, from vespers on Thursday until noon on Sunday, 31 July, the day on which we celebrate the feast of our holy Father Ignatius" (Hubert Cieslik, "The case of Christovao Ferreira," *Monumenta Nipponica,* Vol. 29, nº 1, 1974, p. 14).

372. The Viceroy was the Conde de Baños, Don Juan Francisco de Leiva y de la Cerda (1660–1664), named grandee of Spain in 1691 along with all his descendants (note by C. Le Gobien, *Histoire, book* 1, p. 13).

373. This port, which lies on the Pacific Ocean, is the most renowned in Mexico (note by C. Le Gobien, *Histoire, book* I, p. 14).

(page 10) with fourteen other missionaries of the Society, over whom he was made superior. Trips across the Pacific Ocean are not wont to suffer tempests or setbacks, for the wind is always favorable, and Father San Vitores used this opportunity to give a mission in the ship and instruct the crew in the mysteries and maxims of our holy religion, following the example of St. Francis Xavier, his model and patron.[374]

Before arriving in the Philippines, ships sailed some three-or-four leagues off the Marianas, which were in plain sight.[375] As soon as these islanders saw the fleet from New Spain, they launched their canoes and surrounded the Spanish ships in their little boats loaded with fruit, which they exchanged for knives and other sundries.

Father San Vitores received these poor islanders with a tenderness and kindness that delighted them. But when he learned that this numerous people was mired in the darkness of paganism and that nobody had brought the light of the Gospel to these unfortunate lands, he could not contain his tears. *"How is it possible, Lord"*, he exclaimed, *"that men expose themselves to a thousand dangers each day, and risk spending their lives among barbarians in the most distant countries to acquire passing riches, but nobody is interested in saving the souls that cost Jesus Christ all of his blood?"* Upon saying this, he threw himself at the feet of his crucifix to ask Our Lord not to abandon this poor people and to send them preachers filled with his spirit and moved by its true zeal.[376]

God heard his prayer and led him to understand in the bottom of his heart the words that he had heard during his last illness in Madrid: *Evangelizare pauperibus misi te* [I have destined you to preach the Gospel to the poor]. He had believed until then that God was calling him to Japan, and this is why he was going to the Philippines, harboring no doubts that the Lord himself would provide him with an occasion to enter that empire that had for so long stood closed off from the light of the Gospel. But he then knew (page 11) distinctly that he was destined to the conversion of those islands that he saw in extreme need and total abandonment. If it had been up to him, he would have then and

374. Valladares, *Castilla y Portugal en Asia*, pp. 102-04.

375. Transpacific trade's importance grew after Chinese traders settled near Manila (Parián, Binondo). Traders were limited to a cargo of 250.000 pesos de 8 reales in merchandise from Manila to Acapulco. On the return trip, galleons could bring to the archipelago 500.000 pesos fuertes de plata de 8 reales, out of which came the salaries of government employees (the situado) (Carmen Yuste López, *El comercio de la Nueva España con Filipinas, 1590-1785*. Mexico, DF: Instituto Nacional de Antropología e Historia & Departamento de Investigaciones Históricas, UNAM, 1984, pp. 10; 14).

376. We are to understand here that San Vitores's voice is also that of Father Morales, who uses the martyr to say what a Jesuit procurator would say in the court in order to gain royal favor and economic support for the missions.

there consecrated himself to those poor people, staying with them, but since his superiors had sent him to the Philippines, he believed that it was worthier to go there and then take the necessary measures to pass into the Mariana Islands, to solidly establish the faith in them. Thereafter, all his thoughts were on the conversion of these people, and these thoughts filled him with joy and delight.

He arrived at the Philippines [Lampong] on August 10, 1662, where he was received as an apostle.[377] The zeal that animated him did not allow him a moment of rest. He retired with his companions to carry out the Spiritual Exercises of St. Ignatius, as was the custom in the Society, and to draw the blessings of heaven over the work that the new missionaries were about to begin. Having barely finished them, the missionaries were sent where laborers such as themselves were most needed.

Father Miguel Solana,[378] a man of consummate virtue and extraordinary merit, who had been the provincial of the order in the Philippines, asked Father San Vitores to work with him in the mission of Taytay,[379] to which he had consecrated himself. Although the Tagalog language that was spoken in the Philippines was very difficult, the new missionary learned it in such a short amount of time that Father Solana was greatly surprised. His admiration grew even more when he saw the impression that the words of the new apostle caused in the hearts and spirits of these savage peoples. He looked at him then as an extraordinary man for whom God had special plans.

Father San Vitores went on to Mindoro,[380] one of the largest of the Philippine islands, and although he witnessed the continuous blessings that God showered upon his work, he was constantly preoccupied with the idea of going to the Mariana Islands. He remembered them incessantly, and how they had been the first lands of the Orient (page 12) discovered by Magellan, and how Legazpi had taken possession of them in the name of the king of Spain; and he was deeply saddened by how they had been abandoned because of their extreme poverty. The more he recalled those savages, who lacked all the comforts of life and all that men seek so fervently, the more he felt the need to succor them

377. F. García, S.J., *The Life and Martyrdom of the Venerable Father Diego Luis de San Vitores*, book II, chapter III, p. 96.

378. Miguel Solana was provincial of the Philippines from 1668 to 1670. For more on Father Solana, S.J., see H. de la Costa, S.J., *The Jesuits in the Philippines...*, pp. 410–411; 425; 435–37; 486.

379. It is a great place, six or seven leagues from Manila, the Philippine capital (note by C. Le Gobien, *Histoire, book* I, p. 17). The parish of Taytay was established by Franciscans in 1579; the Jesuits took over it in 1591. By 1630, they had built the church of St. John the Baptist, which is still standing.

380. Le Gobien inserted here the following note: "This lies in front of the city of Manila, separated only by an arm of the sea" (C. Le Gobien, *Histoire, book* I, p. 17). See also F. García, S.J., *The Life and Martyrdom of the Venerable Father Diego Luis de San Vitores*, book II, Chap.s X-XI, pp. 123–35.

and work for their conversion. He thought that he could hear the voice of God constantly repeating to him in the bottom of his heart the following words: "*Remember that I have destined you to preach to the poor and to make my name known to the most forsaken peoples.*[381]"

Preaching in Manila on the Feast of San Juan de Dios [March 8] before a large number of people,[382] these words had come to San Vitores's mind, and they affected him so much that he turned his sermon into a sort of dialogue in which the Marianas barbarians cried to God Our Lord that they were enveloped in the densest darkness and gloom, and that nobody had taken the trouble to bring them into the light of the faith that was preached with so much zeal and success in all other nations. He spoke these words so movingly that his entire audience broke down in tears, and he took this instance as a favorable opportunity to present the plan that he had been thinking of for so long, and he did it with such power and such vitality, that they all showed much interest in this great enterprise.

Animated by this promising beginning, he went to meet the governor[383] and the rest of the king's ministers on whose shoulders these matters rested. He explained to them that the time had come to work on the conversion of those abandoned peoples; that the more miserable they were, the greater the duty to become interested in their wellbeing; and that this pure and disinterested zeal would greatly please God and would at the same time honor the nation. That this was a favorable occasion to confound the malignant slander of heretics, who published everywhere that the Spanish preached the Gospel (page 13) only to take other people's wealth and resources under the pretense of piety; that the affection those islanders had always shown the Spanish, their docility, their kind character, the purity of their customs, and their long-held desire to embrace the Christian religion, deserved some effort be carried out in their favor. That there was no need to fear expenses because these consisted in providing for the needs of a small number of missionaries and the necessary catechists.[384]

---

381. Father Paul V Daly's translation of Book One ends here.

382. See F. García, S.J., *The Life and Martyrdom of the Venerable Father Diego Luis de San Vitores*, book II, chapter IV, pp. 98–101.

383. Don Diego Salcedo, Governor of the Philippines (note by C. Le Gobien, *Histoire, book* I, p. 19).

384. John N. Schumacher argues that these assistants were not really catechists, for there were yet no schools where catechists could be formed (J. N. Schumacher, "Blessed Pedro Calungsod, Martyr: An Historian's Comments on His Philippine Background." *Philippine Studies*, 49 (3), 2001, p. 305). However, one must take into account that, given the lack of personnel faced by Jesuits in these missions, assistants were granted a degree of autonomy in educational and pastoral tasks that was unthinkable in Europe and other places. In any case, after 1680, the students of San Juan de Letrán were very well prepared—there were musicians, "who play the harp very well and fulfill the role as catechists entrusted to them by the Fathers with great

That he was sure the king would approve whatever sacrifices were made for this holy and glorious enterprise; that he knew of his zeal and piety, and he knew the conversion of a nation was more gratifying to him than the conquest of a kingdom, but what would move and interest him most of this enterprise was his accountability to God for the irreparable loss of so many souls rescued by the blood of Jesus Christ, who perished because no missionaries were sent to them.

The king's ministers praised the priest's zeal, but they justified their lack of support for his plans arguing that they had neither ships; nor missionaries; nor money to finance the trip, adding that it was practically impossible to sail from the Philippines to those islands because of the contrary winds that swept those seas; that most vessels that had tried this voyage had foundered; that to solidly establish such a mission it was necessary to have enough funds to cover the expenses that would be continually generated, and a ship each year to take the necessary provisions to the missionaries; that the royal treasury was exhausted, incapable of assuming these costs, so that an enterprise of this importance could not be undertaken. Although at first sight this project seemed quite easy, when one considered it more closely one could not fail to see its many difficulties. Thus, it was best to exhibit a more prudent and less indiscreet zeal, and not persist in carrying out this enterprise (page 14) when there was such a great number of infidels to convert in the Philippines. In sum, that it was necessary [for Father San Vitores] to moderate his zeal and apply himself to cultivating the old missions, instead of undertaking new ones that could not be sustained.

These arguments, supported by the king's ministers, and even by some members of the church who were jealous of the Father or who did not know the matter too well, impressed many. The Father's desire was looked upon as an impossible enterprise that should no longer occupy anybody's thoughts, for the efforts that had been attempted during the last century in order to carry it out had been useless. However, this opposition did nothing but inflame the new apostle's zeal. He was not at all discouraged by the resistance that he encountered wherever he went, stating with resolve and conviction that the time to convert those peoples had finally come, and that what seemed impossible to men was not impossible for God, and so he multiplied his prayers and penances to move God Our Lord to take pity on those poor people.

He was accompanied in his efforts by Don Miguel [Millán de] Poblete [1603-67], Archbishop of Manila, prelate of a rare merit and a truly great apos-

---

diligence" (AHCJC, *Documentos Manuscritos Historia de las Filipinas* (FILPAS), no. 52, 1668–1686, f. 357v).

tolic virtue.[385] However, all of the archbishop's efforts to convince the viceroy and the other Spanish officials were for naught. They all persisted in their first resolution, saying that such a project was impossible. The Father understood that he had to use more energetic and effective means to carry out this project, and he decided to address the king at that time, Philip IV, of whose zeal and good intentions he was well aware.

He therefore sent that prince a memorial that was sure to move him, in these terms:

> *The apostle of the Indies, Francis Xavier, wrote to Father Simon Rodriguez, one of St. Ignatius's first companions and superior of the Jesuits in Portugal, the following (page 15) words: "If I thought that the king would not take offense at the advice of a faithful servant who truly loves him, I would warn him to meditate for fifteen minutes every day upon this divine dictate: What good is it if someone gains the whole world but loses his soul? It is time to show the prince the error of his ways, and warn him that the time of his death is closer than he may think. That fatal time of judgment when the King of kings and Lord of lords will call upon him and say these tremendous words: "Tell me your deeds." Thus, do what you must to fulfill your duties well, and send to the Indies all the necessary resources to increase the faith.*

> *That which inspires me shows me only one way to make religion flourish in the Indies, and it is that the king make it known to all of his representatives in the country that he will only trust those who show zeal in their desire to increase religion; that he order them to work harder on the conversion of the island of Ceylon,[386] and to raise the number of neophytes in the cape of Comorin;[387] that missionaries are sent for across the world; that the workers of the Society of Jesus be used for such a holy ministry as well as all those deemed appropriate; that those neglectful governors who do not carry out their duties must be made to understand that if they do not procure the extension of religion in the Indies, they will be punished upon their return to Lisbon*

385. For more on criollo Miguel Millán de Poblete, the archbishop of Manila between 1653 and 1667, see Alexandre Coello de la Rosa, "El ascenso de los «hijos de la tierra» durante el primer período de gobierno del arzobispo de Manila, Miguel de Poblete (1653-1663)," *Anuario de Historia de la Iglesia*, 33 (2023), pp. 1-47.

386. This island faces the tip of the famous Indian peninsula on the eastern side of the Ganges. This tip is called Cape Comorin, or the Fishery coast, because that is where pearls are fished (note by C. Le Gobien, *Histoire, book* I, p. 24).

387. Cape Comorin is the southernmost tip of the Indian subcontinent.

with the confiscation of all their goods and a lengthy imprisonment.

If the king were to order this, punishing those who did not obey him, by the grace of God we would see many conversions. But if he does not apply himself to this, nothing will be obtained. You know what I think of this. I say nothing of other matters that are irrelevant to this one. I only add that if my desires were to be carried out, neophytes (page 16) would not be exposed to thievery and mistreatment as they have been so far, and this would contribute greatly to the infidels' conversion. However, if the king's orders are not taken into account, and his representatives do not apply them, every effort is useless. I speak from experience and I know why things are done this way, but it is not necessary for me to provide any more explanations. I wish I could see two things in the Indies. The first: that the king gave that order I have talked about, to force his representatives to do their duties. The second: that good preachers were sent to all of the Portuguese strongholds. Nothing would benefit those countries more than the increase of our faith."

These words are all from St. Francis Xavier. He also proposed at that time the means that would be conducive to the increase of our religion in the Indies, particularly in Ternate[388] and Zamboanga, which is a stronghold of Your Majesty in the island of Mindanao,[389] where we know that St. Francis Xavier preached the Gospel. These two fortresses lie abandoned today, causing the loss of many souls and the ruin of religion, which will be lost in Zamboanga entirely if a garrison is not reestablished.

Necessity in those countries is even greater now than it was in the time of St. Francis Xavier. The great island of Borneo and its neighbors are also without resources. But what all persons who are devoted to the glory of God and worthy of Your Majesty's attention

---

388. One of the Mollucas (note by L. de Morales/C. Le Gobien, *Histoire, book* I, p. 26). Today, these are called the Maluku Islands.

389. From the early seventeenth century, Jesuits had been trying to evangelize the island of Mindanao. To dissuade the Dutch from intervening in the Philippines, a fort (*Nuestra Señora del Pilar* de Zamboanga) was built on the western tip of Mindanao in 1634, and from there the Jesuits planned to expand inland, across the "Moorish" lands. The Jesuits stayed until 1663, when the Philippine authorities decided to transfer the soldiers from the Zamboanga fort back to Manila to defend the colony from a hypothetical Chinese invasion. The Jesuits did not return to the fort until 1718 (Miguel Saderra Massó, S.J., *Misiones jesuíticas de Filipinas (1581–1768 y 1859–1924)*, Manila: IF, 1924, p. 35). See also Josep Ma Fradera, "La formación de una colonia. Objetivos metropolitanos y transacciones locales," in Mª Dolores Elizalde, Josep Ma Fradera and Luis Alonso (eds.), *Imperios y naciones en el Pacífico*, vol. I, pp. 89.

deplore the most, is seeing that the Islas de las Velas [Islands of the Lateen Sails][390] *are still submerged in the darkness of idolatry, despite the zeal with which the Catholic kings have had the gospel preached everywhere, and the orders that they have given more than once, that (page 17) the conversion of this abandoned people be undertaken. But what good are these orders if those who must follow them do not do so even when the most favorable opportunities present themselves before them? Alas, these peoples are indeed worthy of pity, for the infinite number of souls that perish there, and for the danger that they are of being polluted by Mohammedanism that is spreading everywhere, forming new proselytes in the Orient each day, to the detriment of Christianity.*

If the voice of the Apostle of the Indies in the past century was heard in the court of Portugal, where his demands were met with great and happy success, what can we not expect of the court of the Catholic king in favor of this poor people redeemed by the blood of Jesus Christ? All of the ministers of our Lord must raise their voice in his name, and this is what the lowliest among them does today. Diego Luis de San Vitores

This memorial was well received in the Spanish court. The king was impressed with its contents. He reflected on St. Francis Xavier's words, to whom he was very devoted, and applied them to himself; and as a consequence, he judged the petition to be fair, and that Father San Vitores had very uncommon characteristics and a particular light from the heavens regarding this matter. This prince[391] died shortly thereafter, a good Christian, as was to be expected given his piety.

The Servant of God was not satisfied with this memorial to the king, and in order to secure the success of an affair that was so important to the glory of God and the health of many souls, he thought it convenient to procure the queen's intercession[392], that she may take these islands under her protection. This virtuous princess already had received news of the matter, and had manifested her sadness that the gospel had not been preached yet in these islands. Conditions were favorable, and the Father made good use of them. He wrote to

---

390. He is talking about the Marianas (note by L. de Morales).

391. Philip IV, king of Spain, died on September 17, 1665 (note by C. Le Gobien, *Histoire, book* I, p. 28).

392. Maria Anna of Austria, wife of Philip IV, king of Spain and mother of Charles II (note by L. de Morales/C. Le Gobien, *Histoire, book* I, p. 28).

Juan Everardo Nithard,[393] who was at that time the Queen's confessor, and who would later be the Archbishop of Palermo and a cardinal (page 18), so that he would direct her interest to this enterprise.

> *I will not remind you,* San Vitores told him, *what traders do each day, the dangers that they expose themselves to, the fatigues that they experience, the vigor with which they accumulate riches that thieves can take from them and of which death will surely rob them. There are better treasures to obtain in this country, and at a better price. It is up to the queen to obtain the better part of them. They are nothing but these abandoned lands where the news of Jesus Christ are yet to be preached, and all that she must do is take them under her protection and interest herself in the conversion of their peoples. Her piety and virtue are so great that she could convince all heretics of their error and win all infidels to Jesus Christ, and given the zeal that she is capable of, I have no doubt that she would. However, if it is not in her hands to convert the entire world, she can at least contribute to convert a great people that has its arms outstretched, and that would henceforth consider her their benefactor. She could indeed send to heaven a multitude of souls of children who die every day in these lands of infidels that would have been saved if only a minister of the Gospel had been there to baptize them.[394] I have often reflected upon this. If a child were dying without being baptized in the queen's palace, and she knew about it, would she not rush to baptize him herself if there were nobody there who could do it? What can we not expect then from this great princess in this occasion, when what occu-*

---

393. Father Juan Everardo Nithard, Grand Inquisitor of Spain, was made cardinal by Pope Clement X on May 16, 1672, and archbishop of Palermo in Sicily in 1675. He died in Rome on February 1, 1681 at the age of 73 (*Histoire, book* I, p. 28). For more on Father Nithard (1607–1672), Queen Maria Anna's influential confessor, spiritual director, and *valido* (favorite), see Julián J. Lozano Navarro, *La Compañía de Jesús y el poder en la España de los Austrias*, Madrid, Cátedra, 2005, pp. 297–324; and Javier Burrieza Sánchez, "Las Glorias del segundo siglo (1622–1700)," in Teófanes Egido (eds.), *Los jesuitas en España y en el mundo hispánico*, Madrid: Fundación Carolina & Marcial Pons Historia, 2004, pp. 169–171.

394. Marianas' Indians could not be considered subjects until they were introduced in the western world by the rite of baptism. Eduardo Subirats argues that "[baptism] established a new sacramental identity, violently and outwardly, but juridically indissoluble (...). Baptism was the first imposition of name and identity. It was the sacramental entry into the order of the verb, that is, of history as a providential order of divine reason" (*El continente vacío. La conquista del Nuevo Mundo y la conciencia moderna*. Madrid: Muchnitk Editores, 1994, p. 208). Baptism represented death as well as rebirth. Immersion evoked a regression to a pre-formal stage, that time before birth where all forms were dissolved, while immersion emulated the cosmogonic act of coming to life. It was a traumatic rite that submerged the individual in a "new life" outside of this world and away from sin. Mircea Eliade adds that baptismal nakedness also has a ritual and metaphysical signification: the abandonment of that, "old dress of corruption and sin that the baptized is rid of by following Christ, that dress that Adam donned after his sin" (M. Eliade, *Lo sagrado y lo profano*, Barcelona: Labor / Punto Omega, [1957] 1985: 116).

*pies us is not the loss of a single child, but the eternal condemnation of
an entire nation? One word from the queen would be enough to save
it. She need simply tell the king[395] to continually send missionaries to
preach the faith. That will be enough to ensure success and procure the
health* (page 19*) of a great people.*

While these letters and memorials reached Spain, Father San Vitores
redoubled his prayers and his austerities for the success of his enterprise. He spent
many nights before the Blessed Sacrament, and there, his heart bled with fervor
and trust as he asked God for the conversion of those islands. He also sought to
attract the wills of those around him, particularly Don Diego de Salcedo, the
Philippines' governor general [1663–1668], but the Servant of God's continual
efforts in fact irritated him more and more. This character openly declared
his opposition to the project and refused to cooperate in any way, so that the
superiors beseeched the Father not to speak of it again for fear that estrangement
with the governor could lead to problems and difficulties for the Society.[396] He
obeyed, but he confidentially told one of his friends that the enterprise would be
carried out despite all the efforts against it. This was in fact verified when, having
read Father San Vitores's memorial, King Philip IV ordered the governor of the
Philippines in a dispatch dated June 2, 1665, to support the missionary's project
and provide him with ships and all the resources that he may need to work in the
conversion of the Marianas. This order was to be executed immediately, and the
governor seemed disposed to obey it. He prepared a ship at the port of Cavite,[397]
and called it San Diego in his own honor, and since he was very interested in the
work being done, it was ready in no time. The boat was supposed to depart for
Mexico and from there go on to the Mariana Islands, when Salcedo changed his
mind and announced that the San Diego was going to sail for Peru.[398]

---

395. He is referring to Charles II, Queen Regent Mariana's son and future King of Spain.

396. The tense relations between Governor Diego de Salcedo and Archbishop Miguel de Poblete are well
known (Leoncio Cabrero (coord.), *Historia General de las Filipinas*, Madrid: Ediciones de Cultura Hispánica
& Agencia Española de Cooperación Internacional (AECI), 2000, pp. 244–46). These relations were probably behind his refusal to support Father San Vitores' plans.

397. This port is three-or-four leagues from Manila (note by L. de Morales/C. Le Gobien, *Histoire, book*
I, p. 32).

398. This is the largest and wealthiest kingdom in southern America (note by C. Le Gobien, *Histoire,
book* I, p. 32). According to the *Real Cédula* of November 23, 1634, direct trade between the Viceroyalties
of New Spain and Peru was strictly forbidden, with the exception of a 200-ton galleon that could transport
a maximum of 200,000 ducats of silver to the port of Acapulco, and return to the port of Callao with merchandise and Chinese and Mexican products. As Borah points out, this prohibition continued throughout
the seventeenth century, but this did not stop some governors, including Diego de Salcedo, from chartering
vessels that carried out illegal trade with the Peruvian Viceroyalty. Woodrow Borah, *Early Colonial Trade and
Navigation between Mexico and Peru*, Berkeley & Los Angeles: University of California Press, 1954, p. 127).

(Page 20) This was a terrible blow for Father San Vitores. Seeing that although his holy zeal had been inflamed, all of his plans were going to fail, he went to Salcedo and threatened him with the ire of the heavens and the indignation of the king if he persisted in his opposition to God's designs. Showing Salcedo a dispatch that the king had honored him with, he asked him to initiate the expedition, giving him no room for excuses. On the other hand, word had spread that God had performed a miracle in his servant's favor. In effect, while the governor's orders [about Peru] were being announced throughout the city, the San Diego was so violently dipped that they feared it could not be raised again. Many saw this as a mere coincidence, but since the Father often publicly stated that the stubbornness of those who opposed the plans of God would bring great misfortunes upon the city, and that the ship would not be able to depart until they changed their ways, everybody was pressuring the governor to revoke his last order. He was finally convinced, and he announced, again, that the ship was going to Mexico and then to the Mariana Islands, where Father San Vitores and his companions would be dropped off so that they could work in the conversion of those peoples.[399]

This news brought universal joy to the city, a joy that became even greater when a second miracle performed by God confirmed the first. For, as if to show that the accident that had befallen the ship had been his work and not the result of a coincidence, the ship returned to its natural position when the order that favored the Father's plans was given. The whole city thanked God for such a marvelous occurrence, and the unanimous opinion was that an enterprise that was so manifestly favored by God could not but succeed.[400]

(Page 21) When everything was ready, Father San Vitores boarded ship in the Cavite port with Father Tomás Cardeñoso on August 7, 1667.[401] All of Manila felt the loss of such a holy man, but nobody as vividly as the archbishop of that great city, who appreciated him as the man in whom he had placed all his trust.

> *I cry*, he said in the letter that he wrote at that moment to Don Jeronimo San Vitores, *I cry tears of blood for the loss that I have just experienced. You cannot imagine how much I loved this fervent*

---

399. Years later (1668), Don Diego de Salcedo left in his will 10.000 pesos to support the missions in the Marianas (AGN, Instituciones coloniales, Real Hacienda, Archivo Histórico de Hacienda, vol. 326, pp. 1683–1896).

400. Here Morales emphasizes Augustinian providentialism as the leitmotiv of the Mariana enterprise.

401. F. García, S.J., *The Life and Martyrdom of the Venerable Father Diego Luis de San Vitores*, book III, chapter III, p. 179.

missionary, how necessary he was for the enhancement of my own virtue and the spiritual wellbeing of all these islands where he brought forth wondrous fruit. But God disposes of men. He has his designs, and I submit to them. Your son is destined to convert a great people. He will be the apostle of the Mariana Islands, which have always been the object of all his desires.

Father Diego de San Vitores arrived in Mexico early the following year. He immediately wrote to the viceroy, who at that time was the Marquis of Mancera[402] to notify him of his arrival and to obtain from him the resources that he needed. The viceroy sent him a courteous reply, but the Father's friends warned him that it was necessary from him to show himself, that he would not attain what he required if he did not come in person. The Father, armed with this warning, presented himself before the viceroy, who received him kindly and promised to protect him. In effect, he had his council meet and put the matter to them. At first, the council seemed inclined to agree, but one of the auditors, men with much authority among them, argued that the treasury was exhausted, in a state that made confronting even the necessary costs impossible. Therefore, even though he only asked for ten thousand escudos to finance an enterprise that would bring much glory to the State and advantages to religion, the Father's demand was rejected.

The Father, surprised at such an unexpected resolution, and animated by the zeal of God's glory, went to meet the (page 22) auditor:

Do you know, he asked, the price for the souls rescued by the very blood of God? Do you know what they are worth? Infinitely more than all of the riches of the Indies. The king, may he rest in glory, was convinced of this, for he frequently said that he would give all the treasures of the New World for the salvation of a single soul. The queen shares these sentiments. Do these beautiful examples not move you to imitate them? Maybe you fear that the necessary expenses to carry out this project will ruin the royal treasury, but if Jesus Christ has prom-ised to reward a hundredfold those who succor a poor man, what can we not expect of the Father of mercy for those who procure the health of an entire people? Believe me, if these souls are allowed to perish for fear of losing riches, [then both riches and souls will be lost], for it is evident that God has given the empire of the Indies to the Catholic Monarchs so that they would extend religion across them and procure

---

402. Don Antonio Sebastián de Toledo Molina y Salazar, Viceroy of New Spain from 1664 to 1673.

the salvation of their peoples. God has done with them what that king in the Old Testament did to Abraham when he said, "give me the souls and take everything else."[403] *And that is the conduct that our princes have followed so far, for they have been less eager to accumulate riches and subjugate peoples to their rule, than to extend the limits of the kingdom of Jesus Christ. The conversion of a numerous people depends on the dictate that you have given. Reflect upon this and keep in mind that you have made yourselves guilty before the sovereign judge of the living and the dead, for the loss of an infinite number of souls that will be condemned merely because of your opposition to God's designs.*

These words, pronounced with such energy and inspiration that they left no room for doubt, vividly impressed the auditor, who, moreover, was an honest man with a good heart, making him change his way of thinking, and he promised the Father to repair the evil that he had caused. (page 23) He kept his word, and with the support of the viceroy's wife who was very pious and was profoundly interested in this matter, he convinced the viceroy to grant the Father what he had asked for the very next day.[404]

This excellent result obtained by the Servant of God filled the whole city of Mexico with joy, for he was greatly esteemed, and they all contributed with such a holy enterprise. The Father, satisfied at having attained what he sought, thanked God, and could think of nothing other than departing. He chose as his companions men of apostolic zeal and proven virtue: Father Tomás Cardeñoso, who had come with him from the Philippines; Father Luis de Medina, an excellent missionary; and Fathers Pedro de Casanova, Luis de Morales, and Lorenzo Bustillo. The last three had not yet finished their studies, but they insisted in their desire to consecrate themselves to this mission until other missionaries could replace them. They went to the port of Acapulco and boarded the ship that was to take them away after having implored for heaven's protection on March 23 of 1668. The vessel was under the command of Admiral Don Bartolomé Muñoz, who would soon have the fortune of dying in Father San Vitores's arms.

This zealous missionary applied himself to the sanctification of the crew

---

403. "Dixit autem rex Sodomorum ad Abram: Da mihi animas, cetera tolle tibi genes" (note by L. de Morales/C. Le Gobien, *Histoire, book* I, p. 37). This is a reference to Genesis, 14, 21–22.

404. Father Francisco Jiménez (or François Guillot), prefect of the celebrated Colegio Máximo de San Pedro y San Pablo, played an important role in this turn of events. It was he, as confessor to both the viceroy and his wife Doña Leonor Carreto, who convinced the viceroy to support the Society in the evangelization of the Marianas and the Californias. Through his intercession, the 10.000 pesos from the Royal Treasury, along with various servants and goods, were assigned to the mission (L. Gutiérrez, *Historia de la iglesia en Filipinas*, p. 261; R. Lévesque, *History of Micronesia*, vol. 9, 1997, p. 438).

and establishing order in the ship, and he succeeded to such a degree that it seemed as if everybody aboard was in fact a member of a religious community. Prayer and other spiritual exercises were practiced with edifying sentiments of devotion, and so were the exhortations and other pious exercises that he had them perform. My sons, he told them, let each one attempt to attract the blessings of God over this new mission with his prayers and the sanctity of his life. Unhappy he who impedes the success of the mission with the smallest sin.

(page 24) Through these words of the holy man, and what is more, his good example in the fulfillment of his duties, the crew was sanctified during this voyage. They were all anxious to arrive at this land of promise, and a novena to the Sacred Family was prayed so that they would reach the islands soon, when on June 15, one of the days of the novena that had been consecrated to St. Anne, they saw the island of Zarpana. The Father then gave the name of great St. Anne to that island. Later they saw the island of Guåhan, and soon they saw more than fifty small boats full of savages that yelled with all their might, mauri, mauri, which in their language [Chamorro] meant: friends, friends. The ship was soon surrounded, and the islanders were invited aboard, but they refused to do so, perhaps because it was getting dark, or because they were naturally wary and they feared a deception. This greatly saddened the Father, who was impatient to talk to them. He tried to gain their trust, but seeing that nothing came of it, he ordered that the litanies to the Virgin be sung, and they had barely started when the savages climbed up the ship's sides and invaded it, mingling with the Spaniards and singing with them.

The Father received them with the joy that was to be expected, and he showered a thousand caresses upon them, and honored the chiefs among them with gifts and affection. He spoke to them of course of the kingdom of God, for during the trip he had learned their language. They listened to him with attention, and were so pleased with what he said, that they invited him to come to their islands. A Spaniard[405] who had lived among them for more than thirty years told the Father that he would be well received and that he would find the inhabitants were well disposed to benefit from his instructions and embrace the faith. He left Fathers Medina and Casanova with this good people, and he promised them that he would visit them often.

(page 25) Since we have had very incomplete news of these islands, where

---

405. This man, simply referred to as Pedro, had survived the shipwreck of the Nuestra Señora de la Concepción galleon. According to Father García's hagiography, Pedro asked Father San Vitores to baptize his daughter, who was given the name Mariana in honor of Queen Regent Maria Anna (F. García, S.J., *The Life and Martyrdom of the Venerable Father Diego Luis de San Vitores*, book II, chapter XVII, p. 160).

religion has taken such great strides in the last few years, I think it is convenient to provide an exact account of them, describing the customs of the peoples that inhabit them.

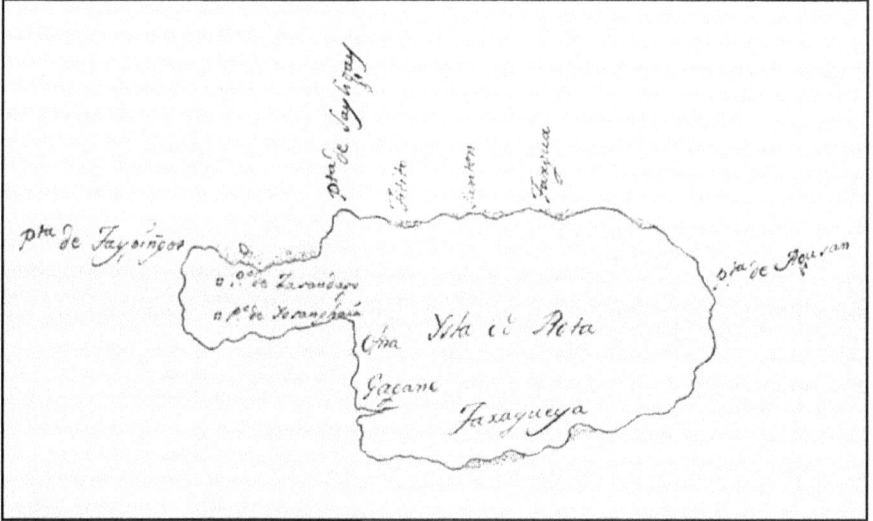

{Figure 7: Map, island of Rota (Zarpana), 1794, by Francisco Antonio Mourelle de la Rúa (1750–1820). Archivo del Museo Naval, MN-55-13}

Book Two[406]

A
mong the numerous islands that comprise the Archipelago of St. Lazarus, only fourteen are well known, those that go from north to south
practically forming a straight line. To ascertain the size and situation of
these islands, one must not look at maps, for they are not drawn with exactitude
[since their names and true locations have been determined only a few years ago.

These islands that have Japan to the North and New Guinea to the
South, lie between the Tropic of Cancer and the equinoctial line, in the extrem-
ity of the Pacific sea and about four hundred leagues from the Philippines. They
extend across the sea for approximately one hundred and fifty leagues, from
Guåhan, which is the largest and southernmost [to Urach, which is the one
closest to the Tropic].

Although they are within the Torrid Zone, their skies are beautiful and
serene; the air that you breathe is pure, and heat is never excessive. Mountains
are always covered in vegetation and crisscrossed by numerous streams that
are spread out across valleys and plains, contributing to making the country
exceedingly pleasant.

(page 26) Before the Spanish came to these islands, their inhabitants lived
in complete liberty, with no other laws than those that each one imposed on
himself. Separated from all other nations by the vast seas that surrounded them,
and locked in their lives as in their own small world, they completely ignored that
there were other lands, and they thought that they were alone in the universe.[407]

Most of the things that we believe are vital to our lives, they lacked. They
had no animals, and they would have had no idea what these were, if it had not
been for some birds, all of the same species, similar to turtledoves, that inhabited the islands.[408] They did not eat them, but they tamed them and taught
them to sing. The most surprising and incredible thing was that they had never
seen fire.[409] This element, which is so necessary, was entirely unknown to them.

---

406. A large part of this chapter's contents are evidently based on the experiences narrated by Father L.
de Morales in his "Relaciones del estado y progresos de la misión de las islas Marianas desde junio de 1681
hasta el 25 de abril de 1684" (RAH, Fondos Jesuitas, Tomo 19, Signatura: 9-3593/26).

407. Father Morales underlines the ethnocentrism of the Chamorro people as a telling element of their
absolute isolation and ignorance.

408. See F. García, S.J., *The Life and Martyrdom of the Venerable Father Diego Luis de San Vitores*, book III,
chapter I, p. 166. In the hagiography of his confrere Luis de Medina, Father García argued that this land "has
only mice, and the only European animals are the dogs and cats who survived the shipwreck of the Spanish
galleon *Nuestra Señora de la Concepción*" (Francisco García, S.J., *Relación de la vida del devotíssimo hijo de
María Santíssima, y dicho Mártir Padre Luis de Medina*, chapter IV, p. 48).

409. [Marcus] Vitruvius, so celebrated for his ten books on architecture [*De Architectura*, 23–27 AC],

They did not know its uses and its qualities, and they had been greatly awed the first time they saw it, when Magellan disembarked in one of the islands and burned fifty houses in punishment for some wrong they had committed. In the beginning, they regarded fire as a sort of animal that attached itself to wood for its nourishment. The first who had dared come close to it was burned, and they warned the others. After that, they only dared look at it from afar, afraid, they said, of being bitten and hurt by the violent breathing of the terrible animal, for such was the (page 27) idea that they formed in their minds of flames and of heat. This belief did not last very long, and when they saw their error they became accustomed to seeing fire, and they used it like us.

It remains unknown when these islands were first populated, and what is the country of origin of their inhabitants.[410] Since one can see in them the same inclinations of the Japanese regarding the nobility's pride, some have suggested that these barbarians came from that country, from which they are separated by only five or six days. Others believe that they came from the Philippines and other neighboring isles because the color of their skin, their language, their customs, and how they govern themselves, are very similar to those of the Tagalog people that inhabited the Philippines before the Spanish took those islands. Maybe they come from one and the other, for perhaps their population is the product of Japanese and Tagalog shipwreck survivors tossed to these coasts by the force of the storms.

Whatever the case may be, the truth is that these islands are very populous. The island of Guåhan alone has around thirty thousand inhabitants, even though it is only forty leagues all around.[411] The isle of Saipan has less people, and the others are populated proportionately [to their size], and all are full of

says that the first men discovered fire by chance when some trees that were violently rubbing against each other erupted into flames because of a storm and caused a great fire. This initially scared the men, but after having come close to the fire and seen that its heat was comfortable and pleasant, they were very careful to conserve it and use it for various ends (note by L de Morales/C. Le Gobien, *Histoire*, book II, p. 44). This belief expressed by Morales/Le Gobien was questioned by various modern naturalists, including German scholar Adelbert Von Chamisso, who ridiculed the Spanish religious fanaticism that he found in James Burney's *Chronological History of the Voyages and Discoveries of the South Sea* (Vol. 3, London, 1813) (Barratt, 1988, p. 29, cited in David Atienza, "The Evangelization of the 'Poor' Mariana Islands and Its Symbolic Use in the West." Manuscript, 2012). I thank David Atienza for the bibliographical reference.

410. Bowers argues that "[Chamorros] probably came from the "Malayan area" (possibly moving through the Philippines) late in the migration periods of Oceania and that primary settlements were in coastal villages with much smaller interior villages, although he makes no reference to the traditional Chamorro caste system" (Neal M. Bowers, "The Mariana, Volcano, and Bonin Islands," in Otis Freeman (eds.), *Geography of the Pacific*, New York: John Wiley & Sons, Inc. 1951, p. 216.

411. Recent studies suggest that this number seems exaggerated. C. Le Gobien includes this same calculation (*Histoire, book* II, p. 46); and in his hagiography of San Vitores, Francisco García raises it to 50,000 (F. García, S.J., *The Life and Martyrdom of the Venerable Father Diego Luis de San Vitores*, book III, chapter I, p. 167).

villages scattered over plains and mountains, some with as many as a hundred or a hundred and fifty huts.

Their dusky skin color is somewhat lighter than that of the Philippines' inhabitants. They are stronger and more robust than Europeans are. Of considerable height and proportionate bodies, even though they only eat roots, fruit, and fish, they are so stout that they seem to be swollen, but their obesity does not make them any less agile. It is not uncommon for some of them to reach the age of one hundred, and the first year that the Gospel was preached to them more than one hundred and twenty who were that age were baptized (page 28), and they were as healthy and strong as if they had been in their fifties.

The causes for their longevity are varied. They grow accustomed to the weather's changing inclinations from infancy. Their diet, which is always the same, by its mere vegetable nature does not excite in them the desire to overeat. The moderate exercise that they carry out when they go fishing and when they tend to their plants and trees, and above all, the peaceful and free life that they have, without concerns, without subjections, without reasons to be upset or worried, makes them enjoy a health that Europeans have never known, regardless of how much they strive to procure it. Most of them reach a very old age without ever having been sick, for it is indeed uncommon to see one in that state, but if one of them does fall ill, they carefully care for and cure him using herbs, whose medicinal properties they know well.

Men go about completely naked, but not so women. The latter regard beauty very differently from European women, but they are similarly concerned with it, and they attempt to attain it by blackening their teeth and whitening their hair. Thus they are continually busy blackening the former with certain herbs and whitening the latter by washing it constantly with water and solutions prepared for that end, and they wear it very long, unlike the men whose heads are entirely shaven except for a small tuft of hair in the middle, like the Japanese.[412]

Their language is analogous to the Tagalog language spoken in the Philippines and neighboring islands and it is pleasant enough, with an easy and harmonious pronunciation. One of its particularities is that it inverts the order of words, and sometimes even the syllables in a word, which gives cause for

---

412. In his 1524 *Relación*, Antonio Pigafetta had already described how the women, whose skin was lighter than that of the men, wore their hair very long and loose, and painted their teeth black and red, and how they covered their pubic area with a type of palm leaf, or *tifis* (Antonio Pigafetta, *Primer viaje alrededor del mundo*, ed. Leoncio Cabrero Fernández. Madrid: Dastin Historia, 2002, pp. 73-74). See also F. García, S.J., *Relación de la vida del devotísimo hijo de María Santíssima*, chapter IV, p. 48, F. García, S.J., *The Life and Martyrdom of the Venerable Father Diego Luis de San Vitores*, book III, chapter I, p. 169; and L. Thompson, "The Native Culture of the Marianas Islands," pp. 10–11.

frequent errors that they enjoy very much.

Although they possess no knowledge of science or fine arts, they have stories full of fables and a certain poetry of which they are quite fond. A poet among them is quite distinguished, and that title alone earns him the respect of the whole nation.[413]

There has never been another people with more ridiculous presumptions or more vain than this one. Submerged in the most profound ignorance ever encountered, and lacking all the comforts of life, they consider themselves the wisest and most discreet people in the world. They view all other peoples with pity, and speak of them with the utmost contempt.[414]

There are three classes among them: the nobility, the plebe, and the middle class.[415] The nobility is excessively proud. One cannot imagine in Europe just how the nobility disdains and dominates the plebe, so much so that the most infamous thing that a noble can do is to marry a girl from the plebe. A family loses its honorable reputation when passion or interest blinds a nobleman to the point of making him commit this act regarded as unworthy of his high birth. [Before they became Christians], an assembly of relatives would gather and consent to cleanse this shameful and ignominious stain with the blood of the man who was guilty of committing it. This shows the extreme to which the nobility was willing to go to conserve its rank and transmit pure blood to its descendants.

These were not the only cases that evidenced the contempt with which nobles view the plebe; they go further, to the extreme of declaring it a crime [for a plebeian] to go near the home or a member of the nobility. If a *chamorris* [*maga'låhi*], which is the name given to this most distinguished part of the nation, if a *chamorris*, I repeat, wants something that a plebeian has, he need

413. In 1683, Jesuit hagiographer Francisco García had already noted that Chamorros admired poetry, and that they regarded poets as "men who worked wonders" (F. García, S.J., *The Life and Martyrdom of the Venerable Father Diego Luis de San Vitores*, book III, chapter I, p. 169).

414. As Glynn Barratt has argued, this statement confirms that the Chamorro were conscious of the existence of other peoples beyond the Marianas (probably the Carolinian islanders) (G. Barratt, *Carolinean Contacts with the Islands of the Marianas: The European Record*, Saipan: Micronesian Archaeological Survey, Division of Historic Preservation, Department of Community and Cultural Affairs, 1988, p. 5).

415. According to the classic study by Laura Thompson (*Native Culture of the Mariana Islands*, 1945), Chamorro society was divided into nobles (*matua* or *matao*), semi-nobles or the second-ranked class (*atcha'ot*) within the *chamorri* caste, and the lower caste (*mangatchang* or *manachang*). Each caste was profoundly endogamous, and contact between classes was restricted through ties and taboos. Members of the matao could be demoted to the atcha'ot caste if they violated the rules of society. However, revisionist studies have suggested that given the lack of social mobility between high and lower castes, Chamorro society was in fact divided in two: the lower caste (*mangatchang* or *manachang*), who were darker complexioned and smaller than the upper castes (*matao* and *atcha'ot*) (George J. Boughton, "Revisionist Interpretation of Precontact Marianas Society," in Donald H. Rubinstein, *Pacific History: Papers from the 8th Pacific History Association Conference*, Mangilao, Guam: University of Guam Press & MARC, 1992, pp. 221-24).

only ask from afar. These notions are so ingrained in them that if a plebeian eats or drinks anything in a nobleman's home, he dishonors that house. These nobles have their chiefs, a hereditary position; however, fathers are not succeeded by their children, but by their brothers or (page 30) nephews [who then take the name of the deceased, or the name of the head of the family].[416] This custom, which seems odd, is so established among them that it causes no disruptions or quarrels.[417]

The nobility that enjoys the most distinction is that of the city of Agadña [Hagåtña], capital of the island of Guåhan, for it is very well located and its waters are excellent, and thus, many noble families settle there, making its nobility the largest in these islands.

The nobles' assemblies are presided by the chief among them. They are respected and listened to, but their words are heeded only when they are considered fitting. Each can take whichever side he wants at any given moment, without any opposition, for these peoples are neither submitted to any authority nor governed by any law whatsoever. However, they have customs that they observe as religiously as if they were true laws.

Although these people are barbarous and coarse, the *chamorris* do not lack a certain education. When they encounter or pass each other, one will greet the other with these words: Ati Arimuo (or Arinmo), which means, "let me kiss your feet."[418] If a noble passes in front of another noble's house, he is invited in and offered food and an herb that they chew in their mouths whose effects are the same as those of tobacco.[419] When they visit somebody or they want to give a gift, they run their hand over the other's belly. This gesture is commonly used among them to show extraordinary affection or courtesy.

They consider it excessively rude to spit in front of a person who merits respect. Their anxiety regarding this is almost superstitious. They rarely spit, and when they do it is with great precautions, and the explanations that they

---

416. This extra information comes from Daly, *History of the Mariana Islands*, p. 4.

417. Chamorro society was organized in clans integrated by extended matrilineal families and avuncular residence in small towns (songsong) made up of 50 to 150 huts (A. de Ledesma, "Noticia de los progresos de nuestra Santa Fe...," pp. 3v-4v.). Women's closest links were those they maintained with their own lineage, so that men depended on their sisters to have heirs. For a more thorough analysis of Chamorro matrilineal organization, see Alice Joseph and Veronica F. Murray, *Chamorros and Carolinians of Saipan. Personality Studies*, Westport. Connecticut: Greenwood Press, Publishers, [1951] 1971, p. 15; Lawrence J. Cunningham, *Ancient Chamorro Kinship Organization*, Agat, Guam: L. Joseph Press, 1984, pp. 2–26; L. Thompson, "The Native Culture of the Marianas Islands," p. 11; and D. A. Farrell, *History of the Northern Mariana Islands*, pp. 88–89.

418. See F. García, S.J., *The Life and Martyrdom of the Venerable Father Diego Luis de San Vitores*, book III, chapter I, p. 169.

419. He is referring to *buyo*, or betel nut, which is still commonly used today (F. García, S.J., *The Life and Martyrdom of the Venerable Father Diego Luis de San Vitores*, book III, chapter I, p. 169).

give about this are not entirely clear.

Their most common employment is fishing. They practice it since infancy and they are in the water so frequently that they swim like fish. The canoes that they use for fishing and for going from one island to the other are very light, and they are good enough to be used even in Europe. They caulk them with a type of pitch and lime that they dissolve in coconut oil. They find this pitch, which they use quite skillfully, in the island of Guåhan.

Their houses are very pleasant. They are made from the wood of the coconut tree and the tree called Maria [Palo María; daog], which is peculiar to these islands.[420] Each home has four apartments that are separated by partitions made of threaded palm leaves similar to straw mats. The roof is made of this same material. These apartments are comfortable, and each is destined to a particular purpose.[421] The first is the bedroom; the second, the dining room; the third, the pantry and storage; and the fourth is destined as a workplace.

Never a people lived in a more complete liberty and a more absolute independence. Each is the master of his actions from the moment they are old enough to reason. Children do not even know what it means to respect their parents, and they acknowledge them as such only as long as they need them. Each takes justice into his own hands regarding problems with others, and if differences arise between villages, they are resolved through war.[422]

They are easily aggravated and are quick to resort to arms, but they make peace just as quickly, and their wars never last very long. When they go into battle they give great shouts, in the manner of savages, but these are used more to rouse themselves than to scare their enemies, for they are not naturally brave. Since they carry no provisions, they spend two or three days without food, focused on their enemy's movements, trying to make them fall into a trap, for they are unrivalled by any other nation in this art. It seems that they only go to battle to surprise one another. They come to blows only if it cannot be helped, and only to avoid the shame of having (page 32) retreated without doing so. It would seem that they fear getting hurt or bloodying the battlefield, for two-or-three men dead or seriously wounded are enough to determine

---

420. See F. García, S.J., *The Life and Martyrdom of the Venerable Father Diego Luis de San Vitores*, book III, chapter I, p. 166.

421. The ancient Chamorro built their houses on stone pillars called latte, under which they would bury their dead (L. Thompson, "The Function of Latte in the Marianas." *Journal of Polynesian Society*, 49, 1950, p. 460).

422. As stated earlier, these arguments are related with some of the ideas on stoicism that regarded austerity and simplicity of "natural men" as positive. The object was none other than to criticize the conduct of the Spanish, considered far from the exemplary demeanor that their civilizing role demanded.

who is victorious. They fear bloodshed so much that they run off and disperse whenever they see it.

The vanquished immediately send ambassadors to the victors, who receive them with the thorough satisfaction that those who are timid and cowardly feel when they face their enemies' humiliation. Since these people are naturally vain and proud, victors exploit their triumph insulting the losers, and mocking them in their feasts by singing satirical songs that allude to them.[423]

They spend very little in weapons. They have neither bows nor arrows nor swords; instead, they use long sticks as lances whose tips are not made of iron, for they do not know this metal, but of human bone taken from an arm or a leg.[424] They are very skillful with these bones, and they sharpen them and poison them so well, that even if only a small flint remains in a body, it will unfailingly bring death amidst the most terrible convulsions and awful torments, and no antidote to this poison has yet been found that can neutralize its effects. They also use stones, which they throw with such proficiency and strength that they can drive them into tree trunks. They do not use defensive weapons, instead trusting their agility and quick movements to avoid their enemies' blows.

They might not be warriors, but they are unrivalled in the art of deception and in the concealment of their true intentions. Not knowing them at all, the Spanish were easily fooled. The islanders appeared to act with charming sincerity and good faith, which led the missionaries to write high praises of them in their letters (page 33) to Europe. But they soon realized that this apparent simplicity was a veil of deception, and that they had to be very careful with these people in order not to be tricked at every step.

Vengeance is one of the passions that most excites them. When they are affronted, their words and gestures do not show that they have taken offense. There is nothing in their appearance: all of the hatred and resentment are stored in their heart, and they control themselves so well that two or three years can go by before they find a favorable occasion to exact revenge. Then, they get even from the violence done to them using everything that the blackest treason and most intense hatred can inspire.

Their inconstancy and levity are incredible: how nothing concerns them as they blindly pursue their caprices and passions, going from one extreme to the other with ease. What they desire with great ardor at one moment, they no

423. See F. García, S.J., *The Life and Martyrdom of the Venerable Father Diego Luis de San Vitores*, book III, chapter I, p. 170.

424. See A. Pigafetta, *Primer viaje alrededor del mundo*, pp. 73–74. See also L. Thompson, "The Native Culture of the Marianas Islands," pp. 18-20.

longer want after a short while. This, which the Spanish have frequently witnessed among them, has been an obstacle for these barbarians' full conversion.

They enjoy having fun, and they joke with one another and go through
a thousand buffooneries to amuse themselves. If they are sober, it is a matter of
necessity and not of virtue, for they meet frequently and give each other fish
and fruit, and a liquor that they obtain from a maritime plant and mashed
coconut. They delight in dancing, running, jumping, and even wrestling, to
exercise and test their strength. They enjoy narrating their ancestors' deeds and
reciting their poets' verses, which are full of fables and extravagant notions.[425]

Women also have their feasts and diversions, for which they festoon
themselves, adorned with seashells and pieces of conch as well as tortoise shells
that they string together and place on their foreheads with flowers. They also
wear belts that are equally decorated with small conches that (page 34) they
value more than pearls and gemstones are valued in Europe. These belts also
hold exquisitely carved small coconuts. In such days, they cover themselves
with cloths made of woven roots, which, being quite coarse, gives them the
appearance of being inside a basket or a cage instead of a dress.

In these assemblies, twelve or thirteen women will stand in a circle,
without moving. They will then sing the fabled verses of their poets with an
intonation and accord that would please Europe itself. Their vocal harmony
is admirable, comparable with the music of Europe's best concerts. They have
small shells in their hands, which they use as castanets. But the most surprising thing is how they accompany their voices and animate their singing with
gestures and expressions that charm all who see and hear them.[426]

---

425. One of the few references that we have today is *Kantan Chamorita*. According to Laura Thompson,
Kantan Chamorita retains significant elements of ancient Chamorro poetry, including the tendency for teasing
and risqué humor (L. Thompson, *Guam and its People*, Wesport, Connecticut: Greenwood Press, Publishers,
[1947] 1969, pp. 274–283). See also L. Thompson, "The Native Culture of the Marianas Islands," p. 24;
C. R. Kim Bailey, "Chamorrita' Songs: A Surviving Legacy of the Mexican Verso?" *Micronesian Resource File
no. 0771*, RFK Memorial Library, University of Guam, pp. 1–9; Michael R. Clement, *The Ancient Origins of
Chamorro Music*, Master's thesis, University of Guam; and Judy Flores, *Art and Identity in the Mariana Islands:
Issues of Reconstructing an Ancient Past*, Ph.D. Diss., University of East Anglia, Norwich, United Kingdom,
1999; Judy Flores, "Kantan Chamorrita Revisitied in the New Millenium," in Helen Reeves Lawrence (eds.),
Traditionalism and Modernity in the Music and Dance of Oceania: Essays in Honour of Barbara B. Smith,
Sydney: Oceania Publications, 2001, pp. 19-31.

426. The Society of Jesus always regarded music as a fundamental evangelization resource. This was
combined with Chamorro women's taste for singing and poetry, and, as Clement argues, the Chamorro
matrilineal creation myth (Michael R. Clement, Micronesia: IV. Mariana Islands. 2. Guam, in *The New
Grove Dictionary of Music and Musicians*, vol. 16, London: MacMillan, 2001, p. 613). The introduction
of new styles, dances, and musical instruments (clarions, hornpipes, bagpipes, drums, lyres) brought over
from Europe, Mexico, and the Philippines, implemented as evangelization strategies had a significant impact
on CHamoru musical culture. In a recent text, David R. M. Irving analyzes music as a mediating element
through which the missionaries tried to "transform the hearts" of the CHamoru people. Festivities and

Polygamy is allowed, but the custom is to have no more than one wife.[427] These have rights in this country that are reserved to husbands elsewhere.[428] Women rule [absolutely] in the home.[429] A wife has all the authority and a husband cannot make use of the smallest thing without her consent.

If a husband does not defer to his wife as she believes he ought to; if he does not conform; or if he is bad tempered, the woman mistreats him or abandons him and her initial liberty is restored. For marriage among these islanders is not indissoluble, and it only lasts for as long as both parties are content with one another. From the moment that discord arises, they separate, and whatever the cause, the woman loses none of her goods, and her children follow her and regard the new husband that she chooses as if he was their real father. Thus, a poor husband is sometimes burdened with the loss of wife and children simply because of the bad (page 35) disposition or quirks of a capricious woman.

Concerning the woman:[430] if she is unfaithful, and her husband has reason to complain, he can take vengeance on the lover to the point of taking his life, but he cannot mistreat his wife under any circumstances, and the most he can do is abandon her. It is not the same when infidelities come from the husband. Then, women take justice into their own hands and punish them, forcing them to fulfill their duties.

When a woman is certain that her husband has relations that are damaging to her, she tells the rest of the women in the village and they all agree upon an hour at which they meet, with their lances, and wearing their husband's hats.

civic-religious celebrations "combined elements from indigenous CHamoru culture (theatrical performances, poetry, and singing in the CHamoru language) with Spanish plays and Mexican dances" (David R. M. Irving, "Jesuits and Music in Guam and the Marianas, 1668-1769," in Yasmin Haskell and Raphaële Garrod (eds.), *Changing Hearts. Performing Jesuit Emotions between Europe, Asia, and the Americas*. Leiden & Boston: Koninklijke Brill NV, 2019, p. 229).

427. Here, the copyist's modern language makes Father Morales's text differ from that of Father Le Gobien: "Men are allowed to take as many women as they like provided that they were not relatives. However, the regular custom is having just one" (C. Le Gobien, *Histoire, book* II, p. 59).

428. Women in Chamorro society had a high degree of independence, especially manifest in the matrilineal and matrilocal structures that are still operative. See Joseph and. Murray, *Chamorros and Carolinians of Saipan*, p. 15; L. Thompson, "The Native Culture of the Marianas Islands," p. 11; and D. A. Farrell, *History of the Northern Mariana Islands*, pp. 88–89.

429. According to Father Gabriel de Aranda, SJ, women "are more esteemed among them than men, because ordinarily they, and not the husbands, rule the house" (Gabriel de Aranda, SJ, *Vida y gloriosa muerte del V. Padre Sebastián de Monroy*, p. 217). On the role of women in the CHamoru society, see Alexandre Coello de la Rosa and Luis J. Abejez, "'Tú no eres quien yo espero': colonización, resistencia y género en las islas Marianas (siglos xvi-xix)", *Mélanges de la Casa de Velázquez*, 52-1 (2022). [Available online] <http://journals.openedition.org/mcv/16338> [Consulted March 2, 2024].

430. There is another difference here, probably also attributable to the copyist. Le Gobien says, "this is not the only disappointment that husbands have to accept. If a woman behaves in an irregular manner, he has the right to complain and may take revenge looking for a mistress" (note by C. Le Gobien, *Histoire, book* II, p. 60).

Disposed like this, they walk in battle formation to the delinquent husband's house and devastate his lands, tearing out plants, stripping the trees of their fruit, and causing an awful havoc, and if the unhappy man has not had the precaution of retiring to a safe place, they attack him and abuse him tremendously.

They have another way of taking revenge. They abandon the house and make it known to their relatives that they can no longer live with their husband. Pleased at the occasion of enriching themselves with the pretext of avenging her, the relatives go to the husband's house, storming and looting it, and take everything that they can find; leaving nothing that can be of use. The husband considers himself fortunate if they content themselves with that and do not go further, destroying the entire house, as some have done.

This power of wives over their husbands is the reason why countless young men do not wish to marry. They hire young women or buy them from their fathers for a few pieces of iron or turtle shells, and take them to public or common houses [*i mangguma' uritao*],[431] where they live with these young women in such a libertine and scandalous fashion that they offend the decency and good customs of those around them (page 36).

They are so afraid of homicide and theft that it was unjust to name the archipelago the Ladrones Islands. Instead, there is so much trust among them that they always leave their doors open, and nobody steals from their neighbor.

They are naturally liberal and inclined to favor others. Spaniards had an opportunity of experiencing this in the famous shipwreck of the vessel La Concepción [1638].[432] These people took in those who were fortunate enough to survive, and procured to alleviate their hardship with all sorts of kindness.

Before the Europeans arrived at their coasts, they imagined themselves to be the only people that existed in the whole world. After meeting the Spanish

---

431. These *I mangguma' uritao*, or "·men's houses," constituted another great obstacle for Chamorro conversion into Catholicism. The sexual freedom between young single men and women that characterized these houses was obviously considered pernicious by the Jesuits for the instauration of a "moral order" in the islands. See Lawrence J. Cunningham, "Pre-Christian Chamorro Courtship and Marriage Practice Clash with Jesuit Teaching," in Lee D. Carter, William L. Wuerch & Rosa Roberto Carter (eds.), *Guam History: Perspectives*, vol. 2, Mangilao, Guam: Richard F. Taitano MARC & University of Guam, 2005, p. 66.

432. In 1664, Admiral Esteban Ramos returned to Manila along with four Philippine natives who had lived in the Marianas for 26 years after having survived the cited shipwreck of the *Nuestra Señora de la Concepción* in 1638 off the coast of Saipan. During their "exile" in the island of Guåhan, they had learned the Chamorro language, and they worked as interpreters in San Vitores' expedition (F. García, S.J., *The Life and Martyrdom of the Venerable Father Diego Luis de San Vitores*, book II, chapter XIV, pp. 144–45. See also Florentino Rodao, "España en el Pacífico," in Javier Galván Guijo (curator), *Islas del Pacífico: el legado español*, Madrid, Ministerio de Educación y Cultura, 1998, p. 28; and D. A. Farrell, *History of the Northern Mariana Islands*, p. 140).

and seeing multiple English and Dutch ships sailing by their islands, they have
had to dismiss this erroneous belief.[433] Since they are fond of fables,[434] their
poets have fabricated all sorts of extravagant notions that they consider true,
because the tales indulge their pride, one of their dominant passions.[435]

These tales said that all nations have their origin in a spot in the island of
Guåhan, where the first man [Puntan] was formed out of that land's earth; that
the earth turned into stone and from that stone sprung the rest of the men that
settled the different countries. Some went to Spain, others to Holland, and others stayed there. Those that left, having settled far from their homeland, soon
forgot its language and its customs.[436] *That is why*, they say, *the other peoples
cannot speak and do not understand us. If they speak it is but with meaningless
words, like lunatics, and they do not understand each other nor know what they are
saying.* This is how far their vanity and presumptuousness go, as they are firmly
convinced that theirs is the only language.

Being this ignorant, they do not believe that the world has existed eternally.
They attribute a beginning to it, and a portion of (page 37) the absurd fables that
they express in bad verses and sing in their assemblies regarding this event.

On the other hand, they do not acknowledge any divinity, and before the
Gospel was preached, they had no ideas related to religion. They had no temples,
no altars, no sacrifices and no priests.[437] All they had were a few charlatans who
claimed to be prophets. These impostors, whom they called *macanas*,[438] had

433. According to the relación written by provincial Father Ledesma, towards the end of 1668 three
Dutch ships approached Humåtak Bay, in Guåhan, as well as the island of Zarpana or Rota (Andrés de
Ledesma, S.J., *Noticia de los progresos de nuestra Santa Fe en las islas Marianas, llamadas antes de los Ladrones,
y del fruto que se han hecho en ellas el padre Diego Luis de Sanvitores, y sus compañeros de la Compañía de Jesús,
desde 15 de mayo de 1669 hasta 28 de abril de 1670, sacadas de las cartas que ha escrito el padre Diego Luis de
Sanvitores y sus compañeros*, 1670, f. 2r).

434. Again, Father Morales highlights the Chamorro love of storytelling.

435. Father Morales seems to suggest that particular human beings or groups are dominated by certain
passions.

436. On the Chamorro gods of creation (Puntan and Fu'una), see William Bingham, "Seeking for the
Origins: the Dao of the Chamorro Creation Myth." *Micronesian Journal of the Humanities and Social Sciences*,
2 (1/2), 2003, pp. 16–22. See also <https://www.guampedia.com/puntan-and-fuuna-gods-of-creation/>
[Available online] [Consulted March 2, 2024].

437. According to Father García, Chamorros' lack of religion made it easier to introduce the Christian
faith (F. García, S.J., *The Life and Martyrdom of the Venerable Father Diego Luis de San Vitores*, book III,
chapter II, p. 174).

438. *Macanas, makåhnas* or *kakahnas* were local shamans who worshiped ancestral spirits, good and bad
(*anites, aniti*). According to Jesuit historian P. Murillo Velarde (Historia de la provincia de Filipinas, f. 330r),
"the Marianas' natives saw so many prodigies, that they refered to [Diego Luis de San Vitores] as Macana,
which means miracle man, for it seemed that he could control the sea and the wind." Anthropologist David
Atienza explains that the term "macana" signifies two different things in contemporary Spanish: the most
widely used is "club" or "truncheon" (and as an analogy, penis) (Gonzalo Fernández de Oviedo, *Sumario de
la Natural Historia de las Indias* [1526]. Estudio, edición y notas de Álvaro Baraibar. Madrid: Universidad

acquired from the *anitis*, that is, the dead, whose skulls they kept at home, the power to rule the elements, heal the sick, change the weather, protect the crops and make sure that abundant fish were caught.[439] Macanas' only object was to take advantage of these people's ignorance and live at their expense, for they had no respect for the skulls that they used. They simply put them in small baskets that they left lying about the house in utter disregard until somebody went to consult them.

Despite the fact that these people do not adore any divinity, they have many superstitions regarding the dead. When somebody is about to die, they place a basket next to his head so as to gather their spirit, and they ask him to get into the basket when it leaves that body, and to stay there, or to return to it whenever he wants to visit them. Others, wanting to treat their dead, rub the deceased with aromatic oil and parade them to their relatives' houses so that they choose which room they like best or which place they will stay in when they come to visit, for they believe in the immortality of the soul. They also acknowledge the existence of a hell[440] located high above the heavens, called *Zazarraguan* or the home of *Chaifi* [*Chayfi*], which is the name of the devil that torments those who fall into his hands. After the Spanish showed them fire, they said that Chaifi has an oven in which (page 38) he burns souls the way we burn iron, hitting them constantly.[441] Their heaven is a place of delights, but they had the silly notion that it was underground, and since their ideas are very

de Navarra & Iberoamericana & Vervuert, 2010, p. 112). But in Argentinian lunfardo and tangos, makana means liar. The Diccionario de la Real Academia Española verifies this claim, revealing that this use of the word is common in colloquial speech in Paraguay, Uruguay, Bolivia and Peru as well as Argentina. In Atienza's opinion, since there is no linguistic explanation as to why a word that meant *porra* (club) would eventually come to mean liar, the meaning "liar" comes from the Chamorro term *macana*, for the Jesuits' most constant complaint regarding Chamorro macanas was that they were liars. Given that there is evidence of this second meaning in eighteenth century texts from the Rio de la Plata Basin and Chile, it is very probable that Jesuits began using the term in their missions after the publication of Francisco García's hagiography of San Vitores (constituting the first Chamorro loanword in the Spanish language!) (David Atienza, 2010, personal communication). Reinforcing Atienza's hypothesis is the fact that macana as *porra* is also a loanword—it comes from the Carib word macana, which described a weapon made of hardwood shaped as a machete or bludgeon that often included a flint blade.

439. In March 1601, Franciscan friar Juan Pobre de Zamora (formerly a soldier in Flanders), and his companion Pedro de Talavera were shipwrecked in the island of Rota. A year later, Father Juan Pobre wrote the first reference to the Chamorro practice of worshipping their ancestors' skulls in his well-known *Relación de la pérdida del galeón San Felipe* (Marjorie G. Driver, *The Account of Fray Juan Pobre's Residence in the Marianas, 1602*. Mangilao, Guam: MARC & University of Guam, [1973, 1988] 1993, p. 22); Augusto V. de Viana, "Filipino Natives in Seventeenth Century Marianas: Their Role in the Establishment of the Spanish Mission in the Islands," Journal of the Humanities and the Social Sciences (Micronesia) 3:1–2 (2004), p. 20).

440. Le Gobien's text had the following extra sentence: They acknowledge the existence of Heaven and Hell, concerning which they have the strangest notions (*Histoire, book* II, p. 65).

441. See F. García, S.J., *The Life and Martyrdom of the Venerable Father Diego Luis de San Vitores*, book III, chapter II, p. 173.

limited, they define its beauty as that given by the palm trees, sugarcanes, and other vegetation that grow there and give fruit of an exquisite taste.[442]

It was not virtue or sin that led to these places. Good or bad deeds were completely irrelevant. Everything depended on the way in which a person died. If they were unfortunate enough to have a violent death, they went to hell, that is, they were locked up in the *Zazarraguan*; but if they had the good fortune of dying of natural causes, they went to heaven, where they could enjoy the fruits and the trees that were so abundant.

These people are convinced that the souls of the dead appear, and whether it is in fact the devil who fools them by taking the shape of their deceased relatives, or whether their exalted imagination represents before them what they hear others speak of, the truth is that they complain that spirits trouble and sometimes scare them terribly. For this reason, when they go to their Anitis, that is, to the souls of their dead ancestors, it is not so much to ask them for graces, but to try to appease them so that graces are not taken. For this reason, they maintain a respectful silence when they fish, and they carry out frequent fasts, fearing that otherwise the Anitis will mistreat them, or scare them in their dreams, by which they set much store.

One would be hard pressed to find a people that expressed their pain and sorrow more vividly. In funerals, they cry bitterly and in such anguish that they touch even the most insensible heart. They eat nothing for a long time, and with their lamentations and long fast, they become so thin that they are unrecognizable. Mourning lasts for seven or eight days, sometimes more, depending on the affection that they felt for the deceased or the favors that they had received from them. During this whole (page 39) time, they cry and sing somber tunes.[443] They also hold a feast [*chenchule*] next to the grave of the deceased,[444] for they always

---

442. As J. B. Diaz argues, "For the Chamoru, heaven was below the earth. It symbolized refuge and peace because everything was still, safe, and unchanging. Hell for the Chamoru was up in the sky, perhaps because Guåhan is a tropical island, susceptible to typhoons and natural disasters. Moreover, the inhabitants were conscious of the constant heat of the sun by day and the innumerable fires of the stars at night. It was easier for them to understand hell as located above because they witnessed the sky changing dramatically and drastically" (J. B. Diaz, *Towards a Theology of the Chamoru*, p. 18).

443. *Techas,* the older women that today lead the rosary prayer in funerals in Guåhan reproduce some of these practices (Lilli Pérez Iyechad, "Reciprocity, Reunification, and Reverence among Grieving Chamorros in Guam: An Ethnographic History of Death Rituals," in Lee D. Carter, William L. Wuerch & Rosa Roberto Carter, *Guam History: Perspectives*, vol. 2, Mangilao, Guam: Richard F. Taitano & MARC, 2005, p. 268).

444. The reciprocal gift-giving, or *chenchule',* is one of the most peculiar and interesting elements of the Marianas' funerary rituals. It includes non-monterary contributions or gifts, such as physical assistance or food, and is reminiscent of the Kwakiutl potlachs (Department of Chamorro Affairs. Division of Research, Publication and Training, *The Official Chamorro-English Dictionary = Ufisiåt Na Diksionårion Chamorro-Engles.* Department of Chamorro Affairs, Division of Research, Publication and Training, 2009, p. 72). See also A. de Ledesma, "Noticias de los progresos de nuestra Santa Fe...," pp. 4r-4v; L. Thompson, "The Native

raise a grave over the place where they bury him, or at the very least, next to it, and they cover it with flowers, palm leaves, conches, and everything that they hold dear. The despair of mothers who have lost their children is inconceivable. Searching for ways to kindle their grief, they cut locks of their children's hair, which they carefully save, and they wear a string around their neck upon which they tie as many knots as the nights that have passed since their child died.[445]

If the deceased is a *chamorris,* or a distinguished woman, then the mourners' pain knows no limits.[446] They enter into a type of furor and desperation, felling their trees, burning their homes, destroying their boats, tearing the sails and hanging the pieces in front of their homes, and covering the roads with palm leaves. If the deceased was skillful at fishing or warfare, two professions that they consider illustrious, they crown his grave with nets or lances to signify his valor or his fishing ability. If he distinguished himself in both, then they thread the lances into the nets, forming a sort of trophy.

All of this is accompanied by wails and other outward expressions of pain that are quite poetic: "*I do not want to live anymore,*" says one of them. "*All that is left for me is tedium and bitterness. The sun that gladdened my existence has been eclipsed, the moon that illuminated me has darkened, the star that guided my way has disappeared. I am sunken in a deep gloom and the abyss of a sea of tears and sorrow.*" As soon as this one has finished, another begins: "*Woe to me, I have lost everything! I will no longer see him who was my joy and the delight of my heart, who animated our warriors, was the honor of our race, the glory of our country, the hero of our nation. He does not exist, he has abandoned us; what will become of us? How can we* (page 40) *go on living?*"

These laments last all day and well into the night, with each one striving to express with their words the vividness of their pain, praising the qualities of the deceased.

This was how these people lived, enveloped in the shadows of darkness for many centuries. And they would have continued to do so if God, who has determined the time in which He will illuminate every man that is born onto

---

Culture of the Marianas Islands," pp. 16–18; and L. Thompson, *Guam and its People,* pp. 262–265.

445. See F. García, S.J., *The Life and Martyrdom of the Venerable Father Diego Luis de San Vitores,* book III, chapter II, p. 174. In the mid-eighteenth century, Jesuit missionary Father Franz Reittemberger kept the locks of women who had belonged to the Congregation *Nuestra Señora de la Luz,* which he founded in 1758. Although the German Jesuit considered them their "daughters," in 1771 he was judged by the Holy Inquisition for sexual solicitation. See A. Coello, "Lights and Shadows: The Inquisitorial Process against the Jesuit Congregation of *Nuestra Señora de la Luz* on the Mariana Islands (1758–1776)." *Journal of Religious History,* vol. 37, No. 2 (June 2013), pp. 206–227.

446. See F. García, S.J., *The Life and Martyrdom of the Venerable Father Diego Luis de San Vitores,* book III, chapter II, p. 174.

this world, had not dissipated these shadows as a result of his kindness and compassion, sending Father San Vitores and his companions to teach them about His kingdom and the road that leads to it, which is Jesus Christ, the Savior of mankind.

BOOK THREE

F ather San Vitores impatiently awaited the return of the two companions that he had sent to the island of Guåhan to become acquainted with the conditions and disposition of its inhabitants. Some of the islanders, having seen the missionaries, went forth armed with their lances. Others followed them, so that in a very short while the coast was lined with barbarians. The multitude, armed with spears and arrows, struck fear in the Spaniards that accompanied the priests, but they recovered their courage when they realized that this entire warlike array was meant to honor the preachers of the Gospel. In effect, they were received with great demonstrations of friendship, and they were led to Agadña, the most important settlement in the island.

The *chamorris* Quipuha [or Kepuha], accompanied by the greater part of the nobility,[447] (page 41) received them in his home. The two Fathers greeted him according to the custom of the country, and expressed their desire to stay among them to instruct them in the law of God and to teach them the way to Heaven. "We thank you very much, Fathers," Quipuha told them affably, "and you bring news that will make the people rejoice, for we have expected you for a long time."[448] The Fathers, gladdened by such a favorable reception, spent the rest of the day and night in Quipuha's home. They raised a great cross on the beach as a sign of their having taken possession of the country in the name of Jesus Christ, and they returned on the morrow to their ship accompanied by the island's chief nobles, who brought provisions for the Spaniards. There, in the name of their compatriots, they invited Father San Vitores to stay and live with them.

The new apostle went later with his companions and with those Spaniards who wished to join him, consecrating themselves to the service of this mission. The moment they set foot upon the land, he raised an altar at the beach, and began the duties of his apostolic life by saying Mass and asking God for the conversion of this country of infidels and for the health of the numerous people who had come from all over the islands to see him and honor him. He then preached in their tongue, which he had already learned. *"My only purpose in coming here,"* he said, *"is to have you know the true God, and to show you the way to Heaven. In order to reach it, you must believe in the mysteries*

---

447. Let us recall that *chamorris* was the title given to the first among the nobles (note by L. de Morales/C. Le Gobien, *Histoire, book* III, p. 72).

448. By converting to Christianity, Quipuha set a very persuasive example for the rest of the natives. See V. M. Diaz, "Pious Sites...," p. 319.

of the Christian religion, practice its maxims, and receive baptism, so as to become the sons of God and heirs of heaven." He then explained in a few, simple words what was indispensable for salvation, and his manner was so moving, that after hearing this first sermon, one thousand five hundred of his listeners converted and asked for Holy Baptism.

Father San Vitores, stirred by the vivid impression and wonderful effects that the grace of Jesus Christ produced in these barbarians, could not hold back tears. He offered God this first harvest, and told the people that the good disposition (page 42) that he saw in them filled him with joy, but that before receiving the precious grace of baptism, it was necessary for them to be convinced of the truths of religion and to be well instructed in them; that they must not grieve for this delay, but should instead apply themselves to learning what was taught them; and that as soon as they were sufficiently instructed, he would grant them the grace that they asked for so fervently. In the meantime, they could have their children baptized, and thus draw the blessings of heaven over them and their families.

The parents, inspired by these words, presented their children that very moment, but he could only baptize twenty, for he was told that the ship in which he had come was about to depart for the Philippines, and that emissaries from a neighboring island as well as from other villages in Guåhan were waiting for him to go with them.[449] He received them gladly and became their friend, promising to visit them frequently, and he kept his word, for as soon as he finished baptizing the children that were presented to him that first day, he went about the island of Guåhan to preach the glad tidings of the kingdom of God.

This island, which has a circumference of about forty leagues, is fertile and pleasant. It has safe, deep-water ports. The main ones are: Ati, to the west; Umatac [Humåtak], where the Dutch who travel these seas sometimes anchor their ships; and Iris and Pigpug [Pipok], which are separated only by a strip of land. All of these ports have an abundance of fresh water thanks to the streams that empty into them; but the best of all is the port of Agadña, for in it the ships are protected from the winds, and anchorage is impeccable in ten-to-eighteen fathoms of water.

Father San Vitores traveled around the island baptizing many children and a number of dying men and women who seemed to be but waiting for his arrival in order to go forth and enjoy the liberty of the children of God in

---

449. Father García says that 23 children were baptized that day (F. García, S.J., *The Life and Martyrdom of the Venerable Father Diego Luis de San Vitores*, book III, chapter II, p. 178).

{Figure 8: Statue of Quipuha in present-day Hagåtña (Guåhan). Photograph taken by Ron J. Castro}

heaven. However, his long absence upset (page 43) the *chamorris* of Agadña. These nobles, jealous of their rank and prerogatives, complained about the missionaries' apparent disregard for these things, and openly murmured against the instruction of the common people, who they believed deserved nothing but contempt.[450]

The Father was informed of this and, not wanting to irritate this proud

450. See F. García, S.J., *The Life and Martyrdom of the Venerable Father Diego Luis de San Vitores*, book III, chapter II, p. 187.

people, whose vanity and susceptibility he knew too well, he decided to return to Agadña immediately. In fact, he did more: he resolved to establish the missionaries' principal residence in Agadña and build a church there.[451] The *chamorris* were pleased with his return and with the deference that he showed them. Quipuha especially marked the occasion by declaring himself the protector of this mission, and granting Father San Vitores land where he could build the church the moment he learned of the priest's intention.[452]

This and the instruction of Agadña's nobility occupied the servant of God nearly the whole winter, and forced him to delay the visit to the neighboring islands that were waiting to receive the missionaries. He sent Father Pedro de Casanova to the island of Zarpana, whose inhabitants had seemed extraordinarily enthusiastic at the idea of having preachers of the Gospel among them.[453] This island is only six-or-seven leagues away from Guåhan. It has two excellent ports, one in the south and the other in the north. The natives call the latter port Socanrago [Sasanlago], while the Spanish refer to it as Puerto de San Pedro.[454] Father Casanova was well received, with gestures of friendship that gave him great hope. He was not mistaken, for in the very first few days of his stay he had the consolation of baptizing three hundred children, and instructed a great number of adults who later received Holy Baptism.

In the meantime, Fathers Cardeñoso and Morales went to the island of Tinian, where their harvest was in no way inferior. Its inhabitants had long wanted to have missionaries. One of their compatriots, called Taga, had planted this desire in them (page 44) telling them of a miraculous apparition that he had been favored with when the vessel called La Concepción had shipwrecked near these islands. This is the story that the two Fathers learned from this good people:

In 1638, the Holy Virgin had appeared to Taga with such an air of sweetness and majesty that he was left speechless. She made herself known to him, and encouraged him to become a Christian and to help the shipwrecked Spaniards. Taga obeyed, assisting the Spaniards, and, having himself instructed, he asked for Holy Baptism and received it from the hands of Don Marcos Fernández de Corcuera, whose family would years later give the Philippines a

451. See F. García, S.J., *The Life and Martyrdom of the Venerable Father Diego Luis de San Vitores*, book III, chapter III, p. 181; see also the phtoographic appendix.

452. D. A. Ballendorf, "From *Latte* to Concrete...," pp. 21–26.

453. See F. García, S.J., *The Life and Martyrdom of the Venerable Father Diego Luis de San Vitores*, book III, chapter V, p. 186.

454. See F. García, S.J., *The Life and Martyrdom of the Venerable Father Diego Luis de San Vitores*, book III, chapter I, p. 167.

{Figure 9: Map of Guåhan. Source: Charles Le Gobien, *Histoire des Isles Marianes nouvellement converties à la religion chrétienne; & de la mort glorieuse des premiers Missionaires qui y ont prêché la Foy* (Paris, 1700)}

Governor.[455] Taga was not satisfied, and he begged the man who had baptized

---

455. He is referring to the shipwreck of galleon *Nuestra Señora de la Concepción* (1638). This episode is described in F. García, S.J., *The Life and Martyrdom of the Venerable Father Diego Luis de San Vitores*, book III, chapter III, pp. 176–180. By showing that there were already Christians in the islands, Morales justified the

him to promise that upon his arrival at the Philippines, he would send preachers to Tinian to teach his compatriots and make them Christian. Corcuera promised to do so, but whether he did not bother to fulfil his promise, or he could not overcome the obstacles that presented themselves, the truth is that no missionaries had come to the island until now, and that is why as soon as they arrived, the natives were willing to listen to them and profit by their instruction.[456] To record the memory of this miracle, the missionaries called this island Buena Vista Mariana. The Holy Virgin showed that it was dear to her and that she had taken it under her protection, for the faith progressed wonderfully in that island in a very short time.

For his part, Father San Vitores worked tirelessly in Agadña with a success that befit the greatness of his zeal, as he prepared a great number of persons of both sexes and all ages and conditions for baptism. Since these barbarians were most impressed by that which dazzled their senses, the Father made sure to inspire in them a profound respect for the mysteries of our holy religion, especially Holy Baptism. He therefore surrounded the act with a great gravity in order to move them more. Everything was ready for celebrating the ceremony with great solemnity, when an obstacle arose that nearly upsets this holy man's designs, ruining his projects.

He had conveyed such a high idea of baptism and the graces that this sacrament bestowed upon those who received it, that the *chamorris*, who looked upon the common people with great scorn, decided that such a precious gift should not be wasted on those who were not worthy of receiving it. *Communicating*, they said, *the graces of Our Lord Jesus Christ to such vile and coarse souls would be like profaning them; and this grace, as the great Father has taught us,* for this is how they called Father San Vitores, *raises man onto God, making him a participant of his divine nature. Therefore, the common people must be excluded, and only people of our rank and quality should be made participants.*

Since these people are very inconstant, the Father thought that that he did not have to pay too much attention to what they said. However, he saw that they were serious, and that regardless of what he said they persisted in their refusal to let the people receive baptism, so that he could not make them see their great error. Thus, with a great pain in his heart, he told this obstinate nobility that if they continued with this bad attitude he would not baptize anybody.

---

need to "recover" them for Christianity.

456. Fernández de Enciso and Fernández de Oviedo mention numerous examples of providential shipwrecks that favored the evangelizing mission of pagan peoples thanks to the intervention of the Holy Virgin. See A. Pagden, "The Savage Critic," pp. 32–45.

"I cannot flatter you," he said. *"In this matter there is no difference between those with noble blood and those without it. God
makes no exceptions, all people are his children, and he destines them
all to the same glory of heaven. Just as he makes the sun shine over all
people and illuminate every nation, he also sent Jesus Christ, his only
son, to Earth to save all men, and he invites them all to believe in the
same truths, to keep the same precepts, and to receive all sacraments
equally. Respect his orders and submit to his will, for you would make
yourselves unworthy of his kindness and his mercy if because of your
pride (page 46) you wanted to exclude from baptism those whom God
does not exclude. Imitate instead the humility of this sovereign Master
who, as great and superior as he was to all men, wanted to make
himself a man and give his life for them. Therefore, far from keeping
the people from embracing the Christian religion, you should invite
them to submit to it, and henceforth look for the way in which you
can distinguish yourselves because of the faithful and exact ways in
which you fulfill the duties that religion imposes.*

This appeal did not move their spirits as much as the Father had hoped.
He decided to convince the *chamorris* of the error of their ways. He spoke in
particular to the most obstinate among them, hoping to make them desist from
their ridiculous pretension.

"I deplore your blindness," he told them after a few days, *"you
insist on a point of honor that would draw the wrath and vengeance
of God, for you ought to know that this sovereign judge will treat you
like demons, who, being the first among the angels, were lost by their
pride and presumptuousness. Bear in mind that if you persist in this,
you will end up in hell, like those rebel angels whose lot it is to moan
there through all eternity."*

God touched the hearts of the haughty nobility and their proud souls
were humbled, and they submitted themselves to whatever the Father wanted.
He made them bury the bones and the skulls of their ancestors, which some of
them kept, misled by the macanas' superstitions[457]. They also had figures carved
out of trees, and the Father had them burn them. Finally, after having prepared
them to receive this first sacrament of the church, he set the date in which they
would be baptized. The ceremony was performed with a fervor that moved all

457. See F. García, S.J., *The Life and Martyrdom of the Venerable Father Diego Luis de San Vitores*, book
III, chapter V, p. 188.

who had come from all over the island to witness the novel spectacle. Quipuha, who had been the first to receive the preachers of the Gospel in his home, and who had served them with (page 47) his advice and supported them with his influence, headed the procession of the nobles, and was the first to receive the Holy Baptism. He was given the name John, in honor of John the Baptist, under whose protection father San Vitores had placed the island of Guåhan, calling it San Juan.

The *chamorris* followed his example, and far from stopping the people from receiving the sacrament, they were the first to invite them and to procure whatever was necessary for them to complete this holy action. The examples of virtue of many of these new Christians were as worthy as those of the first centuries of the Church. Quipuha especially, distinguished himself for his piety as well as for the rank and credit that he enjoyed. This illustrious old man preserved the grace received in baptism and lived thereafter in the careful practice of Christian virtues. He had the joy of dying [1669] with these sentiments, for the edification of all and the great consolation of the Father.

His death served to destroy an old superstition. The *chamorris* had the custom of burying their dead in certain caves destined for this use, and their commitment to this notion proved hard to dispel. *We do not want to be separated from our ancestors who have been buried in these places,* they said, *and we consider it a duty to rejoin them after our deaths.* Father San Vitores decided to abolish this custom so contrary to the laws of Christianity, and have Quipuha buried in the Church with all the honors that are associated with people of his merit and dignified station.[458] The relatives of the deceased opposed this, for they wanted to bury him in his ancestors' grave. The Father convinced them, and he gave Quipuha a funeral that pleased the *chamorris.*[459] They all felt greatly the loss of this protector of religion, and believed that his death had been precious to the eyes of God. What happened to his son a few days later confirmed this belief even more. This young man (page 48) insisted that his father had appeared to him, saying these words: *"My son, rejoice, for I have the delight of being in heaven."* This apparition, the news of which soon spread everywhere, confirmed

---

458. Quipuha died on December 3, 1669. He was assisted by Father Luis de Medina and he was buried in the church, on a Monday, on the Eve of Easter, "going to his house with a trumpet, bearing the standard of our Father St. Ignatius, and St. Francis Xavier (…) and we sang his vigil, as well as his Mass" (Francisco García, S.J., *Relación de la vida del devotíssimo hijo de María Santíssima, y dicho Mártir Padre Luis de Medina,* Cap. IV, p. 64).

459. As Vicente M. Diaz has said, "blessed Diego's victory over the manner of Kepuha's burial is seen to necessitate the destruction of native superstitions surrounding the fate of spirits of the dead. This was the materiality, the staff which, according to C. Le Gobien and Blessed Diego, was used in the "edification" of the new church" (V. M. Diaz, "Pious Sites…," p. 320).

the new Christians in their faith, encouraging them to lead holy lives.

The Father inspired them with such horror for the slightest faults that as soon as they committed a sin they ran to him, saying, "*Father, I have sinned; is it necessary to be baptized again in order to obtain pardon?*" [San Vitores] would answer that he had already taught them that the only remedy in such cases was to ask God for forgiveness and to receive the sacrament of Penance. He then instructed them about what it was, and he took the opportunity to hear their confessions.[460] These neophytes gave them with signs of such grief that Father San Vitores could not contain his own tears. Since they were naked, he gave them clothes to cover themselves; but because what he had brought from Mexico was not enough for the great number of neophytes, which increased each day, he made the decision to have them wear garments made of palm leaves. These were not well liked, for they were considered ridiculous, and indeed, they were. Then, the Servant of God, who wanted them to get used to wearing them, donned one himself, putting it on over his cassock. Once they saw this, none of them refused to wear them again.

While Father San Vitores worked in Agadña, Father Medina traversed the island of Guåhan with a success that gave him the hope that he would soon see Christianity solidly established in the island. In less than three months, during which he visited these island's inhabitants three times, he baptized more than four thousand persons, and he prepared many more to receive this holy sacrament.[461] He was not content with visiting a village once. Instead, he would return often to succor the ill and baptize the children, (page 49) and he was blessed with the consolation of converting at least one person in each visit. It is unimaginable what this servant of God had to endure in these testing and constant marches over rugged mountains and impracticable roads. Impetuous rivers and muddy swamps were the same to him as valleys and plains covered in swordgrass and weeds that made his feet bleed. His zeal was so great that no obstacle could stop him on his quest to win souls for Jesus Christ. When he found a village that was not docile enough, he was not discouraged, and he turned to God and doubled his prayers and rigors, which were already excessive. What happened in Nisichan [Nisihan or Nisi'an], one of Guåhan's principal villages, confirms this.

---

460. See F. García, S.J., *The Life and Martyrdom of the Venerable Father Diego Luis de San Vitores*, book III, chapter V, pp. 188–89.

461. Father García, the author of Father Medina's hagiography, insisted that his confrere had "baptized more than three thousand Marianas' natives, both children and adults, of whom some flew straight to heaven, as if they had been waiting for baptism to open the door for them" (Francisco García, S.J., *Relación de la vida del devotíssimo hijo de María Santíssima, y dicho Mártir Padre Luis de Medina*, chapter IV, p. 55).

From the beginning, its people had shown indifference and even aversion to Christianity. Far from wanting to listen to the Father and profit by his instruction, they had heaped insults upon him and struck him on the face. This mistreatment did not diminish the zeal of God's servant; on the contrary, it inflamed his spirit more and he asked the Lord with renewed fervor for the conversion of those barbarians. On the Feast of St. Francis Xavier [December 3], he felt keenly inspired to entreat the Holy Virgin, and to give that village the name of the saint. He asked his companions to pray for the success of this enterprise, and he himself performed many rituals during the Octave, fasting rigorously and mortifying his body with extraordinary severity.[462] On the day of the Octave [December 10] he said Mass and went to Nisichan [Nisihan or Nisi'an]. His arrival surprised the inhabitants, who thought that he would never return after having been treated so ignominiously. He gathered them, and animated with a truly apostolic zeal, he again announced to them the kingdom of Heaven. The Father had barely begun preaching when (page 50) these people, so indomitable and hardened before, now asked for Holy Baptism. Without wasting any time, the Father began their instruction, baptizing all those who were disposed to receive the sacrament on the day of the Octave of the Immaculate Conception [December 15]. Two things had moved this fervent missionary to ask God so insistently for this village's conversion. One was that, wanting to go to a village where the people were more willing and docile, he had lost his way and found himself for a third time in Nisichan [Nisihan or Nisi'an]; and another, that having had the joy of being mistreated there for the love of Jesus Christ, he believed that it was his duty to apply himself more vigorously to the salvation of those who had procured this great benefit for him.

This holy man had an extraordinary talent for winning souls for God. His zeal made him think of a thousand ways to attract these barbarians to our Holy Faith. While at Agadña during the Christmas holidays, he set up a nativity scene in which the figures, especially the Holy Child, were admirably represented. All of the island's people, attracted by the curiosity of such a novel spectacle, hurried to see the enchanting nativity scene, but he only showed it to those who could recite the Apostles' Creed, the Ten Commandments, the Act of Contrition, and the prayers that he had composed for them. Thanks to this holy artifice, all of them, especially the youth, learned in a moment what would

---

462. In Christian liturgy, Octave is both the eighth day after a feast, reckoning inclusively, which thus falls on the same day of the week as the feast itself; or the eight-day period between these days.

have otherwise taken days of studying.[463]

The Faith was solidly established among the various villages, and there were reasons to hope that all of the islands would soon be entirely Christianized, when a storm was unleashed that threatened to utterly destroy this nascent Church. An idolatrous Chinese (page 51) called Choco was the instrument chosen by the devil for his evil plans. In the year 1638, this man was on his way from Manila to Ternate, when a storm wrecked his ship near the Marianas. Choco fought against the waves and thought he would perish, but finally saved his life. This was a great misfortune for the islanders, who received him with friendliness and treated him well, leading Choco to decide to remain among them.[464]

Since in his own country Choco had been a great follower of the bonzes[465], who are the ministers of idolatry and superstition in China, he thought it would be easy for him to gain ascendance among these savages who had no religion whatsoever. He declared himself a bonze and began to teach the cult of the idols. The people, always avid for novelties, listened to him and became more and more infatuated with his idolatrous doctrines, until the arrival of the missionaries foiled his plans. He feared that he would lose the credit and authority that he had gained among the islanders; and once he saw the great number of conversions carried out by the new apostles everywhere and every day, he knew for sure that he would. Unable to conceal his resentment, he decided to ruin the Fathers' honor and reputation in order to discredit their ministry.

Choco spread the rumor that these foreigners were a perverse and infamous people; that the Spaniards had sent them to these islands to get rid of them; that they were impostors who told fables and lies in order to win their trust; and that their true desire was to fool them and make them the victims of their cruelty; that these vile men brought desolation and death wherever they set foot, and he had been a witness of this in the Philippines; that they could see it for themselves in the great number of children who had died since the foreigners had moved to the islands; that their malice was so [great] that they rejoiced in causing the death of newborn infants by pouring poisoned water over their heads; that they did the same to the ill, applying a poisoned oil on

---

463. See F. García, S.J., *The Life and Martyrdom of the Venerable Father Diego Luis de San Vitores*, book III, chapter VIII, p. 202.

464. See F. García, S.J., *The Life and Martyrdom of the Venerable Father Diego Luis de San Vitores*, book III, chapter VI, p. 190.

465. Bonze is the name given to priests of idolatry in China (note by C. Le Gobien, *Histoire*, book III, p. 90). See the hagiographies of F. García, S.J., *Relación de la vida del devotíssimo hijo de María Santíssima, y dicho Mártir Padre Luis de Medina*, chapter V, pp. 76–77. The term is derived from the Japanese name given to Buddhist monks [bonzou or bonso].

them (page 52); and that they were nothing but enchanters who used their diabolical secrets to attract and trick their followers.

This horrendous depiction made a terrible impression on these credulous souls, and an almost universal change took place. People that had sought out the preachers of the Gospel with great fervor, and that had received them with true affection, now retreated from them in horror. If the Fathers set out to visit the villages and preach the word of God as they had done before, the barbarians would present themselves, spears in hand, to prevent them from entering, treating them as impostors and murderers, and threatening to kill them if they insisted on pressing forward.[466] How different this reaction to those early affectionate and friendly invitations and those sincere and engaging displays of friendship! Their spirits had been stirred [against the Fathers], who were frighteningly discredited.

But what the missionaries deplored the most was that, persuaded by Choco and convinced that the water was poisoned and that baptism would kill their children, mothers hid their sons and daughters from them, or hid themselves in the mountains with their babies in their arms, for fear that they would otherwise be baptized. Thus, the great daily progress of the Faith was suddenly cut short by Choco's imposture and by the credit that these light and voluble people gave to this idolater's calumnies.

The old Christians remained faithful to their duties, and the first among them tried to convince their compatriots of the falsity of the Chinese bonze's words:

> *How could the Fathers commit murder, they would tell them, when they teach us a religion that forbids it, and that orders us to love one another as we love ourselves? When we only see in them sweetness, charity, and an eagerness to serve us and do us good? If they show zeal in baptizing our children and visiting the* (page 53) *sick, it is only because they want to save them and procure for them the glory of Heaven, which is man's happiness for all eternity. How can you be so blind as to give credence to Choco's impostures? Do you not know that he is motivated by passion, and that he berates the Fathers only because they foil his plan to continue instilling in us his extravagant ideas?*

While the more constant and fervent Christians sought to dissuade their compatriots, Father San Vitores undertook to convert Choco in order to uproot

---

466. See F. García, S.J., *Relación de la vida del devotíssimo hijo de María Santíssima, y dicho Mártir Padre Luis de Medina*, chapter V, p. 78.

the evil in its very source, and dissipate the mistrust that this bandit had generated towards the Gospel's preachers.

Only heaven could secure such a conversion. He redoubled his prayers. He increased his fasts and austerities. He implored the assistance of the Holy Virgin, and St. Ignatius and St. Francis Xavier, the patron of this mission. He also prayed to the children who had died after having been baptized, so as to obtain from God the conversion of this infidel and thus have the glory of his Holy Name shine before all these people.

One day, while he was praying with even more fervor than usual, he felt strongly inspired to go and find this enemy of Christianity. Urged by this inner movement, he was about to depart when his companions arrived. They were Father Morales and Father Medina, who had been seriously wounded by the barbarians [in the island of Saipan].[467] Father San Vitores warmly embraced them, for they were the first apostles of these islands who had been judged worthy enough by Jesus Christ to suffer for him, and he told them with a tender but resolute tone:

> *"Fathers, charity demands that I stay here with you, but I feel vividly inspired to work in the conversion of our friend Choco, and I believe that I would be resisting the Holy Ghost if I delayed but one moment more (page 54). Therefore, I entreat you to the care of Father Bustillo, and depart to win for God a soul capable of perverting so many others."*[468]

Since traveling by land would have taken three days, and his desire compelled him to reach Choco that very day, he took to sea, contrary to his custom, to reach Paa,[469] the place of Choco's residence, as soon as was possible. He entered the village singing the Christian Doctrine that he had put into native verse to attract these barbarians, for they loved singing. They all went to hear the Servant of God, and Choco came along with them. Discerning him in the crowd, the Father spoke to him, asking him in front of everybody why he was so bent on discrediting the Christian religion and slandering its ministers. Thus, he began a discussion with Choco regarding the Christian religion and its

467. See F. García, S.J., *The Life and Martyrdom of the Venerable Father Diego Luis de San Vitores*, book III, chapter VI, p. 194.

468. See F. García, S.J., *The Life and Martyrdom of the Venerable Father Diego Luis de San Vitores*, book III, chapter VI, p. 192.

469. Paa was close to Malesso', twenty miles from Hagåtña (F. García, S.J., *The Life and Martyrdom of the Venerable Father Diego Luis de San Vitores*, book III, chapter VI, p. 192). See also F. García, S.J., *Relación de la vida del devotíssimo hijo de María Santíssima, y dicho Mártir Padre Luis de Medina*, chapter IV, p. 58.

mysteries. The dispute became so involved that it lasted many days.

The Father spoke with such forceful reason and such inspiration that the idolater was finally left speechless. Choco publicly confessed that he had spoken of the mysteries of the Christian religion without knowing them and that he was convinced of the truth and sanctity of this religion that the Father had just taught him. To show everybody the sincerity of his words, he dropped at the feet of the servant of God, and asked to be baptized [August 20, 1668]. News of this marvelous conversion spread rapidly. This sudden and unforeseeable change was regarded as a wonderful triumph of the Christian religion. The Father thanked God, and put off Choco's baptism for a few days in order to further test and instruct the new catechumen.

On the appointed day, the neighboring villagers went to witness this surprising conversion. The ceremony (page 55) went well, despite Hell's attempts to perturb it through the demonic possession of two Filipinos that accompanied the Father.[470] Choco seemed pleased, and all the Christians were happy in their belief that the greatest persecutor of their religion was going to become an apostle, but their happiness did not last long. Choco did not persevere, and, disloyal to the grace received, he became an apostate who scandalized the church and continued to persecute it.

Father San Vitores, pleased with his expedition to Paa, where calumny had been publicly defeated, returned to Agadña. There he found out about the trouble in Tinian. This island's inhabitants, naturally inconstant and restless, had blindly accepted all of Choco's extravagant notions. They distanced themselves from the Fathers, and showed a great aversion to religion, going so far as to commit acts of violence: they had seriously wounded Father Morales, as was already stated, and caused Sgt. Lorenzo Castellanos and a Filipino named Gabriel de la Cruz[471] to perish at sea. The missionaries had reason to believe that the situation would get worse, and they warned Father San Vitores of the impending danger. He replied using the words of the prophet, "*Euntes ibant et flebant, mittentes semina sua,*"[472] and he immediately resolved to go to this island

470. According to San Vitores's hagiographer Francisco García, two of his assistants were possessed by the devil that day: one of them, Bautista, ran off to the mountains, while the other attacked the priest and wounded Don Juan de Santa Cruz in the arm. F. García, S.J., *The Life and Martyrdom of the Venerable Father Diego Luis de San Vitores*, book III, chapter VI, pp. 193–94).

471. See F. García, S.J., *The Life and Martyrdom of the Venerable Father Diego Luis de San Vitores*, book III, chapter VII, p. 198.

472. "They weep as they go to plant their seed" (Psalm 126:6) (note by Luis de Morales/C. Le Gobien, *Histoire*, book III, p. 98). See F. García, S.J., *The Life and Martyrdom of the Venerable Father Diego Luis de San Vitores*, book III, chapter VII, p. 198.

to share the dangers and labors of this mission with his brothers. [But they told him] that if the barbarians killed them, this nascent church would perish with him, and therefore it would be more appropriate if someone else went there to try to pacify those people.

> *"Your are mistaken"*, he told them, *"I have less to fear than any other, because I am (page 56) the least worthy to die for Jesus Christ; but if this bliss were to befall me, know that my death will be an advantage to this mission instead of a loss."*

He arrived at Tinian with Father Morales in late October 1668.[473] He spoke to the barbarians, dissipating their fears and taking them out of their error and obstinacy, and he won them over with his sweet and affable ways, so that whenever they saw him they exclaimed, *mauri si Dios*: God is good.[474] After having restored order in this island, and establishing a mission house there, he went to Saipan for a visit. The northernmost isles had yet to hear about Jesus Christ, and he really wanted to go there and announce the kingdom of God to those people, but the needs of the new churches prevented him from going, and he sent in his stead his companion, Father Morales.

God blessed his work. This new apostle discovered the peoples and islands of Inarajan [Inalåhan], Sarigan, Guaguan [Gugan], Pago [Pågu], and Agrigan, and his harvest was so great in such a short time, that he was eager to continue making new discoveries. But the weather became so bad and the sea so rough, that those who accompanied him, even though they were skillful, did not dare risk their lives in their small canoes. He therefore remained in the island for six months, where he had the solace of baptizing more than four thousand persons.[475]

In the meantime, Father San Vitores went about the entire island of Saipan, visiting every single place, whether in the mountains or in the valleys. These marches were quite difficult, and he undertook them wearing a coarse garb made of palm leaves and a hat of the same material. He walked barefoot over impracticable roads, going up craggy mountains in which he was in constant danger of falling to his death, or crossing over the swordgrass that abounds

473. Father García provides the exact date: October 20, 1668 (F. García, S.J., *The Life and Martyrdom of the Venerable Father Diego Luis de San Vitores*, book III, chapter VIII, p. 201).

474. See F. García, S.J., *The Life and Martyrdom of the Venerable Father Diego Luis de San Vitores*, book III, chapter VIII, p. 201.

475. See F. García, S.J., *The Life and Martyrdom of the Venerable Father Diego Luis de San Vitores*, book III, chapter VIII, p. 201.

(page 57) in the valleys and plains of these islands. The only things he carried with him were his breviary, the New Testament, the Imitation of Christ and the book of rules, a great rosary around his neck, and a crucifix in his hand.[476] Thus, he walked and visited the villages.

Before setting out for the road, he would always recite the customary prayers[477] and assume a patron saint to protect the mission that he was about to undertake. Usually, it was the saint whose office the church celebrated that very day. He told the Fathers to do the same always. When he was about to enter a village, he would sing the Christian Doctrine to assemble the people. If a cross was already planted, he adored it, praying a little, and then he went to all of the houses to baptize the children and console the sick and the dying.[478] He spent the entire day in this pious exercise, and despite his hunger or fatigue, he did not rest or take any food until he had finished. Thus he visited the islands of Saipan, Tinian, Aguiguan and Zarpana, from where he returned to Guåhan on the eve of the Feast of the Epiphany [January 6] in 1669.[479]

Since he saw that the children were docile and interested in learning the Christian doctrine and the mysteries of religion, he took particular care in their instruction, and resolved to establish a seminary in Agadña that he named San Juan de Letrán [in mid-1670] to educate the youth in the practice of Christianity.[480] This was a successful project. The children were eager to help in the Church services; they sang the Christian Doctrine each day with enchanting candor; and they went out in the streets ringing a bell to call the other children to catechesis. The oldest and most skilled accompanied the fathers in the missions and served them as catechists and interpreters.

These children won their parents over to God and sanctified (page 58) their families.[481] This is what happened with one of the nobility's leaders: a

---

476. The Breviary [Breviarium] is divided into four parts (Winter, Spring, Summer, and Autumn), and works as a compendium of all the requisite offices to be performed by a Catholic priest, friar, or layman. "*The Imitation of Christ* [De Imitatione Christi], an amply used devotional text, was written by Thomas à Kempis in Latin ca.1418–1427. The Society's *codex legum*, or "book of rules," contained St. Ignatius' Constitutions and various other foundational texts.

477. These prayers were known as the "itinerarium: A form of prayer used by monks and clerics before setting out on a journey, and for that reason usually printed at the end of the Breviary, where it could be conveniently found when required." See Alston, G. C. (1910). "Itinerarium". In *New Advent Catholic Encyclopedia*. [Available online] <http://www.newadvent.org/cathen/08255a.htm> [Consulted March 2, 2024].

478. After these visits, he would erect an altar, say Mass, and provide catechetical instruction for children (note by C. Le Gobien, *Histoire*, book III, p. 101).

479. See F. García, S.J., *The Life and Martyrdom of the Venerable Father Diego Luis de San Vitores*, book III, chapter VIII, p. 204.

480. See F. García, S.J., *The Life and Martyrdom of the Venerable Father Diego Luis de San Vitores*, book III, chapter VIII, p. 205.

481. Catechesis aimed at children, aided by dialogue and song, had revealed itself an effective method

VERDADERO RETRATO DEEL V. P. DIEGO LVIS
Gᵒ Fosman sculp.   DE SANVITORES.   Matriti 1682

{Figure 10: Martyrdom of Diego Luis de San Vitores (1671). Source: *Galerie Illustrée de la Compagnie de Jesus*, By Father Alfred Hamy. Paris: Chez l'Auteur, 14 bis, Rue Lhomond, Tomo 7, 1893} ten or twelve year old boy was fishing with his father, when they saw a guatafe [Tagafe],[482] a fish that is highly valued in these islands. The father, pleased at the prospect of this great catch, invoked his Anitis. The boy, stung by his father's blindness, said: Oh, father of mine! Do not invoke the demons that are the enemies of man, for if you address them you will accomplish nothing!" "Who should I address, then?" asked the *chamorris* brusquely. "God, about whom the great Father has taught us!" said the boy. "Place your trust in Jesus and Mary and speak their names, and our catch will be abundant!" The good man did as his son told him, and that very moment he caught the fish that he desired. He was so impressed that his prayer had granted him what he wanted so immediately, that at that very moment he went to the Church to be instructed in the mysteries of our religion, and in a short while, he was baptized.

Good fruit was harvested in the seminary of San Juan de Letrán, but since it was difficult to support it without the necessary funds, Father San Vitores took great care in procuring them. Since he could only receive them from the Court, he wrote a memorial to the king of Spain, and after providing him with a report on the mission, he spoke in these terms:

> *It is in the interest of the glory of God, and the service of Your Majesty, to establish a seminary in the island of Guåhan for the education of its youth. This establishment is indispensable insofar as the children here are independent of their parents, and they often lose themselves to the most dissolute disorder. In this regard, these islands are worse off than other countries. The devil has set up veritable schools of debauchery, public houses where th young men who do not want to marry live with women that* (page 59) *they buy or rent from their parents, so that even decent islanders are scandalized. These detestable places can only be destroyed by establishments that are contrary to them, where the young are educated in piety and virtue.*

> *The children here are noble and docile, and they can be easily directed towards good. They learn with ease, and in a short time can*

in the indigenous populations in various Jesuit missions elsewhere, such as Peru's Cercado and Juli. The object was to distance the children from their elders' customs and transform them into the propagators of the new faith and their parents' educators. See Alexandre Coello de la Rosa, "La doctrina de Juli a debate (1575–1585)," *Revista de Estudios Extremeños*, Tomo LXIII (2), 2007, pp. 951–990.

482. Friar Juan Pobre of Zamora had already detected this in his famous 1602 *Relación*. On this respect, see Marjorie G. Driver, *The Account of Fray Juan Pobre's Residence in the Marianas, 1602*, p. 22. See also F. García, S.J., *The Life and Martyrdom of the Venerable Father Diego Luis de San Vitores*, book III, chapter IX, p. 205.

be made into good catechists, and later, into excellent preachers and virtuous ecclesiastics, for they have no inclination towards drunkenness such as happens in other countries.

One of the things that St. Francis Xavier cared for the most was the instruction of children, and he looked upon this instruction as one of the greatest advantages of missions. He recommended it to his companions, convinced that religion was fortified with age; and that since these children would later be heads of families, they would become the apostles of their country, and Christianity's most solid support. After the Viceroy of the Indies had applied the four thousand pardaos[483] for the instruction of the youth [of Malabar] which the queen of Portugal collected from the coastal fisheries, the apostle of the Indies beseeched this princess to approve what the Viceroy had done in her name. "For you must be persuaded, my Lady," he told her, "that these children are the greatest and safest guides that Your Majesty will have to lead you into heaven." I can say the same thing of this Seminary, whose foundation I hope will be undertaken by the Queen, that the children there instructed will be Your Majesty's strongest safeguard, and that this house shall be the strongest castle of Your States.

It would also be necessary to establish a similar place for girls, in order to educate them in the exercises of Christian piety, and to keep them from ending up in those houses of dissipation where they lead (page 60) such dissolute lives.[484]

This Memorial was sent to the Queen who governed Spain as regent, given the young age of the king [Charles II]. Since she had taken these islands under her protection, she took Father San Vitores's words to heart, and granted his request, officially decreeing the foundation of the Marianas' seminary for boys on April 18, 1673, and endowing it with an annual assignation of three thousand escudos that the Royal Treasury of Mexico would provide. She also ordered the Marquis of Mancera, Viceroy of New Spain, to contact the Father in order to undertake the establishment of the girls' seminary.[485]

---

483. *Pardao* was a coin used in Portuguese India (Goa) and other eastern possessions (i.e., Ceylon), worth half a rupia. The word "pardao" comes from the Sanskrit word for splendor, *pratāpa*, used as an epithet of kings on the legends of native Indian coins.

484. See F. García, S.J., *The Life and Martyrdom of the Venerable Father Diego Luis de San Vitores*, book III, chapter IX, pp. 206–207.

485. See F. García, S.J., *The Life and Martyrdom of the Venerable Father Diego Luis de San Vitores*, book III, chapter IX, p. 207. The girls' school, Santa Rosa de Lima, was founded on 1674 (ARSI, "Relación de las empresas y sucesos espirituales y temporales de las islas Marianas," Philipinae Historiae, 1663–1734, Tomo

Thus, a house was built in Agadña where the islands' youth has been educated with extraordinary care and solicitude since then. With their good behavior and their dedication to learn the mysteries of our religion, the seminarians have surprised the Spaniards. For it is admirable how children who are free to go where they want, their parents having no authority over them, choose to commit themselves as they do to a discipline that is as rigorous as that of the most observant religious houses in Europe.

The church building of Agadña was completed in early 1669, and Father San Vitores opened it on the Feast of the Purification of the Holy Virgin [February 2].[486] Villagers crowded the church, which in time became famous because of the miraculous cures that took place inside it. Here is one that Father San Vitores relates in one of his letters:

An eight-month old boy from the island of Zarpana was afflicted by what seemed an incurable dropsy. His loving parents brought him to the church of Agadña hoping that his health would be returned to him. They presented him to Father San Vitores, who asked them if he was baptized, which he was not (page 61). However, thanks to the ridiculous ideas spread by Choco, his parents feared that baptism would make the child worse, and so they assured the father that he was baptized, and that they wished to be baptized themselves. The Father instructed them and baptized them. He then applied some relics to the boy, and he prayed and recited the Holy Gospel over him, but to no avail. The boy's illness continued unabated, and indeed, worsened. His frightened parents returned to the church and confessed the fault that they had committed when first presenting the child. The Father baptized him, whereupon the boy was immediately and completely cured. The fame of this miracle spread far and wide across the islands, and was instrumental in destroying Choco's calumnies, giving this people a high regard for this sacrament's virtue.

Thus, God rewarded with new graces the labors of his servants, so that religion was strengthened and this new church flourished each day. Here is what Father San Vitores wrote to the Queen of Spain in a letter dated April 15, 1669:[487]

We offer Your Majesty the first fruit of the Mariana Islands,

---

13, f. 181v).

486. See F. García, S.J., *The Life and Martyrdom of the Venerable Father Diego Luis de San Vitores*, book III, chapter IX, p. 207.

487. Francisco García provides a different date: April 25, 1669 (F. García, S.J., *The Life and Martyrdom of the Venerable Father Diego Luis de San Vitores*, book III, chapter IX, p. 209).

but no gold, pearls, or precious stones, for these islands are known only for their misery, as the Dutch, who have often crossed them, are convinced. We offer Your Majesty that which you value more than these riches, and which is indeed dearer: that is, souls rescued with the precious blood of Jesus Christ, whose grace has been spilled upon these lands of infidels with abundance. For despite the artifices of the devil and Choco's impostures, we have baptized more than thirteen thousand savages this first year, and instructed more than twenty thousand catechumens in the eleven islands discovered. More than one hundred children who ascended to heaven immediately (page 62) after their baptism will guard our sovereign king, watching over the preservation of his person and praying to God for his health and for the increase of his estates. For they owe their eternal happiness to him, since if this mission had been delayed for only one more year, they would have been lost for all eternity.

{Figure 11: "Proas marianas", by Henri Louis Claude de Saulces deFreycinet (1779–1842). Source: Original Copper Engraving from the Volkergallerie. Freycinet Art Collection (c. 1835). Courtesy of Micronesian Area Research Center (MARC)}

Book Four

When Father Morales returned from his voyage to the northern islands, the missionaries assembled to confer on the future of the missions and religion. After having decided upon several matters, they divided the religious care of the newly discovered islands among themselves, and begged Father San Vitores to visit them so that he could know the missions and determine how best to address their needs. Each one then departed for the place that had been assigned to him, and in early July of 1669, Father San Vitores returned to Guåhan.[488]

However, the weather was not propitious for this long journey. The sea gets so rough in that time of year, that large vessels can barely withstand the waves without foundering. The islanders have only small canoes that are more like the debris from a shipwreck than proper ships for sailing, for they consist of three boards joined together over which one must struggle to remain in equilibrium, less the canoe topples to one side or the other. Lacking any shelter, those aboard are at the mercy of the elements, scorched by the rays of the sun or soaked by the splashing seawater or the rain that falls incessantly from the heavens in this time of year (page 63). Moreover, in the midst of all these fatigues, there is no other sustenance but crackers drenched in seawater, which are more likely to upset the stomach than to quell hunger or animate the spirit.

Not a day went by that Father San Vitores did not risk perishing at sea in the rough waters, sailing in such poorly constructed and ill-equipped vessels. He traversed these islands despite the hardship, the dangers that he exposed himself to, and the mistreatment that he received at the hands of these barbarians. What happened in the island of Saipan is a telling example. In this island, Choco had many followers that set the people against the Gospel's preachers, discrediting their ministry. The Father found these islanders entirely predisposed against the missionaries. They did not let him baptize their children or carry out his other apostolic duties. Indeed, they had decided to kill him, and had appointed one of their principal *chamorris* as the executor. Whether out of fear, or simply to amuse himself (as he told his compatriots), he refused this honor, and passed it on to a friend of his who had a passion for extraordinary things and was dominated by curiosity. This man knew that the Father had performed many miracles: that he cured the sick from inveterate illnesses; that he had power even

---

488. See F. García, S.J., *The Life and Martyrdom of the Venerable Father Diego Luis de San Vitores*, book III, chapter X, p. 212.

over death, for he had as of late, brought a child back to life in the island of Zarpana. These miracles had excited in him the desire to witness one. "*Why do you come here?*" he asked the Father curtly. "*What is your purpose?*" The servant of God replied, "*I come to show you the way to heaven and to lead you there, if you subject yourselves docilely to my teachings.*"

The *chamorris* laughed, and invited him to perform a miracle in his presence. Seeing (page 64) that this infidel was acting just like Herod when Our Lord Jesus Christ was presented to him, the Father decided to imitate his Divine Teacher, and spoke not a word to whatever questions he was asked by that man.[489] Irritated by his silence, the barbarian called him a lunatic and a fool, and returned him to his friend. A group of insolent people took him, insulted, and humiliated him along the way. "*Here is the madman you sent our master,*" they told the *chamorris'* friend. "*He is a fool who goes from town to town and island to island agitating the people with his extravagant notions, singing with the children and with other madmen like him, who let themselves be seduced, and follow him. He spared him so that you may amuse yourselves with him.*"

In effect, they did this most cruelly, exposing him to the insults and jibes of an insolent mob that humiliated and vexed him. This did not prevent the Servant of God from harvesting great fruit even there, for despite the jeers, many listened to his words and converted.

He was not content with visiting the islands that had already been discovered, wanting instead to bring the light of the faith to others that were yet unknown. Thus, he went to the islets of Asonsong [Asunción; August 15, 1669] and Maug [August 17, 1669], where the inhabitants were so docile and inclined to profit from his instruction, that he baptized almost every one of them.[490] He fervently wished to have some missionaries stay and cultivate these new churches, but since they were all occupied, he had to content himself with leaving two catechists to baptize the children and watch over the faith in the new Christians.

On his return, he stopped in the island of Anatajan, where he visited some of the villages, while one of his catechists, a man named Lorenzo, visited the others. Lorenzo was so zealous carrying out the Father's mandate, that he was graced with the crown of martyrdom.[491] While he was baptizing a girl, he was

---

489. Lucas 23: 6–12 (Note by Luis de Morales).

490. See F. García, S.J., *The Life and Martyrdom of the Venerable Father Diego Luis de San Vitores*, book III, chapter X, p. 213.

491. See F. García, S.J., *The Life and Martyrdom of the Venerable Father Diego Luis de San Vitores*, book III, chapter X, p. 214.

assaulted by a group of enraged savages, who had been persuaded by the mother of a deceased child that his death was caused (page 65) by his recent baptism. They threw themselves at Lorenzo, pierced him with their spears, gouged out his eyes, and threw his body down a cliff. This Servant of God, who became the first martyr of the Marianas, was from the Malabar Coast. He had survived the famous shipwreck of the Concepción [1638] that had crashed against the rocks of the Mariana Islands, and had lived among these savages ever since then, learning to speak their language perfectly. Since he was very pious, he had joined the Fathers, serving them as catechist and interpreter and accompanying them in their trips. Animated by a fervent desire to save souls, he shared all their labors.

Heaven was moved by the death of this fervent Christian, for a few hours after he was martyrized, a horrible storm broke out over Anatajan and struck fear into its inhabitants. The thunder and lightning were fierce, but what most frightened them was a thick and flaming material that they had never seen before falling all around them. Terrified, they sought out Father San Vitores and threw themselves at his feet, begging for his forgiveness at the death that they had just caused. They believed that Lorenzo's soul had gone to the Philippines demanding vengeance, and that the governor, indignant at their cruel deed, had sent his artillery against them. The atmosphere was thick with a dense, black smoke that spread as far as the islands of Saipan and Tinian.

Fathers Medina and Casanova worked zealously in this last island, and met with great success despite the terrible discord caused by an ongoing feud between the two main population centers, Marpo and Sungharon.[492] The nobility in this country is proud and insolent, and some *chamorris* from each of these towns had insulted each other. This private quarrel soon became a general conflict, and the entire island was divided in two camps. They took to their (page 66) weapons, went out, and were about to clash, when, warned of the danger that this new church was facing, Father San Vitores arrived and without hesitation, placed himself in the middle of the field holding the crucifix in his hands. He called them to peace, he begged them, he threatened them, but to no avail: their spirits were too excited. These barbarians did not listen to the words spoken by the Servant of God, and hoping to rid themselves of the mediator who was upsetting their plans, they started throwing rocks at him. But they were greatly surprised when they saw that this holy man stood motionless in the

---

492. See F. García, S.J., *Relación de la vida del devotíssimo hijo de María Santíssima, y dicho Mártir Padre Luis de Medina*, chapter V, pp. 78–85; and F. García, S.J., *The Life and Martyrdom of the Venerable Father Diego Luis de San Vitores*, book III, chapter X, pp. 215–17.

midst of the hail of stones, and that instead of hurting him, the stones turned to dust the moment they touched him or the crucifix that he held in his hands.

Yet, not even a miracle as spectacular as this was capable of quelling these furious barbarians, who were still bent on fighting each other. I have already stated that hand-to-hand combat was a last resort; an extreme arrived at only after spending all imaginable means to ensnare their enemies. Seeing that this war could ruin one of the archipelago's most flourishing churches, Father San Vitores decided to return to Guåhan in order to bring help [November 15, 1669].[493] Once there, he gathered his catechists and the Spaniards that usually accompanied him in his apostolic endeavors and told them that it was necessary to take up arms and march against the infidels, not to enrich themselves with the spoils of war, but to deliver the barbarians from the devil's tyranny, for the glory of God and the good of the souls. They obeyed, going with him to Tinian [November 25, 1669], and although Captain Don Juan de Santa Cruz had only ten Filipinos under his command, they marched with as much ardor and boldness as if they had been ten thousand.[494]

Father San Vitores, having implored Heaven's favor, placed himself between the two armies that were still facing each other (page 67), and once again urged them to make peace, to no avail. The Marpoese, the more agitated of the two, began throwing rocks at him, and again they turned to dust without harming him. Then, seeing that the barbarians refused to listen to appeals or threats, he signaled for Captain Don Juan to advance with this small company, ordering the captain to stay with him between the two enemy [camps] and fortify this position as best he could. Saints have resources that other men lack. It was impossible for ten men armed only with three muskets and a small piece of field artillery to impose themselves on two armies. Mocking them, the barbarians attempted to overtake them, but the Spaniards, filled with their trust in God and assisted by San Vitores's prayers, did not lose heart, and instead threatened the barbarians to attack the party that refused to make peace. Each side named delegates to confer with the missionaries, and Father San Vitores went to Sungharon and Father Medina to Marpo. They tried to convince the people that it was in their best interest to avoid the consequences of a terrible war that would bring ruin to their homes and the loss of their families, and

493. See F. García, S.J., *The Life and Martyrdom of the Venerable Father Diego Luis de San Vitores*, book III, chapter X, p. 216.

494. This was the famous *Escuadrón Mariano*, the Marianas Squadron, commanded by Captain Don Juan de Santa Cruz. See F. García, S.J., *The Life and Martyrdom of the Venerable Father Diego Luis de San Vitores*, book III, chapter XI, pp. 218–20).

exhorted them to undertake peacefully what the Spaniards would otherwise force upon them. The Marpoese, who had started the quarrel, were persuaded, and they sent some of their *chamorris* to Sungharon to negotiate the conditions of the peace. Having reached an agreement, the Marpo *chamorris* were returning to their compatriots when they fell into an ambush prepared by some renegade enemies who disagreed with the pact that had been made. This infuriated the Marpoese, and they refused to consider peace afterwards, despite the missionaries' words and entreaties.

Deeming that it was necessary to let time pass in order for their inflamed spirits to calm down, Father San Vitores went about the island of Tinian and sent Father Medina to visit Saipan. Upon his return (page 68), they both redoubled their peace-making offices. The two parties finally listened to the missionaries' propositions and accepted them, and on January 24, 1670, they happily arrived at a peace agreement.[495] Its main stipulations were that the past was to be forgotten; that two churches would be built, one in Marpo and the other in Sungharon; and that the two armies would march to a given place where reconciliation would take place.

Father Medina was at the head of Marpo's army, which marched in an orderly fashion carrying the flag of the Holy Virgin and the mission's patron saints. Father San Vitores, carrying a cross in his hands, led those of Sungharon. Both armies approached each other, adored the cross amidst displays of sincere sorrow, and exchanged presents of rice, fruit, and tortoise shells, which are tokens of friendship among these people.[496] The Marpoese presented such a large shell that it was consecrated to Our Lady of Guadalupe, in the island of Tinian. To commemorate these events, the field where San Vitores was stoned was named De la Santa Cruz, and eventually a shrine was built there dedicated to Our Lady of Peace.

With tranquility restored in Tinian, Father Medina returned to the island of Saipan [January 27, 1670] to complete the visit that he had started.[497] This Servant of God longed for martyrdom, and the closer he came to that

---

495. See F. García, S.J., *The Life and Martyrdom of the Venerable Father Diego Luis de San Vitores*, book III, chapter XI, p. 220. The Mariana Islands were the only area in Oceania to have rice (Lawrence J. Cunningham, "The Ancient Chamorros of Guam", in Lee D. Carter, William L. Wuerch & Rosa Roberto Carter (eds.), *Guam History: Perspectives*, Vol. 1, Mangilao, Guam: Richard F. Taitano & Micronesian Area Research Center (MARC) & University of Guam, 1998, p. 14).

496. See F. García, S.J., *The Life and Martyrdom of the Venerable Father Diego Luis de San Vitores*, book III, chapter XI, pp. 220.

497. F. García, S.J., *Relación de la vida del devotíssimo hijo de María Santíssima, y dicho Mártir Padre Luis de Medina*, chapter VI, p. 98.

blissful moment, the more his heart became inflamed with the desire to give his life for Jesus Christ. He had had ardent desires for this since his youth, and he had consecrated himself to the mission of the Indies with the hope that God would grant him the grace of being a martyr. He begged Our Lord God for this grace each day, and held this intention while he recited the beautiful prayer composed for this purpose by Father Carlo Spinola,[498] one of the most illustrious martyrs of Japan, who had (page 69) the joy of being burned alive in the city of Nagasaki on September 10, 1622.[499]

On his first trip, the Father had already visited most of Saipan, but he still had to go to the villages that were most opposed to the missionaries. It is incredible how great an impression that Choco's impostures had caused in the spirits of these barbarians, and how much hatred they now harbored against our Holy Faith. This impostor had a great influence in this island, for he had married there, and his wife's relatives were decidedly in his favor, and they helped him procure the absolute discredit of the Gospel's preachers in order to make them odious to the people, inventing and spreading infamous calumnies against them each day.

As soon as Father Medina arrived, an insolent mob began to follow him, heaping insults and indignities on him. The inhabitants of Raurau [Laulau] also received him with an unfavorable disposition. Parents hid their children for fear that he would baptize them, believing that baptism made them sick and caused their death. He went on to Tatasu, another village where the more docile people listened to him and profited from his instruction. He finally went toward Sugrian Mountain, whose rude and ferocious inhabitants could not stand the missionaries. When the women heard that he had come, they fled with their children in their arms for fear that just by coming close to them, the Father would kill them, for this was the idea that they had formed of the missionaries. When the Servant of God found out that they had gone to sea to hide from his sight, he followed, and finding a newborn boy in Tipo beach, he asked if he could baptize him. Some of the Sugrian mountainfolk, waiting but for an occasion to insult him, began to murmur and agitate the residents of the coast against the Father. *"Do you not know,"* they said, *"that man's wretched designs? He comes here to take our children and cause their death* (page 70) *by*

---

498. C. Le Gobien added the following note: "His life has been written in Italian by Father Fabio Ambroise Spinola, and in French by Father Pierre Joseph d'Orleans, both Jesuits" (*Histoire*, book IV, p. 122).

499. He was burned to death over a slow fire in Nishizaka, in the so-called Great Martyrdom of 1622. For more on Carlo Spinola (1565–1622), see *DHCJ*, vol. IV, pp. 3623–3624; *Varones Ilustres de la Compañía de Jesús*, Tomo I, pp. 372–407.

pouring poisoned water over their heads. He is a madman who preaches extravagant notions and adores a false God that he has created himself." To this, they added even more blasphemies.[500]

The Father, grief-stricken, tried to appease the mountainfolk, telling them that he would not baptize the child, for he was healthy, unless his parents asked him to. However, his words fell on deaf ears. The barbarians became more and more infuriated, until they resolved to kill the Servant of God. Not knowing their plan, the priest and his two companions took the road up the mountain to renew the visit.[501] The barbarians followed him, insulting him all the way to the village of Cao, where as soon as he arrived he went from home to home announcing the kingdom of God and asking if there were any children at risk of dying. He heard the cries of a child in one of the houses, and thinking that the child was perhaps ill, he tried to go inside, but a mob of more than thirty furious barbarians stopped him as they pronounced more blasphemies against God and his ministers. One of the most audacious struck him with a spear from behind, but despite the violence of the blow, the Servant of God did not fall. He tried to press on, [but] the barbarians did not allow it as they fell upon him, wounding him everywhere. He fell, covered in his own blood, but he then arose as best he could, and holding in his hands the small crucifix that he carried around his neck, he addressed the Lord with such affection and fervor that the savages retreated from him, fearing, as they revealed in their later confession, that the God whom the Father was addressing so vehemently would avenge his death.[502]. However, the holy martyr was possessed of very different sentiments. He called to them, exhorting them to convert and repudiate their errors, and after having told them that he forgave them for his death from the bottom of his heart, he continued to speak to Our Lord. One of these soulless scoundrels, tired of the long soliloquy, lanced the holy man's neck, causing his immediate death. This transpired on the day of January 21, 1670.[503] His death seemed so extraordinary (page 71) to these barbarians, that when they spoke of it they used the word *Macana*, which in their language means marvelous,[504] because they believed that after he fell the first time, pierced by so many lances, he had come back to life,

---

500. See F. García, S.J., *Relación de la vida del devotíssimo hijo de María Santíssima, y dicho Mártir Padre Luis de Medina*, chapter VI, pp. 99–100.

501. Father Morales underscores the predestination of Father Medina's martyrdom.

502. See also F. García, S.J., *Relación de la vida del devotíssimo hijo de María Santíssima, y dicho Mártir Padre Luis de Medina*, Cap. VI, ff. 103–104.

503. See F. García, S.J., *Relación de la vida del devotíssimo hijo de María Santíssima, y dicho Mártir Padre Luis de Medina*, chapter VI, p. 105.

504. For an analysis of the term "macana," see footnote 392, in Book Two.

or else he could not have spoken the way that he did.[505]

This zealous missionary had been born in Malaga on February 3, 1638.[506] His mother, a virtuous woman, instilled in him a great devotion to the Blessed Virgin from his early childhood, and he remained her faithful servant throughout his life, receiving many extraordinary blessings from this Lady. The innocence and purity of his customs made others regard him as a saint. He had fervently asked to be admitted into the Society, and Father Francisco Franco,[507] provincial of Andalusia at the time, saw such sanctity in him that he admitted him despite his poor health and the fact that he had difficulty speaking clearly. Some Fathers had tried to make him see that it was wrong to admit this young man because his weak constitution would prevent him from properly fulfilling the Society's duties. "*You are mistaken, my Fathers,*" he had told them, "*let us receive him: his duty will be to become a saint.*"[508] Thus, Luis de Medina was received into the Society on April 30, 1656.

The young novice perfectly fulfilled the duties of his vocation, and soon rose to a high degree of perfection in the practice of all virtues, distinguishing himself particularly in humility and mortification of the flesh. God was thus preparing him to be one of the apostles of the New World, and indeed, all that he thought about during those first few years was sanctifying himself in order to become worthy of such a noble ministry. As he grew older, his desire to work in the salvation of souls also grew, and it was soon so ardent that he consecrated himself to the missions using the following words:

> (page 72) *Eternal and omnipotent God, although I am unworthy, I, Luis de Medina, appear before Thy divine presence moved by the desire to serve Thee more fervently, and to amend my sins, and to work and suffer more for Thee, oh, my God! And above all, to show Thee how much I love Thee, I vow to Thy Divine Majesty before the Blessed Virgin my mother and all the celestial court, to go to the Indies, wherever the Father General of the Society chooses to send me, as I have*

505. See F. García, S.J., *The Life and Martyrdom of the Venerable Father Diego Luis de San Vitores,* book V, chapters V–VI, pp. 365–84.

506. Father Garcia says of this: "The same year that a Spanish ship, the Concepción, was lost in the Mariana Islands," arguing that this seeming temporal coincidence was a sign of providence (F. García, S.J., *Relación de la vida del devotíssimo hijo de María Santíssima, y dicho Mártir Padre Luis de Medina,* f. 2).

507. He was also rector of the College of Zaragoza (1637) and provincial (1640) and vice-provincial (1660–1661) of Toledo (C. Sommervogel, SJ, *Bibliothèque de la Compagnie de Jésus,* T. III, 1892, p. 937).

508. X. Baró, "'Recibámosle para santo': las hagiografías sobre Luis de Medina (1637–1670), protomártir de las islas Marianas," in J. Martínez Millán, H. Pizarro Llorente, y E. Jiménez Pablo (eds.), *Los jesuitas: religión, política y educación (siglos XVI-XVIII),* 3 Vols. Madrid: Universidad Pontificia Comillas, 2012, Vol. 3, pp. 1503–1522.

obligingly requested him to send me there where the need for laborers is greatest, if this be for the greater glory of God. However, if it would please Thy Divine Majesty more, and would be more advantageous to my spiritual salvation, that I remain in Spain instead of going to the Indies, in that case [I] would stay here peacefully, persuaded that this is the will of God. In testimony of which I sign this on the day of the Assumption of Our Lady, August 15, 1664. Luis de Medina

He sent this vow to the Father General, who granted what he asked, destining him to the mission of the Philippine Islands. The superiors of the province, saddened by the loss of a man capable of rendering great services for God and the Church in Spain, wrote to the Father General asking him to revoke his orders. This afflicted Father Medina greatly, for his desire to work in the conversion of infidels had greatly increased since he had obtained permission to go to the Indies. He redoubled his prayers and penitence to obtain this grace from God, that no obstacles be placed before his vocation, but seeing that his superiors insisted on retaining him, he wrote the following letter to Father Cristobal Perez,[509] at that time provincial of Andalusia:

My Reverend Father:

The reticence that you show in confirming the permission granted to me by the Father General so that I may (page 73) *go to the Philippine Islands has given me qualms of conscience, and to free myself from these as well as to inform you clearly on what regards this matter, I will expose the motives that urge me to go to the Philippines.*

One year before entering the Society, I fell ill, which was an obstacle for my admission. The refusal that met my demand afflicted me more than the malady itself. Grief-stricken, I asked the Blessed Virgin before one of her images on the Feast of Our Lady of the Snow to obtain from her Holy Son Jesus Christ my complete recovery, so that I could join the Society and go to the Indies to preach the Gospel. God heard my prayers, and to the amazement of my doctors, I was cured, and was blessed with the joy of being admitted into the Society. During my novitiate, Our Lord God inspired within me an even greater desire to consecrate myself to the mission in the Indies, but, not wanting to deceive myself, and to know more surely the will of

---

509. For more on Father Cristóbal Pérez (1601–1647), see Sommervogel, S.J., *Bibliothèque de la Compagnie de Jésus*, Tomo VI, p. 516.

God, I delayed my departure until after I finished my studies, too see if my feelings were still the same. Since then, this desire has been more vehement each day, and five years ago on the Feast of St. Francis Xavier, I felt such a great urge to go to the Indies, that I could not resist asking God through the intercession of the Blessed Virgin Mary and this great saint, to let me know if it was His will that I consecrate myself to this mission, and if it were not, to extinguish the desires that I felt. I continued asking the same thing for three years, and with this intention in mind performed multiple penances. Finally, on the Feast of St. Ignatius, in my thanksgiving after the Eucharist, I prayed to God our Lord to let me know his designs regarding my person. I then heard in the bottom of my heart, the voice of St. Ignatius telling me that I should take a vow to go to the Indies when I finished my studies, that this was (page 74) God's will. I did not do it immediately, however, but on the feast of Our Lady of the Snow, again during my thanksgiving after the Eucharist, while I was commending myself to the Blessed Virgin Mary. I cannot say how, but I can assure you, that I heard the Holy Virgin inside of me, saying: "Take a vow to go to the Indies, because this is the will of my Son, who wants you to go, and this is why on this day he restored your health through my intercession." I still delayed in obeying this inspiration, until I was completely certain that it indeed came from Heaven. I continued my prayers and penance until the feast of the Assumption, and upon that day, when my soul was inundated with sweet, spiritual solace, I took a vow to go to the Indies when I completed my studies. Since then, I feel such an ardent desire to fulfill my vow that my greatest joy comes from thinking that I will soon be in those lands of infidels working for the salvation of their souls.

I give you all these details, my Father, because I believe in my conscience, that I am obligated to it, so that, being informed of the reasons that impel me to go to the Missions, you will then resolve what will be for the greater glory of God. If after all that I tell you, you decide that I should not use the permission granted to me by the Father General, I declare that on the Day of Judgment, when God asks me to account for what I have manifested to you, I will answer to his Divine Majesty with whatever you may reply to this letter and nothing more. And I beg you to please reply to this letter, so that I may find tranquility. I am writing the same to the persons that have consulted me on this regard, so that, well informed, you may resolve what you judge to be for the greater glory of God[510] (page 75). *My Reverend Father, Thy very humble and obedient servant, Luis de Medina. Montilla, April 27, 1666.*

Upon receiving this letter, rather than opposing such a manifest vocation, the provincial granted this Servant of God what he had so longed for. However, the galleons would not depart for New Spain until the following year. Luis de Medina embarked in one of them on July 17, 1667, with fourteen confreres, over whom he was made superior, and he arrived in Mexico at the time when Father San Vitores was preparing to take the light of the Gospel to the Mariana Islands. Father Medina joined this new apostle and went with him to the islands where he was blessed with the joy of being the first of the Society to spill his blood for the faith. One of his companions on that fateful day, Hipólito de la Cruz, shared with him the joy of receiving the crown of martyrdom [January 21, 1670].[511] The barbarians had attacked him with their spears together with Father Medina, and he died right then and there. The Father's other companion [Agustín de la Cruz] managed to escape to Tinian, where he gave the news of these deaths that were so precious in the eyes of the Lord.[512]

It was decided that their bodies should be retrieved, but they were in the hands of the Sugrian mountainfolk, who were willing to attack anybody who came near them.[513] Captain Don Juan de Santa Cruz decided to take charge, and immediately went to the island of Saipan with nine of his men [April 24, 1670]. He did not stop at Opian, where he disembarked, but went on with his people to Raurau. There, he made it known to the inhabitants of Cao and Sugrian that he and his men were going to lay waste their fields and burn their villages if they did not bring him the bodies of the holy martyrs. The barbarians obeyed, but not daring to show themselves before the Spaniards, they deposited the (page 76) precious relics on a nearby hill. Don Juan and his Christians took them, and reverently carried them to the beach, singing the Christian Doctrine in their language.[514]

Upon seeing the Spaniards, most of the savages responsible for the deaths

---

510. Ad Maiorem Dei Gloriam (note by Luis de Morales)

511. See F. García, S.J., *Relación de la vida del devotíssimo hijo de María Santíssima, y dicho Mártir Padre Luis de Medina*, chapter VI, p. 106; and F. García, S.J., *The Life and Martyrdom of the Venerable Father Diego Luis de San Vitores*, book III, chapter XI, p. 221.

512. See F. García, S.J., *Relación de la vida del devotíssimo hijo de María Santíssima, y dicho Mártir Padre Luis de Medina*, chapter VI, pp. 106–07.

513. Juan de Santa Cruz, *Vida y martirio del padre Luis de Medina*, p. 3.

514. See F. García, S.J., *Relación de la vida del devotíssimo hijo de María Santíssima, y dicho Mártir Padre Luis de Medina*, chapter VI, pp. 108–10.

of the father and his companion repented for their deed, and threw themselves at Don Juan's feet in the midst of great sorrow and fear. This excellent captain, not content with having recovered the holy martyrs' bodies from the hands of the barbarians, demanded to see the place where the sacrifice had been consummated, whereupon he erected two commemorative crosses [April 26, 1670].[515] He also gave orders to arrest the two men [Poyo and Daon][516] who had first struck the Father and his companion, and even though there were one hundred and fifty of them, and only ten Spaniards, the barbarians dared not oppose this measure.[517] These two rascals confessed that they had killed the father and his companion out of hatred towards the faith, angry that their children had been baptized and their Anitis taken from them, but they added that they had been incited by the inhabitants of Raurau and Sugrian to carry out the detestable deed.

Don Juan López,[518] bishop of Nombre de Dios in Cebu Island,[519] upon whom the Marianas depended for spiritual matters, ordered a juridical investigation into the death of the two martyrs, and the facts that I present here come from those inquiries.[520] Their bodies were taken to the Church of Our Lady of Guadalupe in Tinian, and from there, they were transferred on Father San Vitores's orders to Agadña, the capital of Guåhan, where they were received with all honors.

Father San Vitores had already returned to said island, where he had found its inhabitants (page 77) in a dreadful state. Drought was killing their fruits and crops, and they were threatened with famine. They had already gone to their Anitis hoping that they would bring rain. The Father was saddened when he heard this, and he assembled all of the nearby peoples and reproached them their

---

515. See Francisco García, S.J., *Relación de la vida del devotíssimo hijo de María Santíssima, y dicho Mártir Padre Luis de Medina*, chapter VI, p. 111.

516. See F. García, S.J., *Relación de la vida del devotíssimo hijo de María Santíssima, y dicho Mártir Padre Luis de Medina*, chapter VI, p. 112.

517. Father Morales alludes to a characteristic situation of "unequal war" that proves that God was indubitably on the Jesuits' side. For more on this, see Coello, "Colonialismo y santidad en las islas Marianas: los soldados de Gedeón," pp. 17–44.

518. See F. García, S.J., *Relación de la vida del devotíssimo hijo de María Santíssima, y dicho Mártir Padre Luis de Medina*, chapter VI, p. 115. Relations between friar Juan López and the Society of Jesus were tense, as his sentence against Father Jerónimo de Ortega for having administered the sacraments without license to do so shows (J. de Ortega, *Defensa por la provincia de la Compañía de Jesús de las Islas Filipinas contra una sentencia que Juan López, obispo de Cebú, fulminó por decir administró los Santos Sacramentos sin jurisdicción* (s.n., 1671). On July 20, 1676, the seat of Cebu was vacant because Bishop López was promoted to the Archbishopric of Manila (Remedios Contreras, *Catálogo de los Fondos Americanistas de la Colección Salazar y Castro*, RAH, Madrid, 1979, p. 151).

519. This is one of the Philippine Islands (Note by C. Le Gobien, *Histoire*, book IV, p. 135).

520. He is referring to the information on Juan de Santa Cruz in "Vida y martirio del padre Luis de Medina" and "Mártires de la Provincia de Filipinas de la Compañía de Jesús," that are kept in Catalonia'Arxiu Històric de la Companyia de Jesús.

inconstancy and infidelity, and exhorted them to place all of their trust in God.

"*You seek*," he said, "*the help of demons whose only desire is to deceive mankind and drag it down to hell with them. Address the Creator of heaven and earth, who has formed you from nothing. He is your Father, he loves you, and he will provide for you. Pray to him with confidence, and he will grant your requests.*"

He then made them recite a prayer that he had composed in their language for such public needs, and assured them that the next day they would feel its effects with the abundant rain that would fall. His prediction came true, and this palpable miracle consoled these good neophytes, reaffirming them in their faith, while leading many infidels to convert and ask for the holy baptism. The *macanas* were angered, for they saw that they were losing their authority and their credit; that nobody came to them; and that they were looked upon as mere impostors who entertained the people with their vain superstitions. Therefore, they decided to rid the islands of the preachers of the Gospel and the rest of the Spaniards. They expounded that the contempt expressed toward the Anitis would attract their wrath, and that very soon there would be no fruit on the trees, no crops in the fields, and no fish in the ocean; that diseases were going to spread everywhere; and [that they] would be punished with all sorts of plagues if they did not obey the mandate to expel these strangers that seduced the people and took their freedom. They were supported by Hurao, one of the most influential *chamorris* of Guåhan. This astute man worked covertly against the fathers, and since he was more skillful than his compatriots (page 78), he acquired great authority among the people and even the nobility, and he was held as an oracle, his words blindly taken for truths.

Hurao was a great obstacle in the way of religion, for although until then he had kept up appearances in his treatment toward the missionaries, deep down he was their avowed enemy. Father San Vitores had tried to win him over and convert him to Christianity, rendering him services and looking to please him by doing him favors, but to no avail. Instead, Hurao had learned to hate him more, becoming more insolent. He was always looking for ways to stir up trouble against the Spaniards who thwarted his influence, and here is the occasion that he found to carry out his plans:

A young Spaniard named Peralta had gone to cut some wood to make more crosses of the kind that were placed in Christian houses.[521] A native who coveted his dress killed him, took his clothes, and escaped. Peralta's companions

---

521. ARSI, "Historica narratio illorum (1668–1673), Filipinas Tomo 13, fol. 95–110, transcribed in R. Lévesque, *History of Micronesia*, vol. 6, 1995, p. 47.

thought that they should pursue the authors of this crime to punish them and make justice, and sergeant major Don Juan de Santiago [Bozo][522] ordered the arrest of some of Agadña's residents who were considered suspects. These men were imprisoned and interrogated, but since nothing came out of these inquiries, they were released and sent home. This process infuriated the barbarians, and they agitated one another to avenge affronts that seemed more insufferable than death itself. The Spaniards tried to explain that this was a just procedure, and that it benefited everybody, for it preserved the peace and prevented conflicts and battles, for if the guilty party was not brought to justice, everybody would be constantly exposed to his violence and abuses, whereas the proceeding to secure his death would save innocent lives.

Agadña residents were not convinced by this reasoning, and they looked upon the arrest of their compatriots as a shameful affront against their people. Hurao and his faction incited this sentiment, secretly encouraging the people in it. He joined Chaco, this avowed enemy of the Christian religion, who, like him, wished to be rid of the missionaries and to expel the Spaniards from the country. The death of Guafac [Huasac or Cha'fa'e, 1671], one of the most important *chamorris*, gave them an opportunity to carry out their evil designs. The Spaniards went to detain a Marianas' native whom they suspected had killed Peralta. Guafac and his companions stopped them, and seized the suspected criminal from the Spaniards' hands. In the ensuing fracas, a soldier with more daring than prudence killed Guafac, and with this, a war signal seemed to have been sent.[523] Hurao declared himself openly against the Spanish, and he used all means available to make his enemies odious in his compatriots' eyes, inciting them to take arms against the Spanish and drive them from the country.

The Europeans would have done better, he told them, to remain in their own land. We had no need of them to be happy. Satisfied with what our country provided us, we lived without desiring anything else. The knowledge that they have given us has only served to increase our necessities and excite our ambition. They think our nakedness is wrong, but if clothing were necessary nature itself would have

---

522. This is a mistake, for Captain Don Juan de Santiago Bozo did not arrive in the Marianas until May of 1672. He must be referring to Captain and Sergeant Major Juan de Santa Cruz. This mistake was also made by F. García in *The Life and Martyrdom of the Venerable Father Diego Luis de San Vitores*, book III, chapter XIV, p. 236.

523. Indeed, this first general uprising is referred to historically as the First Great Chamorro War (11/09/1671). See F. García, S.J., *The Life and Martyrdom of Diego Luis de San Vitores of the Society of Jesus*, book III, chapter XIV, p. 236. A more detailed study of these events can be found in A. Coello, "Colonialismo y santidad en las islas Marianas: la sangre de los mártires," pp. 707–745.

*provided us with it. Why should we burden ourselves with clothes,
a superfluous thing that limits the freedom of our arms and our legs,
because they allege that we must cover our bodies? They treat us as a
coarse people and regard us as savages. Must we believe them? Do we
not see* (page 80) *that under the pretext of instructing and civilizing
us, what they do is corrupt us? That they have made us lose the primitive
simplicity with which we lived, taking from us our liberty, which
is dearer than life itself? They want to convince us that they bring us
happiness, and many among us have been blind enough to believe
them. They would believe them no longer if they realized that all our
miseries and maladies appeared with the arrival of these foreigners,
who came to disturb our peace. Did we know any of the insects that
cruelly pursue us, before their arrival on these islands? Did we know
mice, flies, mosquitoes, and all these other pests whose only purpose
is to torment us? Those are the gifts that they have brought us in the
floating houses in which they sail! Did we know rheumatism and
colds.*[524] *What diseases we had, we also had the means to cure them,
but they have given us their ills without teaching us their remedies.*[525]
*Were iron and the other bagatelles that they have brought, which have
no intrinsic value, were they worth seeing ourselves surrounded by so
many calamities?*

*They reproach us our poverty, our ignorance, and our lack of
industry, but if we are as poor as they say we are, what is it that they
seek among us? Believe me, if they had no need of us, they would not
expose themselves to so many perils or work so hard to settle in our
midst. Why do they insist on subjecting us to their laws and customs,
forcing us to abandon the precious gift of liberty that our parents
bequeathed us? In sum, why give us sorrow in exchange of a good that
we can only hope to enjoy after our deaths?*

(page 81) *They regard our histories as fables and fiction. Do we
not have the same right to regard as equally fictitious what they teach*

---

524. It is hard to believe that these islanders were not subject to rheumatism and colds, and that they had
no insects before the Spanish arrival, but it is indeed true that they attributed all these things to the Spanish,
and frequently reproached them for it (Note by Luis de Morales/C. Le Gobien, *Histoire*, book IV, p. 142).
Despite Morales/Le Gobien's tongue-in-cheek note, the Spanish quite probably brought insects and pests that
were unknown in the islands, just as they brought new diseases that the Chamorro remedies could not cure.

525. The Chamorro had doctors, called *suruhåno* (or *suruhåna*), who used herbs, massage and other
"natural" medicines to cure both physical and spiritual ills. See P. McMakin, "Suruhanos: A Traditional Curer
on the Island of Guam." *Micronesica*, 14:1, 1978, pp. 13–67. Cunningham, "The Ancient Chamorros of
Guam", pp. 35–36.

*us and preach as incontestable truth? They abuse our good faith and
our simplicity. All of their arts are directed towards fooling us, and all
of their science tends only to make us wretched. If we have ever been
as ignorant and blind as they would have us believe, it was when we
overlooked their pernicious desires until it was too late, and when we
allowed them to settle among us. Let us not lose courage in the face
of our many misfortunes; they are only a handful of men and we can
be rid of them easily. Even though we do not have those murderous
weapons that spread terror and death all over, we can finish them off
because we greatly outnumber them. We are stronger than we think,
and we can soon free ourselves of these foreigners, and regain our
primitive liberty.*

These speeches, which he delivered before the *chamorris,* while his com-
missaries repeated them to the people, had the result that he expected. Many
natives took up arms and prepared to attack the Spanish, who became aware
of this only too late. Since they had only come to these islands to help the missionaries and work with them in the conversion of these peoples, they had not
built fortifications nor taken any other precautions to keep themselves safe from
any bad surprises from such a malicious people. Yet, they had realized from the
start that they faced an indomitable and deceitful nation. Thus, they found
themselves in a bind when (page 82) they learned that there were more than
two thousand men ready to fight them.[526] Indeed, if the islanders had attacked
immediately, the Spaniards could not have resisted, but the barbarians bid their
time, and the Spaniards were able to fortify themselves.

A palisade was built around the church and the missionaries' residence,
with two small towers that they referred to as forts. They called the one next
to the mountain the Fort of St. Francis Xavier, and the one facing the sea, the
Fort of Saint Mary. There they raised a flag with *the image of the Holy Virgin
that had been blessed by the Archbishop of Manila.*[527] Each of these forts was then

---

526. See Francisco García, S.J., *The Life and Martyrdom of Diego Luis de San Vitores of the Society of Jesus,*
book III, chapter XIII, p. 236.

527. Father Morales seems to confirm that the statue that many thought was the image of Santa Marian
Kamalen (or Santa Marian Camalin), which had been aboard the galleon *Nuestra Señora del Pilar de Zaragoza*
that shipwrecked on June 2, 1690, in Cocos Island, near Humåtak Bay, was in fact an image consecrated
by the Manila archbishop and brought to the Marianas by Father San Vitores in 1668. At present, it is held
in the church Dulce Nombre de Maria, founded on February 2, 1669, having survived typhoons, wars, and
Guåhan's near total destruction during World War II. This image holds a place of privilege in Chamorros'
religious life, and its popular veneration has turned it into a key symbol of Guåhan's Catholic legacy (Marilyn
A. Jorgensen, *Expressive Manifestations of Santa Marian Camalin as Key Symbol in Guamanian Culture,* Ph.
D. Diss., University of Texas at Austin, 1984, pp. 1–24). The emphasis in the text was added by the editor.

equipped with two old cannons that they had found on the beach. Captain Don Juan de la Cruz reviewed his troops, which were constituted by thirty-one men, twelve of whom were Spaniards armed with muskets, and nineteen were Filipinos armed with bows and arrows, and he had them spread out in given locations.[528] Since their strength was not in their numbers or weapons, but in their trust in God, they made sure to confess and take communion in order to procure the Lord's blessings. Father San Vitores exhorted them fervently, telling them that they should entrust everything to Heaven, and assuring them that they had nothing to fear, for since they were fighting for God, God would fight for them.

Aware that Hurao had aroused the whole island and that his authority and intrigues alone were enough to raise an army, the Spanish held a council to see what measures they should take against him. They decided that it was necessary to seize this *chamorris*, thinking that this would terrorize the barbarians and force them to sue for peace. The enterprise was difficult and risky, but it was carried out just as it was planned, although the consequences were not those that had been intended.

Hurao's capture greatly upset his friends and relatives (page 83). They did all that they could to obtain his freedom: they begged, they cried, they made promises, trying to convince the Spaniards that they were doing everything in their power to pacify their compatriots and thus secure the liberation that they desired so strongly. Instead, they were secretly animating them to take up arms against the Spaniards and avenge Hurao's capture. Perhaps moved by seeing the soldiers exposed to the privations and rigors of war, Father San Vitores was deceived by their hypocrisy, and he decided to make peace.

> *Suing for peace now,* his companions told him, *will lead to our defeat. Do you not know, by Jove, these barbarians' character? What will they say when they see us going towards them? They will regard our conduct as a sign of weakness and discouragement, and that will raise their spirits, making them more insolent, so that once Hurao is free, they will renew hostilities with even more fervor than before. It is they who must sue for peace, and this is what our military honor and dignity demand.* "I would concede to this," replied Father San Vitores, "if we had come here as conquerors, but our object when coming to these islands was to convert their inhabitants, to make them know

---

528. See F. García, S.J., *The Life and Martyrdom of the Venerable Father Diego Luis de San Vitores,* book III, chapter XV, p. 238.

*and love Jesus Christ. Can we obtain such an object through war? Let
us sacrifice, then, this small point of honor for the greater glory of God
and the salvation of souls."*

The respect and deference that they felt towards the Father made them
blindly accept his line of thinking. They sued for peace, sending emissaries provided with goods and tortoise shells[529] to meet with the Marianas' rebels, as was
customary among these peoples. The barbarians received them with contempt
and ridicule, and they composed satirical verses in which they scoffed at the
Spaniards' timidity and cowardice. Father San Vitores thought that if he spoke
to the infidels personally, they would be more docile and reasonable. But as he
approached them, crucifix in hand, he was worse received than the first emissaries (page 84), for the natives showered him with stones and threatened him
with their lances. Not stopping there, a mob of two thousand natives furiously
set out to attack the Spanish positions. This was on September 11, 1670.[530]
The resistance that they encountered did not discourage them, and animated
by Choco, who headed the group, they continued their assaults day and night,
throwing an incredible number of stones at the Spanish. This first offensive
lasted for eight days, but, seeing that they lost many of their own men in these
repeated [attacks] in which their bodies were bare, they looked for a way to
protect themselves from the Spanish firearms. Choco had them build wooden
palisades or walls attached to wheels, which they then took before them as far
as was necessary to stay undercover while they threw their rocks and firebrands.
The Spanish defended themselves vigorously, and carried out successful sorties.
The *macanas* were becoming desperate with the unexpected resistance, and
encouraged their men, sustaining them with the hope of a swift victory. They
made them place their *Anitis* on top of their lances, and assured them that
they could advance fearlessly, for those lifeless heads would be their shields and
safeguards. The barbarians heeded their advice, and trustingly advanced closer
than usual to the Spanish trenches, but they soon learned that their *macanas*
were impostors and that their *Anitis* did not stop musket balls.

The Spanish were exhausted, their situation dire. They were reduced
to a small number of men locked up behind a hastily constructed palisade,

529. As stated earlier, tortoise shells were tokens of friendship, not just in Guåhan but also in many
islands across the Pacific. See F. García, S.J., *The Life and Martyrdom of the Venerable Father Diego Luis de San
Vitores*, book III, chapter XI, pp. 220–21.

530. In his *Histoire*, C. Le Gobien gives the same date (book IV, p. 148). Father Francisco García, S.J.,
mistakenly gives another, September 11, 1671 (F. García, S.J., *The Life and Martyrdom of the Venerable Father
Diego Luis de San Vitores*, book III, chapter XV, p. 239).

surrounded by enemies on all sides, without hope of receiving any assistance. Their trust in God and in the prayers of Father San Vitores was so great, that they fought as if they were certain of their imminent victory. The barbarians centered their attacks (page 85) on the church, trying to destroy it by throwing firebrands that would ignite its palm leaf covering. Nevertheless, what they could not accomplish was carried out by a terrible hurricane[531] that passed through the island, demolishing most of the inhabitants' homes [September 30, 1670]. The church and the missionaries' residence suffered the same fate.[532] This gave the barbarians new hope, and they began preparing for a last effort and a general assault. Seeing this, the Spanish prepared on their part to receive them, and in effect, they did so with such courage that the great number of casualties that they caused the barbarians terrified and disheartened them. The next day, they sent Hurao's two confidants as emissaries, and they conceded everything that was asked of them, with the condition that Hurao be freed. Captain Juan de Santa Cruz, who knew the character of these people perfectly, wanted an unconditional peace, but the Marianas envoys asked for Hurao's liberation so insistently, that Father San Vitores begged the captain to grant it. Deference to this Servant of the Lord prevailed, Hurao was liberated, and peace was concluded. However, it lasted as long as it took this [Hurao] *chamorris* to get home, for once there, his friends and relatives, who until then had maintained strict neutrality for fear that Hurao might be harmed, took up arms and started the war again.

The furor and rage displayed by the barbarians on this occasion was astounding. Bent on overtaking the Spaniards' position by sheer use of force, they fought for thirteen consecutive days, carrying out extraordinary efforts to reach their goal with unremitting attacks day and night amidst the most alarming screams. The Spanish, tired of this continuous fighting, decided to make a general sortie against those furious barbarians, and met with such success, that the barbarians fled (page 86). The Spanish proceeded to destroy their trenches, smash the skulls left there, and devastate everything. The barbarians were so terrified that they were forced to humble themselves and sue for peace that very day [October 19, 1670], for which they sent Quipuha, a relative of the *chamorris* of the same name who had been the protector and supporter of religion, and was by then deceased. The next day [October 20, 1670], they accepted the

---

531. This is a violent tempest accompanied by maelstroms (note by C. Le Gobien, *Histoire*, book IV, p. 150).

532. See F. García, S.J., *The Life and Martyrdom of the Venerable Father Diego Luis de San Vitores*, book III, chapter XV, pp. 240–41.

conditions of the peace, the main ones being that henceforth they would all go to Mass on Holy Days and Sundays, and to the explanations of the Christian Doctrine; and that parents would punctually send their children to Catechism.

Thus ended this war [on October 21, 1670] that had lasted forty days, during which the Spanish had felt God's protection in a very special way. For although they were constantly exposed to the enemies' attacks, not one of them perished, and only one of them was hurt, Don Antonio Alexalde, commandant of the Fort of St. Francis Xavier, who was struck in the stomach by an enemy stone. He had fallen on the ground immediately, and believing him dead, his companions had prayed for the help of the apostle of the Indies and St. Theresa, whose feast it was that day [October 15], and he recovered in a few hours.[533]

Thereafter the barbarians treated the Spaniards with more deference and respect. Their resistance had frightened them, and the hurricane that had destroyed everything was regarded by many as a punishment sent by the Christian God, who had also preserved the Spanish from death, despite their being constantly exposed to danger. Many villages no longer believed in their *macanas*, considering them impostors who had convinced them to enter into a disastrous war, and they praised Father San Vitores, who had always favored them and behaved as their father and protector, even when they were bent on procuring his death.

---

533. See F. García, S.J., *The Life and Martyrdom of the Venerable Father Diego Luis de San Vitores*, book III, chapter XV, pp. 243–44.

BOOK FIVE

Father San Vitores, seeing that tranquility was restored in the island of Guåhan with the peace that had just been made and certain that there were many children and sick people who needed his succor, set out eight days after the treaty had been concluded. His companions warned him that he was exposing himself to evident peril; that spirits still ran high; and that it was necessary to let time pass in order for them to calm down; that having made peace with the Spaniards only out of fear, the natives could try to avenge themselves on his person if they saw him alone and in their power. However, these reasons were not enough to stop him or make him alter his resolve. Animated by the zeal of the glory of God and the salvation of souls, he left Agadña with two catechists and a Filipino and headed for the mountain of Chuchugu [Chuchugu'].[534] His guide assured him that the inhabitants of that mountain, who were naturally defiant, would not suffer their presence, and that it might be best for him to not go alone, for in no case should he deliver himself into the hands of people who hated the Spanish and who were known for their treachery and violence. Father San Vitores was undeterred, and he sent his companions to visit the valley while he went up to Chuchugu, a hard and laborious task, for this mountain was nearly inaccessible. Upon arriving at the village, he had the joy of baptizing eight children. The other Fathers went about Guåhan consoling the Christian natives and repairing the disorders caused by the war, after which they went to visit the neighboring islands [November 17, 1672].

God had sent new workers to reap more abundant harvests, and these were Fathers Francisco Solano,[535] Alonso López,[536] Diego Noriega,[537] and Fran-

---

534. Today's *Chochogo* is in the valley of La Cañada, northeast of Hagåtña (F. García, S.J., *The Life and Martyrdom of the Venerable Father Diego Luis de San Vitores*, book III, chapter XVI, p. 245).

535. Born in the walled Spanish city of Plasencia (Extremadura) on October 4, 1635, Francisco Solano joined the Society of Jesus on September 1, 1655. Upon San Vitores's death he was named the Superior of the Marianas mission, but he died soon after, in 1672 (ARSI, "Primus Catalogus Anni Personarum Anni 1671." Philippinae Cat. Trien. 1649–1696, Vol. 2-II, f. 352v; "Historica narratio illorum" (1668–1673), ARSI, Filipinas Tomo 13, ff. 95–110, c. transcribed in R. Lévesque, *History of Micronesia*, vol. 6, 1995, pp. 66–67).

536. Alonso (or Alexius) López was born in Plasencia (Extremadura), on July 16, 1646. He joined the Society of Jesus on September 30, 1662, and was a cartographer and missionary to the Chamorros in the Mariana Islands from 1672 to 1675. López wrote the Annual Letter that spanned the period from June 1674 to June 1675 (ARSI, "Primus Catalogus Anni Personarum Anni 1672." Philippinae Cat. Trien. 1649–1696, vol. 2-II, f. 363v). See also F. X. Hezel, *From Conquest to Colonization*, p. 90.

537. Very little is known about Father Diego Noriega, except that because of his delicate health, his superiors had ordered him to continue on to Manila, but he had refused, dying in Hagåtña of tuberculosis on January 13, 1672 (F. García, S.J., *The Life and Martyrdom of Diego Luis de San Vitores of the Society of Jesus*, book III, chapter XVI, pp. 229; 246).

cisco Ezquerra,[538] the last of whom later had the joy of giving his life for Jesus Christ. All four arrived on the galleon *Nuestra Señora del Socorro* [July 9, 1671], bringing with them the [papal] Brief and presents that Pope Clement IX had sent to this new Christianity. Feeling fortified by this additional help, Father San Vitores sent Fathers Casanova, Morales, and Bustillos to the Philippines to finish their studies. The Christian chamorrris, outraged at how the infidels insulted the neophytes and killed the missionaries, took the opportunity to offer their respects to the Governor of the Philippines and the Viceroy of Mexico and place themselves under the protection of the Spanish Crown. Three fervent Christian nobles were chosen for this important commission: Don Ignacio Osi, Don Pedro Guiram [Guirán], and Don Matias Yay[539]. These three men left the island of Guåhan on July 13, 1671, and after a peaceful voyage at sea reached Manila on the day that the Church celebrates the feast of St. Ignatius, founder of the Society of Jesus [July 31, 1671][540]. Since they had never before left their islands, they were greatly impressed by the size of the city of Manila, the beauty of its buildings, the great number of people that inhabited it, and the abundance of food and articles of every sort that could be acquired everywhere. In short, everything that they saw in the Philippine capital enthralled them. Nevertheless, what impressed them the most was the sumptuousness of the churches, the solemnity with which religious functions were carried out, and the people's devotion. They met with the governor of the Philippines, who promised to protect them and support them against those that opposed the establishment of the Christian faith in the Mariana Islands.

They wanted to continue their trip to Mexico, but various accidents

---

538. Father Francisco Ezquerra had been born in Manila on October 4, 1644, to General Don Juan de Ezquerra and Doña Lucía Sarmiento. He joined the Society of Jesus on January 17, 1661 (ARSI, "Primus Catalogus Anni Personarum Anni 1672." Philippinae Cat. Trien. 1649–1696, Tomo 2-II, f. 363r). His brother, Father Juan Ezquerra, was also a Jesuit, and his uncle, provincial Father Domingo Ezquerra (1601–1670), was the one who sent Father San Vitores to the Marianas. Upon Father Solano's death, Father Ezquerra was named superior of the mission, but on February 2, 1674, at the age of 30, he died also, killed by a group of Marianas' natives on a beach in Guåhan (AHCJC, Carpeta "EI.b-9/5/2. Martirios, naufragios, &c.," in EI/b-9/5/1-7. "Martirios y varones ilustres." AHCJC. Loose leaves. See also P. Murillo Velarde, *Historia de la provincia de Filipinas*, pp. 334–336). See also C. R. Boxer, "Two Jesuit letters on the Mariana Mission, written to the Duchess of Aveiro (1676 and 1689)." *Philippine Studies*, 26, 1978, pp. 37–42.

539. There is not much information about those noblemen: Don Matías Yay and Pedro Guiram were brothers and fervent CHamoru Christian converts. The first one remained in Manila from 1671 to 1674, while the second one was lost off the coast of New Spain along with Father Juan de Landa (1617-74) in terrible weather. As Father Francisco García pointed out, "Don Pedro ended his embassy in a happier manner, as we may hope, since God rewarded his zealous desire to bring new light to his country by showing him the eternal brightness of heaven" (F. García, S.J., *The Life and Martyrdom of Diego Luis de San Vitores of the Society of Jesus*, book III, chapter XIII, p. 231).

540. They left on the galleon *Nuestra Señora del Socorro* (F. García, S.J., *The Life and Martyrdom of Diego Luis de San Vitores of the Society of Jesus*, book III, chapter XIII, pp. 229–233).

retained them in the Philippines longer that they had hoped[541]. It was not [until] after the death of Don Pedro Guiram[542], who met his end as a good Christian, when Don Ignacio Osi and Don Matias Yay embarked for New Spain in early 1675.[543] The Archbishop of Mexico, Augustinian friar (page 89) Payo Enríquez de Rivera [y Manrique], [1622–1684], who was at the same time the viceroy of New Spain, received them with great kindness. They postrated themselves at his feet, declaring themselves faithful to the king of Spain as their Lord and Sovereign, and stating that they had come in the name of all of the Christians in the Mariana Islands to beg him to send them a governor and a garrison that could suppress the violence of the infidels and defend them against their enemies.[544] The Archbishop made them rise, and granted what they asked him in the name of the king. Gladdened by this great reception and by the positive resolution that their request had met, they returned to the Marianas, where many things had taken place during their absence.[545]

Father San Vitores had received on the month of July of 1671 the aforementioned reinforcements. He had sent Father Ezquerra to the island of Zarpana, and Father López to those of Aguisan [or Aguijan], Tinian, and Saipan, which had not been visited since Father Medina's death. These missionaries worked hard and met with success, and Father López had the solace of establishing a seminary for boys similar to Agadña's San Juan de Letrán in the town of Sungharon, Tinian.

Father San Vitores returned to Agadña at the end of 1671, where he prepared for the feast of the Lord's Nativity with fasts, penances, and an eight-

---

541. According to Father F. García, "Don Ignacio and Don Pedro sailed aboard the galleon *San Telmo* in the year 1672 with Father Juan de Landa, who was on his way to Rome as procurator of the Province of the Philippines. When the ship was forced by contrary winds to return to Manila, they reembarked in 1673 on the Buen Socorro. Twenty leagues out from Manila, there being no wind, the ship anchored near an island and don Ignacio went ashore. Meanwhile the wind began to blow and the ship put out to sea leaving don Ignacio. He returned to Manila in a small boat, taking a month for the voyage" (F. García, S.J., *The Life and Martyrdom of Diego Luis de San Vitores of the Society of Jesus*, book III, chapter XIII, p. 231).

542. We do not know if Guiram's body was buried in the Philippines or if it was brought back to Guam.

543. According to F. García, the galleon sailed from Cavite on June 5, 1674 and arrived at Acapulco on January 13, 1675. Don Ignacio Osi and Don Matías Yay reached Mexico City on January 31, 1675 (F. García, S.J., *The Life and Martyrdom of Diego Luis de San Vitores of the Society of Jesus*, book III, chapter XIII, p. 232).

544. They were asking for the establishment of a military government, which implied the designation of a captain and a garrison of soldiers as well as the construction of a permanent fortification (F. García, S.J., *The Life and Martyrdom of the Venerable Father Diego Luis de San Vitores*, book III, chapter XI, pp. 232–33).

545. This was a habitual practice. In 1684, Jaramillo went to Manila with three Marianas' boys, one of whom was the son of Matapang, Father San Vitores's assassin. The object was to show religious and lay sectors alike the advances of evangelization in the Mariana Islands, as well as impress the natives with the "wonders" of the capital (Antonio Jaramillo's letter to King Charles II, Manila, June 29, 1684, transcribed in R. Lévesque, *History of Micronesia*, vol. 8, 1996, p. 134). See also F. García, S.J., *The Life and Martyrdom of Diego Luis de San Vitores of the Society of Jesus*, book III, chapter XIII, pp. 232–33.

day retreat. After having closed Father Noriega's eyes in death on January 13, 1672, he renewed his visits to the missions. God our Lord continued blessing his labors and those of his companions. The number of Christians in the islands was considerable, and it kept growing each day, which led Father San Vitores to divide the island of Guåhan into four districts and to raise a church in each of these, in order to care for the new Christians more effectively, and for the ease and comfort of the people. The division was so perfectly accomplished that each district had forty villages. He built the first church in Nisichan [Nisihan or Nisi'an], and he took care of it personally because its inhabitants had always been reluctant to listen to the word of God, and they were rude and hard to govern. Father Cardeñoso took care of the second church, built in Pigpug [Pipok]. The church of Pagat [Pågat] was under the custody of the catechists, while the fourth, which was to be established in Merizo [Malesso'],[546] was to be attended by Father Ezquerra, who was working in the island of Zarpana at the time.

Returning to Guåhan immediately, Father Ezquerra sent Diego Bazán to see Father San Vitores. He was a youth of rare virtue whose zeal for the glory of God would soon win him the crown of martyrdom even at his young age. Bazán worked as a catechist in the town where Quipuha, the man who had signed the peace with the Spanish, lived. Although he had friendly relations with the Spanish and he was a Christian, Quipuha was involved in a scandalous relationship that was a discredit to religion. When he found out, Bazán repeatedly tried to get him to mend his ways and live as a true Christian, but Quipuha was too attached to his bad inclinations to pay heed to such saintly advice, and, blinded by the passion that consumed him, he resolved to get rid of the zealous youth who was like a bothersome censor. For a long time he thought of how to carry out his criminal intent, and finally, not daring to do it himself, he engaged two acquaintances, men who were hopelessly astray that he convinced with gifts and presents to do as he wished. The two murderers set out, and on the road met Bazán, who was on the aforementioned errand for Father Ezquerra. They approached him pretending to be his friends, and then sprang upon him, so that while one of them stabbed him with a knife, the other speared him with his lance. He fell dead upon the ground and the two men left him there. This happened on March 31, 1672.[547]

---

546. Malesso' was one of Hagåtña's main ports. Towards the end of the seventeenth century, the governor had a "house" or palace built there for him (Marjorie G. Driver and Francis X. Hezel, *El Palacio. The Spanish Palace in Agaña, 1669–1898*, Mangilao, Guam: Richard F. Taitano & MARC, 2004, pp. 10–14).

547. See F. García, S.J., *The Life and Martyrdom of the Venerable Father Diego Luis de San Vitores*, book

The way in which this young man had been called to the missions was providential. One day, Father San Vitores was walking across the great plaza in Mexico city while he was waiting to embark (page 91) on his trip to the Mariana Islands, when Bazán, who at the time was only fourteen years old, caught his attention because of the air of innocence and vivacity that enveloped him. He went towards him and asked, *"My son, do you want to come with me to be a martyr?"* To which Bazán replied without hesitation that he truly desired it, and from that moment, with his parents' permission, he followed the Father. Since he had an excellent disposition and a bright mind, he profited greatly from the holy advice and instructions of the Servant of God, and in a very short time he was an able catechist, entirely dead to self. His usual employment was accompanying the Fathers in their visits, although sometimes he acted as a missionary himself, and all his works were manifestly blessed by God.

On the other hand, since he had talent, prudence, and bravery, and he burned with zeal for the glory of God, he was capable of carrying out the most perilous enterprises, and the Jesuits charged him with various such tasks. Of proven virtue, he ardently desired to suffer affronts and disdain, so that he was immensely pleased whenever the barbarians insulted or mistreated him, believing himself blessed when he could suffer in the name of Christ. Thus, he had been sanctifying himself as if in preparation for the martyrdom that God had destined for him, and which he suffered at the age of eighteen.

As [soon as] Father San Vitores learned of his glorious death, he sent his companions Nicolás de Figueredo and Damián Bernal to warn Father Solano and the other missionaries to be on guard. Miguel Rancher[548] joined these two catechists with the idea of visiting the church of Agadña. While passing near the place where Bazán had been murdered and talking of his death, they were beset by an ambush of twenty men from Chuchugu and Mazan, who attacked them. Rancher, who was unarmed, was (page 92) killed immediately, but the other two defended themselves desperately and managed to kill the barbarians' leader, whereupon the barbarians ran away in fear.[549] Figueredo, with a thigh wound, and Bernal, with a head wound, were capable of reaching the neighboring mountain, where they split up to improve their chances of survival. The first

---

III, chapter XVI, p. 248.

548. Le Gobien also mentions a Miguel Rancher (*Histoire*, f. 169). But Father García erroneously refers to a *Manuel Rangel* (F. García, S.J., *The Life and Martyrdom of the Venerable Father Diego Luis de San Vitores*, book III, chapter XVI, p. 249).

549. See F. García, S.J., *The Life and Martyrdom of the Venerable Father Diego Luis de San Vitores*, book III, chapter XVI, p. 249.

went to the village of Ipao, where a Marianas' native embraced him as a sign of friendship, and then killed him with a poisoned dart. Bernal took the road to the village of Funhon, and he was as unfortunate as his companion, for another savage, who had been a friend of his, with the pretext of examining his sword, used it to strike a terrible blow against his head that left him dead at his feet.

These barbarous crimes greatly upset the missionaries, but the death of these virtuous Christians seemed to be God's preparation for that of the apostle of these islands and founder of the mission. In effect, since his last spiritual exercises Father San Vitores longed for martyrdom and for the joyous moment in which he would join his God. He left Nisichan [Nisihan or Nisi'an] on April 1, 1672 to return to Agadña, accompanied only by a Filipino named Pedro Calungsor [Calungsod].[550] They were both headed to the village of Tumon [Tomhom] to baptize the daughter of a Christian named Matapang [Matå'pang]. The Father had instructed and baptized this man after having cured him from a mortal wound, more by the efficacy of his prayers than by the remedies that he had applied, but Matapang, unfaithful to the grace he had received, was perverted.[551] Wanting to help this wretched man return to the way of righteousness, the Father asked Matapang to allow him to baptize his daughter, who was said to be in danger of death. The barbarian, looking at him with fierce eyes and adding impiety to ungratefulness, said: *Come to my house, impostor: there you will find a skull. Baptize her if you wish: I consent to it,* while he showered him with insults and indignities and threatened to kill him if he did not leave immediately.

Feeling more distress for the deplorable state of this unfortunate man than for his own peril, the Servant of God (page 93) did all he could to calm him. *"Let me baptize your daughter,"* he said, *"you know that you are baptized yourself. You can kill me afterwards if you want, for I would gladly lose the life of my body in order to procure the life of the soul for this child."* Nevertheless, seeing that his words only irritated the barbarian even more, he went to gather some children in order to explain the Christian Doctrine to them. The Father invited Matapang to listen, but his only answer was a horrible blasphemy. This

---

550. Pedro Calungsor was a Filipino survivor of the 1638 Concepción shipwreck. For a brief biography on this man whose services as translator and auxiliary to Father San Vitores were quite significant, see J. N. Schumacher, "Blessed Pedro Calungsod, Martyr," pp. 287–336; Resil B. Mojares, "The Epiphany of Pedro Calungsod, Seventeenth-Century Visayan Martyr," in Alfred W. McCoy (eds.), *Lives at the Margin. Biography of Filipinos. Obscure, Ordinary and Heroic.* Quezon City-Madison: Ateneo de Manila University Press & University of Wisconsin Press, 2000, pp. 34–61.

551. See F. García, S.J., *The Life and Martyrdom of the Venerable Father Diego Luis de San Vitores*, book III, chapter XVII, p. 251.

man said that he was tired of God and that nothing mattered to him less than his doctrine. Moved by his hatred towards the Father, this wretch resolved to kill him, and asked one of his friends, called Hirao, to help him. Although Hirao was not Christian, he was horrified at the prospect of committing such a heinous act, and he tried to convince Matapang to desist from it.

"*Would you be so ungrateful,*" he told him, "*as to kill a man from whom you have only received favors, who has procured peace for us, and who has always held our interests at heart? A man who cured your wound and thus saved your life?*" Matapang derided his friend's words, and reproached him his timidity and lack of valor, saying scornfully: "*I have been wrong in addressing you; you are too brave, and must be reserved for greater occasions. I am enough to carry this out, and have no need of you.*"

Slighted by this mockery, Hirao gave in and consented to his friend's plan. While these two evil men went to arm themselves, Father San Vitores entered Matapang's home and baptized the girl. The barbarian became even more furious when he learned of this, and he attacked the missionary's companion because he encountered him first. Although the young man skillfully avoided many of the blows cruelly thrown against him, he was finally wounded by this enemy. Hirao arrived then, and gave him one final blow on the head, leaving him lifeless on the ground. Such was the recompense that this virtuous catechist obtained for (page 94) having served the Fathers during four years with such zeal: the crown of martyrdom.

Father San Vitores realized that his time had come. He addressed the two murderers, crucifix in hand, exclaiming, "*Know that God is the sovereign Lord of all nations, and he is the only Lord that the island of Guåhan must worship,*" adding, "*Ah, Matapang! May God have mercy on you!*"[552] He had barely pronounced these words when the two men sprang upon him. Hirao struck him on the head while Matapang drove a spear through his body. This happened on the Saturday before Passion Sunday, between seven and eight in the morning, on April 2 of 1672.

Thus died the apostle of the Marianas, at the age of 45, after having had the joy of establishing the faith in thirteen islands where the name of Jesus Christ had never been pronounced; after having founded eight churches, established three seminaries for the education of the young of each sex, and having baptized more than 50,000 islanders.[553] He was a man of intelligence, solid doctrine,

---

552. See F. García, S.J., *The Life and Martyrdom of the Venerable Father Diego Luis de San Vitores*, book III, chapter XVII, p. 252.

553. According to other demographic data, this number seems quite exaggerated.

and greatness of spirit, capable of carrying out the most difficult enterprises. Nothing seemed too difficult for him; contradictions and persecution could not deter him, because he was convinced that an apostolic man finds as many obstacles in his path of glorifying God as steps he takes.

Although he was of medium height, he had a majestic and grave appearance; a clear forehead, his eyes vivid and penetrating, ruddy lips and cheeks, an aquiline nose, his face well proportioned, perhaps a bit round, with skin whiter than most Spaniards. His apostolic work and the penances that he imposed upon himself had so extenuated him that in the last years of his life he seemed more like a (page 95) walking corpse[554] than a living being, he was so pale and emaciated.

Not content with taking his life, Matapang pillaged San Vitores's corpse, finding a rough cilice and an iron cincture upon him. Taking the small crucifix that the Father wore around his neck, Matapang destroyed it as he vomited a thousand blasphemies and yelled with all his strength: "*See here He whom the Spaniards acknowledge as their God and Lord!*"[555] This wicked man, who had been taught the mysteries of the Christian religion that he now disavowed, wanted to prevent the Christians from rendering to the Servant of God the honors due to those who die as martyrs for Jesus Christ. And so, he burned the ground and threw ash over the places where the holy man's blood had been spilled, and with Hirao's assistance, he dragged the bodies to the beach, where they tied heavy rocks onto their feet before throwing them into the sea to erase all trace of them[556].

Yet God, who is pleased to honor his Servants, soon made Father San Vitores's sanctity clearly manifest by the great number of miraculous cures obtained through his intercession. Don Damián de Esplana [soon to be governor of the Marianas] was the first to feel the influence that the martyr saint had in Heaven. He was gravely ill, and he vowed that if he recovered his health, he would erect a chapel on the place where the Father had been martyrized. His

---

554. Le Gobien speaks of an "animated corpse" (*Histoire*, Libro V, p. 168).

555. F. García, S.J., *The Life and Martyrdom of the Venerable Father Diego Luis de San Vitores*, book III, chapter XVII, p. 253.

556. Right after San Vitores's killing, a strange thing happened. As F. García pointed out, "the body of Father San Vitores, having sunk, came twice to the surface, and the hands seized the outrigger of the canoe, which the natives use here to counterbalance the sail. Matapang, who was alarmed, twice loosened the hands with a stick, but the body rose a third time and grasped at the stern of the small craft, where Matapang was stationed. By now he was so frightened and aghast that he did not know what to do. He wanted to throw himself into the sea, believing that the father was going to climb into the boat. At last he got back his nerve and struck one heavy blow at the head with one of the oars. Then he quickly rowed away from the scene, leaving the sacred body buried into the sea, to the perpetual grief of those whom he left behind" (F. García, S.J., *The Life and Martyrdom of the Venerable Father Diego Luis de San Vitores*, book III, chapter XVII, p. 253).

prayer was heard, and so the chapel was edified and a cross was raised in the place where the two martyrs had shed their blood.

When news of this precious death reached the Philippines and Spain, public displays of the impression that it caused were seen in Manila as well as in Madrid. A *Te Deum Laudamus* was sung in the Cathedral of Manila, and all of the church bells of the city were rung. In Madrid, Father Vazquez, a Jesuit, pronounced a funeral oration in the church of the Imperial College before the whole court and the most important people of the city.

(page 96) God had armed Father San Vitores since his youth with extraordinary graces and an ardent desire for martyrdom. He spoke of it often, expressing how much he wished to shed his blood for Jesus Christ, and so while he was a Philosophy professor at Alcala, the students of this famous University, certain that God would listen to his prayers, would say whenever they saw him: "*There goes Father San Vitores, who will one day be a martyr.*"

There seemed to be no other man animated by as much zeal for the salvation of his brothers' souls as he, and he would often say that if he could only win one soul for Heaven, he would deem himself rewarded for all his toils and sorrows. One day he said to a companion, "*If during a shipwreck I had the good fortune of finding a plank with which to save myself, and at that moment, somebody told me that they were in mortal sin, and that they had not received absolution, I would gladly give them the plank, so that they may reach land and have enough time to reconcile with God.*"

He worked incessantly in the conversion of the Mariana Islands with the idea of later going to Japan to reestablish the faith, or to austral lands where Jesus Christ was still unknown. His charity was as vast as his zeal, and people were so certain of this that they placed their needs before him trustingly, and he had the ability of gladdening their spirits, as well as giving them all he had. Although he worked incessantly, he only ate insipid roots and herbs, refusing to eat bread or meat or drink wine since he began working for the conversion of souls in the Marianas.[557]

---

557. Historians have recently argued about the centrality of food in the first European colonial expansion (Rebecca Earle, *The Body of the Conquistador: Food, Race and the Colonial Experience in Spanish America, 1492-1700*. Cambridge: Cambridge UP, 2014; Heather M. Martel, *Dirty Things: Bread, Maize, Women, and Christian Identity in Sixteenth-Century America*, in K. Albala and T. Eden (eds.), *Food and Faith in Christian Culture*, 147-70. New York: Columbia UP, 2011; Martel, "Ferocious Appetites: Hunger, Nakedness, and Identity in the Sixteenth-Century American Encounters," in C. Kosso and A. Scott (eds.), *Poverty and Prosperity in the Middle Ages and Renaissance*. Turnhout: Brepols, 2012, pp. vii–xxiv). In her doctoral dissertation, Verónica Peña Filiu combined written and archaeological sources to analyze the changes and continuities that were produced in the diet of the Marianas' inhabitants during the Jesuit period (1668-1769). The process of evangelizing and occupying the archipelago entailed the introduction of cattle raising and new methods of

His ardor for suffering was extraordinary; not one day went by when he did not impose great penances upon himself that would awe anybody who learned of them (page 97)[558]. He was sustained and encouraged by a single consideration in the midst of his great labors: "*How sweet it is to suffer for God*", he would often say, "*we have no cause for complaints, for any of our sufferings pale in comparison to that which Jesus Christ suffered for us.*"

Since God our Lord granted him such an abundance of extraordinary graces, he was heard exclaiming, like St. Francis Xavier: "*Oh my God! Why do you fill my soul with so many graces and spiritual consolations! Stop, my Lord, stop!*" His trust in God knew no limits. In all his necessities, he invariably went to Him, and God granted him anything that he needed. Here is an example of this. When he arrived at the Mariana Islands, he found that its people were lacking in all nearly everything that is necessary for life, and, pitying them, he wrote a memoir listing in it the most indispensable items that would cover the most basic needs of his beloved neophytes. He sent it to Mexico, to a scholar friend of his called Cristóbal Javier Vidal whose only wealth was his integrity.[559] He wrote to him asking him to ask for these things as alms, and to take from what he received payment for whatever expenses he underwent in procuring what was requested in the memoir. The scholar, who had spent some time with the Father and felt a great veneration for him, obeyed him immediately, and asked another scholar, called Juan de Garate [y Francia, 1630-80], to assist him. When they saw a man approaching them, and asked him for alms for the mission in the Marianas, the stranger said, "*I am very glad that you address me. I have been charged with giving one hundred pesos for a work of charity; I did not yet know how to employ them, but I will give them to you, for there is no better use for*

agriculture, carried out in small ranches—known as *lanchos* (or *lanchus*)—to produce new foodstuffs (wheat, grapes, legumes, beef, pork). Changes in dietary and culinary practices were multidirectional, however, with local, Iberian, and American foods comprising the fare of the archipelago's inhabitants (Verónica Peña Filiu, *Alimentación y colonialismo en las islas Marianas (Pacífico occidental). Introducciones, adaptaciones y transformaciones alimentarias durante la misión jesuita (1668-1769)*, Ph. D. Dissertation, Departament d'Humanitats, Universitat Pompeu Fabra, 2019).

558. According to Father F. García, "[San Vitores's] hardships were enormous, but the last four, in the Marianas, were the worst, but always greater still was the joy of suffering them for Christ. Consequently, in all his labors and fatigues, he repeated with inexplicable happiness: "Let it all be for the love of God. How little it is, compared with what the good Lord deserves, with what he suffered. How much more we should desire, O, my good Lord, to suffer for love of you!" (F. García, S.J., *The Life and Martyrdom of the Venerable Father Diego Luis de San Vitores*, book III, chapter XVII, p. 299).

559. See F. García, S.J., *The Life and Martyrdom of the Venerable Father Diego Luis de San Vitores*, book IV, chapter X, pp. 310–11. Cristóbal Javier Vidal Figueroa led the knights of the congregation of St. Francis Xavier. He was the brother of Jesuit Father Joseph Vidal Figueroa, rector of the College of San Ildefonso and henceforth procurator of the Marianas (L. Gutiérrez, *Historia de la iglesia en Filipinas*, p. 261; R. Lévesque, *History of Micronesia*, vol. 9, 1997, p. 438).

them than the conversion of infidels."

Among the many graces that God gave his (page 98) Servant, the gift of prophecy was the most extraordinary. Don Diego Salcedo, that Philippine governor who opposed the Mariana Islands enterprise, and who was later persuaded of the sanctity of Father San Vitores, became his friend and asked him to obtain from God the grace of going through Purgatory in his earthly life. Upon hearing this petition, the Father was shortly taken aback, and, looking at him closely, he said: *"Are you willing to receive from God's hands whatever He may choose to send you, sorrows related to your goods, your honor, and your person?"* Having replied to Don Diego that indeed he was ready to face any suffering, the Servant of God spoke again filled with divine inspiration: *"Rejoice, for you will suffer much, but you will reach that which you desire."* This prophecy was fulfilled. Don Diego embarked towards New Spain and, once there, he lost his fortune, was disgraced before everybody, and his enemies betrayed him to the Tribunal of the Holy Office, which locked him up in a dungeon where he was covered in chains. In the midst of these misfortunes, Salcedo did not lose faith, and a letter sent to him by Father San Vitores renewed his strength. He understood that everything that was happening to him was the means God was using to sanctify him and have him expiate his sins; and so, blessing the hands that tormented him, he submitted himself with perfect resignation to God's will. He was taken to Mexico for his trial, and one night, St. Brigid [of Sweden], to whom he was especially devoted, appeared before him along with Father San Vitores, and they encouraged him to bear his suffering in good spirits, for on October 24 of [1668] his sorrows would end. And they did, for Salcedo died that day, full of confidence in God's mercy, and happy to see that his wishes had been fulfilled thanks to Father San Vitores' intercession.[560] Soon after his death, the falsity of the accusations that had been laid against him by his enemies was made manifest. He was declared innocent, and the General Tribunal (page 99) of the Inquisition gave his brothers an official document that cleared the memory of the deceased.

Brother Mateo [de] Cuenca, who had come to the Philippines with Father San Vitores, also experienced the truth of his predictions. Speaking with him one day, he opened his heart to the missionary and manifested his innermost sorrows and his fear that he would not persevere in his vocation. *"You are right*

560. As evidence of his repentance, the former governor left 10,000 pesos to the Marianas mission in his will (1668) (AGN, Instituciones coloniales, Real Hacienda, Archivo Histórico de Hacienda, vol. 326, ff. 1683–1896).

to fear this," the Father told him, *"but be consoled, for I assure you that you will die a Jesuit, and that you will see me a few days before your death."* In effect, Cuenca eventually left the Society and served as a parish priest in Ahun, in the Philippines. However, repentant for not having been faithful to what God had asked of him, he fell gravely ill and manifested his desire to confess. For some reason, the only other priest in the area could not go to him and, alarmed, the sick man had himself transferred to Iloilo, where there was a Jesuit residence.[561] He embarked on a small boat, but contrary wind conditions prevented him from leaving port. Afflicted by this obstacle and fearing that he would die without confession, he took out a relic of Father San Vitores and begged him to obtain from God the opportunity to confess, reminding him of the promise that he had made some time earlier. Just then, the Father appeared before him, filling him with hope. The wind suddenly became favorable, Cuenca reached Iloilo, and after having received the holy sacraments, he addressed Father [Pedro] del Valle and asked for permission to reenter the Society. Thus, after being readmitted, he had the solace of dying amongst his brothers on August 27, 1677.[562]

What happened to Don Jeronimo de San Vitores, our saint's father, was equally surprising. This man had fervently wanted his son to assist him in his deathbed, and he could not support him in his decision to consecrate himself to the mission in the Indies, which he opposed with all the means in his power. The Father [San Vitores], tired (page 100) of this resistance, told him one day: *"Do not stand in the way, father of mine, of God's designs, for I promise you that in the hour of your death, you will have the solace of my assistance."* Thus, the Servant of God left with his father's blessings, for the latter felt reassured by this promise. It came to pass that a few hours before his death, Don Jeronimo was consoled and fortified by his son, who appeared to him in his last passage. Wanting to manifest the joy that possessed him, he was heard to exclaim up to three times: "Oh, Diego, holy son of mine! You are doing me so much good!" He then kissed the crucifix many times, and died amidst a great display of piety on December 20, 1675, at the age of 80. He had always been a religious man of integrity and probity, and had been one of the most skilled ministers of the Spanish Catholic monarch, to whom he had rendered great services in difficult times.[563]

---

561. He was in the island of Panay, in the Visayas.

562. Father F. García erroneously refers to Father del Valle as Pedro Vello (F. García, S.J., *The Life and Martyrdom of the Venerable Father Diego Luis de San Vitores*, Libro IV, Capítulo XIII, p. 322).

563. His oldest son, Don José de San Vitores, Marquise of la Rambla and Viscount of Cabra, gentleman and member of the Council of the Treasury, married the daughter of the Count of Priego, and had various sons by her, the first of whom married the daughter of the Count of Gracias (Note by Luis de Morales/C. Le Gobien, *Histoire*, book VI, p. 178).

I will say nothing of the infinite number of persons who have been cured by applying relics of the holy martyr, or through his intercession. One example is enough. Seeing himself gravely ill and with no hope of recovery, Father Pedro de Montes, [564] of the Society of Jesus, and superior at the time of the Silang residence in the Philippines,[565] commended himself to Father San Vitores, promising to consecrate himself to the mission of the Mariana Islands if he was cured. He immediately noted a change in the course of his illness, and within a few days, his health was entirely restored.

Father Francisco Herrera also said that he was cured from a severe throat malady by applying a relic (page 101) of Father San Vitores. That same priest, having been called to assist Doña Beatriz de Cascos in her deathbed, found this lady, who lived near the city of Badajoz in Spain, given up as a hopeless case by the doctors. She had been given the last rites, and was waiting to expire, when Father Herrera applied the relic of the Servant of God that he had used on himself, and encouraged her to commend herself to Father San Vitores to recover her health through this holy saint's intercession. She confidently did so, and from that moment on, noticed that she felt better, and within a few days, she was well, making Don Jacinto Lobato, the head doctor of the army at Extremadura, whose fame and ability were known to all, say that such a recovery had been a miracle indeed[566].

I conclude with what happened to Don Antonio de Saravia y Villar.[567] This man had embarked to the Marianas in 1680 to occupy the post of governor, an appointment that he had solicited moved by the honest desire to procure the wellbeing of the mission and to work for the conversion of its

564. This is not the same Father Pedro de Montes (1560–1610), a missionary who died in Hapan (DHCJ, vol. III, pp. 2731-2732). See also *Descalzo Yuste, La Compañía de Jesús en Filipinas*, p. 349.

565. The Silang mission had two subordinate towns: Indang and Maragongdong. It was located at a crossroads where several routes from the provinces of Batangas and Cavite converged towards Manila. Together with the parish of San Miguel, Silang was one of the Jesuit missions in the Tagalog area (Descalzo Yuste, *La Compañía de Jesús en Filipinas*, p. 178).

566. See also F. García, S.J., *The Life and Martyrdom of the Venerable Father Diego Luis de San Vitores*, book III, chapter I, p. 325.

567. Don Antonio de Saravia was the first governor and captain general of the Mariana Islands, named by the Royal Decree of November 13, 1680. On June 13, 1681 Governor Saravia reached Guåhan from Acapulco in the galleon *San Antonio* with his retainers Juan Moreno and Antonio Sotera ("Expediente de información y licencia de pasajero a Indias de Antonio Saravia, gobernador y capitán general de las islas Marianas," AGI, Contratación, 5443, N. 1, R. 5, ff. 1-10v). Saravia was an experienced military man who had served for 30 years in Sicily (AGI, Filipinas 13, f. 97, c. in M. G. Driver, *Cross, Sword and Silver*, p. 20; M. G. Driver, "Notes and Documents. Quiroga's Letter to King Phillip V, 26 May 1720." *The Journal of Pacific History*, 27:1, 1992, p. 99). See also F. García, S.J., *The Life and Martyrdom of the Venerable Father Diego Luis de San Vitores*, book III, chapter I, pp. 325–26.

inhabitants, more than by ambition or self-interest.[568] In Mexico, he became ill, contracting dysentery, and was soon reduced to a dire condition. Declared hopeless by his doctors, he called Father Balthasar de Mansilla,[569] procurator of the mission, who had visited him frequently during his illness. *"You had told me, my father,"* the governor said, *"that you had a portrait of Father San Vitores; I beg you to lend it to me, for I hope in God that this Servant of God will cure me."* The priest brought him the portrait, and almost as soon as the sick man saw it, his pain ceased, and in a few days, his health was fully restored. Thus, he continued his trip to the Marianas, where with his zeal and virtue he did much good to those islands.

The death of Father San Vitores gravely afflicted the Christians and left the missionaries in a difficult situation. They found themselves without support, exposed to the furor (page 102) of the infidels and bereft of a man who gained and enjoyed everybody's trust. The residents of Chuchugu supported Matapang's murder of the Father, and they animated their compatriots to rise against the Spanish, and soon the entire island was up in arms. The villages of the northern coast, more ferocious and indomitable than the rest, showed a clear desire to rid the islands of the European presence. Those in the south, more peaceful and affectionate towards the missionaries, did not join their compatriots, but remained strictly neutral and simply waited to see what would happen.

The Spaniards' situation was dire: they were few, and their numbers diminished each day, which made them fear, with good reason, that they would succumb at the hands of this multitude of enemies. But God, who never abandons those who trust him, sent them the succor that they needed. The ship that went annually from New Spain to the Philippines arrived at the port of Umatac, where the Islands' Christian residents had impatiently waited for it. The barbarians tried to keep its arrival secret, fearing that their treasons and crimes would be punished, but a faithful Christian told Father Francisco Solano, who immediately went to Umatac, where he was consoled by the presence of Admiral Don Leandro Coello and Captain Don Antonio Nieto, friends of San Vitores,[570] whose death they mourned. Being pious and virtuous men,

---

568. In the "Relación del estado y progresos de la misión de las islas Marianas desde junio pasado de 1681 hasta el de 1682," Father Morales described Governor Saravia with similar terms of praise, highlighting his exemplary virtue and good disposition (transcribed in R. Lévesque, *History of Micronesia*, vol. 8, 1996, pp. 66–72).

569. Born in Villagarcía on March 4, 1638, Baltasar de Mansilla was admitted into the Society on April 11, 1654. He was superior in the College of Silang and Procuraror General of the province in Mexico (C. Sommervogel, *Bibliothèque de la Compagnie de Jesus*, T. V, 1894, p. 506).

570. They were two of the admirals loyal to the Society of Jesus who had come in the Manila galleon. See

they spoke with much praise of this holy man, and manifested to Father Solano that it was necessary to continue with the work that this great man had begun. Many of those who accompanied them offered to stay in the island to defend the Christians from the insults of their enemies and work in the propagation of the faith. But their indiscreet zeal took them too far; so much so, that it endangered this flourishing Christianity. Here is how it happened:

(page 103) A Spanish soldier met Hurao, that famous *chamorris* of whom we already talked about, who was walking with a friend of his. After talking for a while, they touched upon the subject of the last war, in which Hurao had had such an important role. Irritated by the Spaniard, the *chamorris* insulted him. He in turn wounded him with his sword, and if the other native had not fled immediately, he would have suffered the same fate. The missionaries lamented this incident, for they were sure that there would be violent consequences. Another incident that happened that same day gave them even greater cause for consternation.

A young woman from Zarpana, scared at the sight of armed Spanish soldiers, ran off, and the Marianas' native who was accompanying her, and who was engaged to marry her, ran after her with all his might. Two of the recently arrived soldiers, unfamiliar with the customs of these people, thought that the couple was trying to escape because they had committed some crime. They chased them, but unable to catch up with them, one of the soldiers fired his musket with such bad luck that he killed the girl and seriously wounded the man.

This incident terrified the population, producing great excitement and anger against the Spanish. "*Here is what these people have brought us,*" they said. "*They have only come here to hurt us and lead us astray. The Fathers kill our children with their poisoned water, and the sick with the oil that they anoint them with; and the soldiers that accompany them exterminate us without pity with their deadly weapons.*"

The missionaries could not submit the Spanish soldiers to their authority, and fearing that with their violence, they would ruin their Christianization efforts, they gathered them to exhort them to moderate their impulses and be patient:

> You have come here, they said, with the desire (page 104) to serve God and assist in the conversion of these peoples, and so far, we have looked upon you as the defenders of this new Christianity.

A. Coello, "Colonialismo y santidad en las islas Marianas: los soldados de Gedeón," pp. 17–44.

However, your conduct and your attitude is contrary to this! Are not the barbarians right when they say that you only tend to their destruction? How can they trust us if you vex them with your untoward behaviors? What is most deplorable is that you are offending God by hampering our missionary visits with your disorder. Where would we be if the Lord, who is our strength and our support, abandoned us? Because, what can twenty or thirty men do against more than thirty thousand barbarians bent on destroying us?

These words produced the desired results, and the soldiers were more docile and circumspect from then on.[571] Don Juan de Santiago,[572] who commanded them, resolved to raise a fort in Agadña that would keep them safe from the barbarians' attacks. He designed the project, and work was begun immediately. However, the necessary materials were in the neighboring mountain, and the barbarians ambushed them and scattered the workers, who barely escaped with their lives.

This hostile act led Don Juan to resolve that they should frighten the barbarians with such a forceful attack, that they would be too terrified to bother them again. For this, he chose twenty men and went to the town of Tumon, where Father San Vitores had been martyrized. When the natives saw this march, alarm spread across the island, and most residents fled their homes. Upon their arrival at Tumon, the soldiers found no one, which satisfied Don Juan, whose only object was to frighten these people and thus prevent new attacks. Thus, he contented himself with setting fire to a dozen homes, among them (page 105) that of Matapang, who had so cruelly killed Father San Vitores.

Vexed by this act, the barbarians thought only of exterminating the Spaniards with a surprise attack in which their superior numbers would bring victory. There were only two roads to return to Agadña from Tumon: the one taken by the Spanish on their way there, and the other, which went along the beach. The enemies obstructed the first road, making it impracticable by spreading a great number of felled trees over it. The Spaniards had to take the other road, and they did so without hesitation. Knowing that they would, the barbarians had covered the road with sharpened bone points drenched in poison. Don Juan,

---

571. The soldiers' violence was one of the main obstacles for the evangelization of the Marianas' natives. Many soldiers were former convicts who arrived in the archipelago from Manila or New Spain. To avoid it, Father Luis de Morales asked the king to make sure that "Philippine governors do not send any exiled residents or soldiers to the Mariana Islands" ("Memorial de 1685," transcribed in R. Lévesque, *History of Micronesia*, vol. 8, 1996, p. 411).

572. It should read Don Juan de Santa Cruz.

who was not only skillful but experienced in these cases, immediately realized the barbarians' stratagem, so that instead of taking the road, he had his soldiers wade waist-deep in the water and thus march on to Agadña. This disconcerted the enemies, who would from time to time show themselves in the nearby mountains. Matapang, more daring than the rest, came to insult them in his canoe, yelling at the top of his voice: *"I am Matapang, and you are a bunch of cowards!"* The only answers he received were a dozen shots. Since they did not touch him, he became bolder, coming closer to the shore to shoot them with arrows, but a solder fired his musket at him and wounded him on the arm.

The soldiers continued their difficult march, and believed themselves to be out of danger, when a mob of barbarians that had been hiding behind some rocks suddenly assailed them. Amidst terrifying screams, these men showered them with stones, but these fell around the Spaniards without touching them. Don Juan ordered his men to face the barbarians, but a large number of canoes that had remained hidden from their sight behind the same rocks beset them. The enemies encouraged one another, throwing their stones with amazing speed and skill, and immediately diving under the water to take cover from the Spanish gunfire, (page 106) which caused so many casualties. This attack was intense but brief. Don Juan was able to return with his people to Agadña, but he was seriously wounded along with three of his soldiers. These men died a few days later, assisted by Father Solano, in the midst of a horrifying, but truly Christian, suffering.

The expedition had alarmed the barbarians, and the inhabitants of three villages [Aniguag, Asan and Tupingan] went to Agadña the following day to sue for peace. This was granted to them with the condition that they send their children to catechism; that they destroy the houses of vice that perverted them; and that they not prevent the Christians among them from going to Mass or the Divine Office.

While these things were taking place in Guåhan, religious work in the other islands met with success, to the great consolation of these evangelical laborers. The church of Tinian flourished, and the number of Christians increased each day thanks to the cares of Father Alonso López and four excellent catechists that worked under his direction. This fervent missionary had the great joy of winning for God one of the most important *chamorris* of that country, Caiza, turning him into a zealous protector of religion whose services were very useful to the Fathers in delicate situations. Father López, concerned about the lack of news from Guåhan, asked various Tinian acquaintances to go to Agadña to find out what was happening. The members of this commission did not return when they were expected, and their families, fearing that some misfortune had befallen them, asked a group of Saipan islanders who had come from Guåhan for news about their relatives. These malicious men replied that the Agadña missionaries had killed them. This was all that it took for the residents of Tinian to become riled up, and Father López was blamed for the (page 107) messengers' supposed death. The fury of these men's parents was such, that they would have undoubtedly sacrificed Father López if Caiza had not contained them with his authority and his credit. The envoys arrived some days later, and they did much more than disprove the calumnies of the Saipanese; they also spoke to their compatriots of the warm welcome and good treatment with which the Agadña missionaries had received them, and they spoke especially favorably of Father Solano, who had succeeded Father San Vitores as superior of the mission.

Father Solano was indeed a highly virtuous man who joined a rare prudence with a greatness of soul that made him seek out the most difficult enterprises in order to suffer for Jesus Christ. He had been born in Jarandilla, in Spain. He had gone to Rome accompanying the Duke of Infantado, ambassador of the Catholic Monarch in that court, where by his own means he obtained a considerable benefice. He later retired to his own country, where he fell gravely ill. Father San Vitores, who at that time was in Jarandilla, assisted him in his illness, and inspired in him such disgust for the world and its vanities that Father Solano resolved to renounce all of the century's desires and enter the Society of Jesus, which he did on the first day of September of 1657. Since then, all he ever thought of was to render himself worthy to work in the conversion of infidels, and he asked so insistently to be sent to the Indies that his superiors gave in to his request and sent him to the Philippines. He had traveled to Mexico with Father San Vitores, who proposed him as a model of virtue and sanctity to his companions. This is what Father San Vitores had said of him in a letter that he wrote while still in the Philippines to Father Juan Gabriel Guillen, S.J.:

> *Even though many,* he said, *offer to work in the conversion of infidels, it is preferable to send only those who perfectly embody the spirit of St. Ignatius and St. Francis Xavier; that is, those who have* (page 108) *a profound humility, absolute obedience, and angelic purity. In sum, that their virtue is as proven as that of Father Francisco Solano. If you sent several men like this fervent missionary, the greater glory of God and the conversion of these islands would be secured.*

When he arrived at the Philippines, Father Solano was sent to Negros Island, where he worked for three years. From there he went to Mindanao, the most demanding and onerous mission in the Philippines. After having spent some time there, he was called to Manila and given an important task. While there, he learned the language of the Mariana Islands with the hope of consecrating himself as a missionary there, which he did when in the year 1670 he went there with Father Francisco Ezquerra [and two other missionaries].

Father San Vitores, who recognized the ardent zeal and heroic valor in Father Solano, destined him to the island of Zarpana, which needed a man with precisely those virtues. However, the troubles that arose in the island of Guåhan kept him from going there, and he stayed in Agadña where he cared for its flourishing Christianity, the most numerous in these islands. He rendered great services during the siege that was suffered there, standing out because of his charity to the soldiers who looked upon him as a father. The barbarians had great respect and veneration for him; his words, filled with spirit and unction, impressed them so, that those who listened to him were immediately convinced of all that he said. He was bent on the conversion of Quipuha, whom we have already talked about, but to no avail. This man was very beloved to the missionaries, because of the departed *chamorris* Quipuha who was the first among his compatriots to receive them in his home, and who contributed much with his authority and example to the conversion of the islanders. This younger relative of his had had the joy of receiving (page 109) baptism, but unfaithful to his duties, he had abandoned his wife and taken another, scandalizing everybody. Blinded by his passion, all of the exhortations of Father San Vitores and the words of the fervent catechist Bazán had done nothing but irritate him. Father Solano pitied him, for the blindness that left his soul in such a deplorable state, imperiled his salvation and threatened to condemn him:

"*That is my problem,*" Quipuha responded coldly. "*I would rather go to hell than separate myself from somebody who is so dear to me.*" "*Ah, you wretch!*" Father Solano exclaimed with a tone capable of striking fear in him, "*be wary of what you say, for you will die tomorrow.*" Quipuha scoffed at the Servant of God and his threats, and on the following day, he set out from his home accompanied by the person whom he loved so blindly, when, after only a few steps, he fell dead on the ground in her presence.

Such a manifest punishment of God produced marvelous effects among the neophytes, who looked at Father Solano as if he were a saint. He really was, for his only passion was to make God known and loved, and his one desire was to give his life for Jesus Christ. He let himself get carried away by the ardor of his zeal, the effect of which soon took its toll, for he fell ill a few days after Father San Vitores' death. His malady was a continual exercise of virtue, and the sweetness, patience, and ardent desire to possess God that he evidenced were admirable. He finally died with these sentiments on June 13, 1682, after receiving all the Sacraments of the Church.

Book Six

The barbarians took advantage of Father Solano's death, whose ardent zeal and exemplary conduct had contained them. They started vexing the Spaniards, albeit not openly, for the sight of the latter's firearms kept them at bay. Nevertheless, they would throw rocks at them while concealed in the forests and mountains; or they would hide behind trees to ambush and kill them. Sometimes, large groups would gather and come to Agadña to harass the Spanish. A small forest that they used as refuge was cut down, but they soon recovered from this loss by making a long ditch four leagues away from the village, to keep the missionaries from visiting the towns in those parts.

This ill disposition greatly afflicted the fathers, who tried to win them over with kindness and gifts, but these ferocious men became bolder and more intractable each day. Yet the indomitable character of these peoples was not the only source of grief suffered by the missionaries. The Spanish soldiers also gave them cause for vexation, multiplying their labors, whether because of the imprudence of their valor, or the irregularity of their conduct. However, since the Christians had asked for their presence, and their help was necessary to protect them from the violence of the infidels, there was no remedy but to put up with them. They were assembled every week, encouraged to fulfill their duties and live as true Christians. Father Ezquerra, who substituted Father Solano at the head of this mission, had them perform [spiritual] exercises (page 111) following the method of St. Ignatius. Many of them indeed sanctified themselves, and later experienced the joy of giving their lives in the defense of the Faith. During the brief period in which he was superior, this Father regulated the mission's affairs quite prudently, and the foundation of Fuña [Fu'una], which allowed the communication with the Spanish ships that came each year to the Philippines, was due to his efforts.

Fuña is in the western coast of Guåhan, and it owes its name to a nearby rock. These superstitious people look upon this rock as a wonder of the world, and it would indeed be so if what is said about it were true. We have already said that before the Spanish arrived, these people thought that there were no other countries but their own, nor any other men but those of this nation. Since they ignored their origin, they fabricated one to their liking, stating that the first man had been formed from that rock, which was therefore to be regarded as the cradle of the human race.[573]

This famous rock is located in a kind of gulf, close to a peninsula that stretches out into the sea towards the west. It is an elevated place, a good vantage point from where ships can be seen at a great distance, and since it is accessible through only one side, it is easy to defend. The inhabitants of Fuña learned of the missionaries' desire to settle in their village, and they were so happy that, to facilitate this project, they themselves worked in the construction of the church. It was dedicated to St. Joseph, for this great saint was chosen as protector of this mission.

The [news] received from Spain gave the Fathers great hopes for the continuation of their work of conversion in these islands, for they spoke of the powerful support of the Queen.[574] This pious princess had ordered the Viceroy of (page 112) New Spain and the Governor of the Philippines to provide the missionaries all that was necessary, and to send new missionaries to a field where the harvest was abundant and the fruit considerable. She had granted the two hundred Filipinos that had been requested and the small ship that was needed to sail from one island to another and to make new discoveries; ornaments were sent for all the new churches, and the missionaries were informed that she would spare nothing towards the conversion of a people that she had taken under her protection.

Although the Queen's orders were final, this occasion revealed that what the sovereigns ordered was without effect when its execution was entrusted to subjects who were bent on disobeying. The Viceroy of New Spain fulfilled the Queen's dispositions and showed the zeal that animated him regarding the propagation of the Faith. However, this was not the case with the Philippine governor [Don Diego de Salcedo], a stubborn man who was against the evangelical workers, and had decided to ruin the mission in the Marianas. To carry out his evil purpose without incriminating himself, he had given secret orders to the ship captains that went from New Spain to the Philippines not to touch the island of Guåhan, leaving the missionaries without succor and at the mercy of the barbarians, so that they would either perish, or be forced to abandon the islands. Thus, the Queen's orders vexed him at first, but he soon resolved to deny the provision of the two hundred Filipinos that this princess had so

---

573. According to the Annual Letter of 1673–74, the town of Fuña or Fu'una was famous for "a rock that they like to show, and from which [the Marianas' natives] believe all men came from. The church of San José was built there" (transcribed in R. Lévesque, *History of Micronesia*, vol. 6, 1995, pp. 201–204).

574. Maria Ana of Austria, mother of Charles III, King of Spain (note by Luis de Morales/C. Le Gobien, *Histoire*, book VI, p. 198).

generously granted under various pretexts. Regarding the boat that was to serve as a transport for the missionaries, he could not delay its construction, since he had received categorical orders as well as the money to carry them out, but he sought to make sure that it was useless. To this effect, he himself designed it and encouraged the shipbuilder to follow his plan, but the latter said that he could not build a boat (page 113) with such irregular proportions, because it would expose those aboard it to deadly peril. Moreover, building such a craft would discredit him completely. The governor insisted, and his orders had to be followed, so that the boat was built without a bridge, masts or ropes, which means it was merely a launch. The governor had it carried aboard the *San Antonio* galleon, persuaded that it would be useless to those in the Marianas.

Fathers Bouwens and Bustillo, who were going from the Philippines to New Spain in order to go from there to the Marianas, informed the Archbishop of Mexico, who had succeeded the Marquise of Mancera as Viceroy, of what had taken place. The archbishop consulted the sailors, and ordered that experts visit the boat and give their opinion; the sailors said that using such a craft was risking death. However, the captain of the *San Antonio,* who was loyal to the Philippine governor, found means to stop the visit ordered by the archbishop, so that the Queen's orders were left without effect, and the missionaries were deprived of a vessel that was absolutely necessary for them to visit the different islands and make new discoveries.

The ship that brought the court's orders had reached Agadña on May 22, 1673, and a few days later, it had continued its voyage on to the Philippines. The General Don Juan Duran de Monfort, who was aboard the ship, was very friendly towards the missionaries, and he even left them a horse to make their trips more comfortable. This horse caused widespread surprise. The savages, who had never seen one, did not tire of contemplating it. News spread soon that a beautiful animal of extraordinary strength had arrived, and people from all the various islands came to see it. Its shape, its speed, its arrears, etc., filled those who looked at it with awe. (page 114) They were most surprised by the bit in its mouth, for they thought that it was its food, and they could not understand how he chewed and digested iron. The animal's long tail thrilled them, and they regarded its mane as the most beautiful thing in the world; they spared nothing in order to get some of its [hairs]. They would come close to the horse, pet, it, and feed it coconut in order to befriend him, hoping that he would allow them to pull some of the hairs from its tail, which led to some of them being kicked by the animal.[575] Urritaos, the debauched young men we have spoken about,[576] used the hairs to decorate their *tunas*, a kind of painted baton with a hole in one end that serves as their token, or more exactly, as their symbol of debauchery. They usually adorned them with three rings made of tree bark, and thick filaments that fell in the form of plumes. When they saw the horse, and were able to pull some of its hairs, they used them instead, forming plumes of inestimable value among them.

General Monfort's horse was certainly not useless in the establishment at Fuña, where two missionaries settled to zealously instruct its inhabitants and those of the neighboring area. They had the consolation of baptizing some four hundred children in their first expedition, and forming a great number of catechumens who were excited to learn the Christian Doctrine and sing hymns. Thereafter, the echo of the praise to God resounded in these mountains and valleys, substituting the profane compositions of other times.

The blessings of heaven that descended upon this new mission were followed by the peace that the savages sued for when it was least expected, for they had rejected all of the peace proposals made until then, and they had even insulted the messengers (page 115) sent to present them. This happened on November 13, a day consecrated to honor the memory of the blessed Stanislaus Kostka,[577] novice of the Society of Jesus; and the savages ratified the peace, not by giving turtle shells, as was their custom, but by bringing their children for baptism. The conditions demanded for the peace were that they would observe the Commandments of God; that they would procure baptism for their chil-

---

575. This description corresponds to that found in the Annual Letter of 1673–74: "It was no small wonder for these people to see a brute that was ever unknown to them, and more when they contemplated its speed and the strength of its limbs, so that its fame spread throughout these Isles and moved them to make various visits, and take strands from his mane as if they were great gifts, proudly carrying them in their limestone cups" (transcribed in R. Lévesque, *History of Micronesia*, vol. 6, pp. 204–205). Louis-Claude de Saulces de Freycinet would later reproduce this same episode in his *Voyage autour du monde exécuté sur les corvettes de S. M. l'Uranie et la Physicienne pendant l'anné 1819* (L.-C. de S. de Freycinet, *An Account of the Corvette L'Uraine's Sojourn at the Mariana Islands, 1819*, Suplemented with the Journal of Rose de Freycinet, translation by Glynn Barratt. Saipan, CNMI Division of Historic Preservation, 2003, p. 33).

576. *Urritaos* [single young men] engaged in sexual relations with single young women in a sort of institutionalized prostitution that the Jesuits tried to eradicate by all possible means. See F. García, S.J., *The Life and Martyrdom of the Venerable Father Diego Luis de San Vitores*, book III, chapter I, p. 171.

577. Stanislaus Kostka (1550–1568) had died while still a novice in the Society (he had entered at 17, a year before dying). Beatified in 1605, he was canonized in 1726 by Pope Benedict XIII. This Pope (1724–1730) inaugurated a new model of sanctity in the eighteenth century characterized by young novices who had joined the Society of Jesus without their parents' approval. Most of them were frail and died young, and their purity and chastity were extolled (R. Po-Chia Hsia, *The World of Catholic Renewal 1540–1770*, Cambridge: Cambridge University Press, 1998, p. 137; Fernando R. de la Flor, "La "fábrica" de los nuevos santos: el proyecto hagiográfico jesuita a la altura de 1730," in Pablo Fernández Albadalejo (ed.), *Fénix de España. Modernidad y cultura propia en la España del siglo XVIII (1737–1766)*, Madrid: Marcial Pons, 2006, pp. 215–235).

dren; and that they would attend catechism and any other instructions that the missionaries decided to give them.[578]

This peace produced very good results. Many neophytes whose faith was still weak, and who had been seduced by the infidels, were reconciled with the church; a great number of children were baptized; the entire island of Guåhan was visited, and superstitious practices were abolished, so much so that various saintly customs were established and are still followed today. In the places that he visited, Father Ezquerra found various young Christian women and girls who had resisted the wiles and snares used to try to make them lose their innocence, with a valor worthy of the first centuries of the church.

This zealous father went to the mission of San José de Fuña at the beginning of 1674, and he visited the nearby mountains, baptizing four-to-five hundred people that he had first instructed and prepared.[579] On February 1, he spent the night in the village of Ati, which the Spanish call Port San Antonio. The following day was the feast of the Purification of the Blessed Virgin Mary, for which he needed to return to Fuña to say Mass and give communion to the six catechists who accompanied him. They left at dawn, and had barely gone one league when they found four Marianas' natives carrying a woman in labor pains to a nearby village. Since she was Christian (page 116) and her condition was critical, the Father heard her confession, and was about to anoint her according to her wishes, when the men who accompanied her, convinced that the oil used for the unction would cause her death, rudely prevented the Father from doing so.

The Servant of God tried to explain the efficacy of this sacrament and the benefits that it provided to those who received it. However, the barbarians refused to listen, and instead called to their compatriots with the intention of killing the priest and his companions, of whom only two were left, for the other two had fled upon perceiving the barbarians' furor. One of these, Sebastián de Ribera, was reached by a spear on his back as he escaped, and he jumped into the sea trying to save himself, but there he was killed off. He was a young man from Manila who had been in the Marianas for a year and had rendered great services to the missionaries.[580] The other, Francisco González, escaped to the neighboring mountain and hid among the bushes. The barbarians looked for him everywhere, and since they could not find him, they set the forest on fire

---

578. Annual Letter of 1673–74, transcribed in R. Lévesque, *History of Micronesia*, vol. 5, 1995, p. 206.

579. Annual Letter of 1673–74, transcribed in R. Lévesque, *History of Micronesia*, vol. 5, 1995, p. 207.

580. See F. García, S.J., *The Life and Martyrdom of the Venerable Father Diego Luis de San Vitores*, book V, chapter X, pp. 409–10.

to force him out, but González was able to escape the flames by climbing over precipices and hard-to-reach promontories.

Seeing himself surrounded by murderers, the Father burnt the holy oils so that the impious barbarians could not profane them, and he prepared himself to die along with the other two companions that were with him. He had barely finished taking their confession, when the barbarians assaulted them with terrible blows. The first to die was the eldest of the two catechists, a man named Luis de Vera Picaso [or Picazo].[581] Born in Manila, this man had become disgusted with the world and had consecrated himself to the Marianas' mission in order to work for the glory of God and the salvation of souls, and he usually accompanied the Fathers and served as catechist (page 117). He was in Ritidian [Litekyan] when Father San Vitores was martyrized and, finding that the enemy had cut off all the paths to Agadña, he was forced to remain concealed for fifteen days inside a crevice, where he would have starved to death if some Christians who went by had not taken him to Agadña. He had the consolation of dying in the arms of Father Ezquerra, who would soon perish at the hands of the barbarians himself.[582]

This holy missionary fell to the earth covered in blood, the barbarians having struck several blows to his head. After taking his clothes, they left his body lying at the beach, unable to tear from his hands the crucifix and an image of the Holy Virgin that he held tightly. His other two catechists, Pedro de Alejo [from Puebla de los Ángeles, Mexico] and Matías Altamirano [from Oaxaca], who had set out from Port San Antonio a bit later, arrived at that moment, and were horrified to see two of their companions dead, and the Father exhaling his last breaths. They jumped into a canoe that they happened to find near the shore, with the hope of escaping from the barbarians that pursued them. However, since the canoe had neither oars nor sails, and they did not know how to operate it, it overturned and they fell in the water, where the furious barbarians attacked them with rocks until they both perished. This was a true loss to the mission, for these two men rendered great services to it. Pedro de Alejo acted as a procurator, for he provided the Fathers with food and other necessities. He was invaluable, for he did the work of four men. On his part, Matías Altamirano was a skilled surgeon: his talent as a catechist and his charitable

---

581. "Lista de mártires sacrificados en las islas Marianas, 1671–1685," transcribed in R. Lévesque, *History of Micronesia*, vol. 8, 1996, p. 184.

582. See F. García, S.J., *The Life and Martyrdom of the Venerable Father Diego Luis de San Vitores*, book V, chapter X, p. 410.

treatment of the sick made his death regrettable to all.[583]

While the barbarians were killing these fervent catechists, Father Ezquerra raised his heart to (page 118) God, kissing his crucifix and imploring the assistance of the Holy Virgin. He saw a young man from Fuña whom he had instructed in the past, and called to him, saying, "*While I have lived, I have been your father; now that I die, I still am, and I always will be.*" Another Christian, moved by pity, came to him, and said, "*Father, why have they treated you like this? What is the cause of your death?*" "*There is no other cause,*" said the Father, "*but having wanted to do you good, to baptize your children and to teach all of you the way to heaven.*" Since the Christian, moved by anger, began speaking of vengeance, the father said, "*Oh, my son! Do not let these sentiments into your soul; it is best for you to retreat from here, lest the barbarians mistreat you, too.*" The poor Christian obeyed him, crying for the Servant of God, who lay covered in blood, badly bruised, in terrible pain, and naked on the burning sand that added to his torment.[584]

After having killed Alejo and Altamirano, the savages returned contentedly, but upon discovering that Father Ezquerra was still alive, they struck him repeatedly until they finally killed him. This happened on February 2, 1674.

Seeing that one of the father's companions had escaped to the mountains, those furious men pursued him as if he were a wild beast, and upon finding him they killed him with their spears. His name was Marcos de Segura, and he was a young man of a sunny disposition who had gained the neophytes' trust through his sweetness and piety.

Of all of Father Ezquerra's companions, only Francisco González had survived, climbing over rocks and precipices for over two hours after having abandoned the forest as he tried to gain the shore. However, he had the misfortune of falling in the hands of one of his enemies, (page 119) who went to him thinking that he was Father Ezquerra. González assured him that he was not the Father, and, trying to save his life, said that the priest was behind him, but the barbarian began striking him with a sort of scimitar. Since González tried to use his arm to fend off the blows, the enemy became even more enraged, and holding the other's hand, struck him new blows until he had caused him to fall on the ground, where, after hitting him some more, he left him for dead. In that state, González still found the strength to crawl to the woods, where

583. See F. García, S.J., *The Life and Martyrdom of the Venerable Father Diego Luis de San Vitores*, book V, chapter X, p. 411.

584. Annual Letter of 1673–74, transcribed in R. Lévesque, *History of Micronesia*, vol. 6, 1995, pp. 208–212.

he laboriously made himself reach a friendly village after losing much blood. There, he was charitably received by the Christians, who took him to Agadña, where he was able to relate part of what I have just written regarding Father Ezquerra's death.[585]

This holy missionary, who died in the prime of his life, was the son of General Don Juan de Ezquerra, known for his services and offices; and Doña Luisa de Sarmiento, daughter of one of the most illustrious Spanish families. At the age of sixteen, he had joined the Society of Jesus, in which he already had a brother and an uncle who distinguished themselves for their virtue and merit. The vivacity of his intelligence and the innocence of his customs had won everybody's affection, but after consecrating himself to God, the austerity of his life and his profound humility made everyone regard him as a saint. One cannot but feel admiration when reading what the author of a summary of his life has said of him.[586] When he was told that he treated his body with too much rigor, he responded amiably: it is necessary to prepare oneself for the labors of the apostolic life.

He had insistently asked to be sent to the Marianas, with the hopes of suffering greatly and obtaining the crown of martyrdom. His superiors had heeded his insistent pleas, but (page 120) since the delicacy of his conscience bordered on the scrupulous, he feared that he had acted according to his own will instead of looking for divine inspiration. These reflections troubled him, and so he placed himself at his superiors' disposal, so that they did with him what they believed was most appropriate, protesting that he was completely indifferent and that he would accept whatever destiny they chose for him from the hands of God. In this regard, he wrote to the provincial Father, saying:

> *I am convinced that separating oneself from the will of one's superiors is like separating oneself from the will of God; thus, one deceives oneself most miserably, and it is even worse to go against obedience by bending the superior's will to one's own. And so, my Father, if my voyage to the Mariana Islands is not the will and desire of my superiors, and it has been my insistence rather than their desire what has compelled them to grant me the grace that I asked of going to this mission, I beg you to not take my request into account, and to even forget that I ever made it, suspending my voyage, for I want to*

---

585. Annual Letter of 1673–74, transcribed in R. Lévesque, *History of Micronesia*, vol. 6, 1995, p. 213.

586. This author is Francisco García, S.J., who wrote in Spain the life of Father San Vitores and of his companions (note by Luis de Morales/C. Le Gobien, *Histoire*, book VI, p. 212).

be completely indifferent to whatever charge you give me. In this way, I will work without scruples wherever I am sent, and place myself without reservation in the hands of obedience, as a dead body or an old man's cane, assured that I will be following the will of God.

Later he adds:

> *I know that I am unfit for such an important work as is that of the missions, unless it is through obedience that I come to it. For this reason, I place myself in the hands of my superiors, so that they do what they will with me, sending me to wherever it may please them. For I am so resolved to do their will, which I believe is that of God, that if they want me to spend the rest of my life serving in the kitchen, I will do so without repugnance and without any sorrow. And if I am refused (page 121) permission to go to the missions, far from being vexed, I will look upon it as a grace with which God wants me to know how he wants me to serve him. Since you informed me that I was to go to the Mariana Islands, I have asked God every day at Holy Mass, while holding the sacred Host in my hands before communion, that if this trip were not for his greater glory, nor for the salvation of my soul and the souls of the infidels, may there be obstacles to impede its realization and change your way of thinking. Thus, my Reverend Father, if you do not judge it proper that I go to the Marianas, you merely have to notify me, and that will be enough for me to stop thinking about it and to never mention this business again. I would only regret one thing, and it would be having made myself unworthy of such a holy ministry, for this is the only reason that I believe would keep this from being God's will, convinced as I am that my infidelities are a great obstacle to his designs for me.*

The Father Provincial was impressed by this holy man's virtue, and ordered him to go to the Mariana Islands to work converting those peoples who so needed it. Words cannot express the joy that Father Ezquerra felt when he saw himself having to obey being destined to do the work that was the object of all his desires. Even though he had great talents for the missions, his humility did not let him see them, for he considered himself unfit for everything. Yet, he was convinced that God, who called him to this holy ministry, would also grant him the grace that he needed to make up for his weakness and incapacities. This idea sustained him in all his apostolic labors, and it gave him solace until the end of his life. In the islands, the loss of this zealous father was to be repaired with the arrival of the new missionaries that were expected. The ship that brought them reached Agadña on June 16, 1674. Father [Peter] Coomans,[587] superior of the mission, (page 122) went to receive them, but the moment he set foot aboard the ship, a great gale pushed the vessel into the open sea, forcing him to continue on the voyage to the Philippines. This unexpected setback greatly troubled the missionaries, whose needs for life's essentials were by then dire. Nevertheless, God afflicts his servants only to reward them amply later, and he therefore crowned their work with more success, if that were possible, than that of previous years, even though there were so few of them left.

Having placed himself in charge of the islands' troops, Captain Don Damián de Esplana, who was ashore when the ship suddenly departed, contributed greatly to the progress of religion with his conduct.[588] His first task was to keep the soldiers from giving themselves to laziness, and he had them cut down a forest near the town of Agadña that the enemies often used for their ambushes. Whether it was because the barbarians were angered to see these woods disappear, or whether because they were inconstant and fickle, they began harassing the Spanish. Don Damián resolved to deal them such a blow, that these people, who saw themselves as the most terrible and fierce, would be thoroughly intimidated. Nevertheless, to be loyal, he first sent various envoys to tell them to respect the treaties that had been reached and to simply observe the two promises that they had made: namely, that they would fulfill their Christian duties and that they would not impede the missionaries from exercising their ministry.

The barbarians thought that these messages were a sign of weakness and cowardice, which made them bolder, and they continued harassing the Spaniards with even more insolence and daring. Chuchugu, a difficult-to-reach village, was the (page 123) place that Don Damián decided to attack, and it was indeed a large risk to try to take a village that was high up in a mountain.

---

587. Peter Coomans, also known as Pedro Cumano, was born in Antwerp on January 30, 1638. He joined the Society on September 19, 1656. At the time of his death, in 1685, he was a graduate and professed of the four vows (2/2/1675) (ARSI, "Primus Catalogus Anni Personarum Anni 1684." Philippinae Cat. Trien. 1649–1696, Tomo 2-II, f. 422v).

588. Father Morales does not hide his favorable opinion of Sergeant Major Esplana, who until then was a great support for the missionaries' evangelical labor. During the rebellion of 1684, the Jesuits, especially Father Gerardo Bouwens, criticized his conduct (A. Coello, "Colonialismo y santidad en las islas Marianas: los soldados de Gedeón," pp. 17–44). By then, Father Morales was no longer in the Mariana Islands. In a 1685 *Memorial* he suggests to the King that "confirming said Sergeant Major Don Damián de Esplana in the post of Governor, and Sergeant Major Don Joseph de Quiroga as his successor" would be beneficial (transcribed in R. Lévesque, *History of Micronesia*, vol. 8, 1996, p. 414).

But having placed this enterprise under the protection of St. Michael, patron of these islands, the captain left Agadña with thirty men on the night of July 26 [of 1674], protected more by the protection of heaven than by the number and valor of his soldiers.

After difficult feats, the expedition reached a narrow pass at the entrance of the town of Chuchugu. This pass was surrounded by three heights that were occupied by the enemies, who were ready to defend themselves at all cost. As soon as they saw the Spaniards, they attacked, and the latter immediately responded with a volley of musket fire. The barbarians kept their positions, and animating each other for combat, showered the Spaniards with stones. Since the Spaniards were fighting inside a gulch against enemies that were above them, they hindered each other's movements and, not knowing how to get out, Father Alonso López, who was with them, raised his voice to implore St. Michael's assistance. At that moment, the combat changed completely, for the soldiers, suddenly animated as if by a supernatural fervor, began climbing up the rocks that surrounded them without any fear for the arrows and stones that rained over them. Terrified at this display of bravery and intrepidness, the barbarians fled and dispersed in a moment, so that the Spanish entered freely and razed the town to ashes. Despite being tough, this combat was not bloody, for only one barbarian lost his life, and only after Father Lopez had the consolation of baptizing him.[589]

Heartened by such a complete victory, Don Damián de Esplana carried out some incursions in the area of Fuuna [or Ga'una] (page 124) and Fuhon, burning some villages to intimidate their residents and to persuade them to fulfill their duties. These triumphs produced the desired effect, for the barbarians became more docile and stopped harassing the Spaniards.

With such a favorable situation for the progress of religion, public schools were established in Agadña for the instruction of girls and boys, and special care was given to them, for the missionaries' greatest hope rested upon these children. The Church of [San Francisco Javier, in] Ritidian [Litekyan] was completed, and a new one was built in Tarragui [Tarague or Talågi, the Church of San Miguel de Tarragui]. There had never before been such fervor among the natives, and the people from the neighboring villages came en masse to be instructed, so that the missionaries could not find the time to teach them all. The younger folk distinguished themselves, for in less than two months

---

589. See the Annual Letter of 1674–75 written by Father Alonso López, Hagåtña, June 6, 1675 (transcribed in R. Lévesque, *History of Micronesia*, vol. 6, 1995, pp. 283–293).

they had learned the Christian Doctrine and the religious hymns. Those who were older tried to follow their example to such a degree, that meets were set up in which the parishioners of Ritidian [Litekyan] and Tarragui [Tarague or Talågi] competed to see who responded better to questions proposed in public conferences held to that effect. They went as follows:

A place, date, and subject were decided upon, and the parishioners arrived in a procession headed by their banner-bearers. Two rows of children dressed in white followed, the girls on one side and the boys on the other, with crowns of flowers on their heads and palm leaves in their hands. Men and women came after, also in rows, singing in two choirs, with an admirable disposition, the religious hymns that were a compendium of the Christian Doctrine. The procession [of the parishioners] of the village where the conference was to be held met the other, and once they were all assembled, they began proposing questions related to catechism or (page 125) the mysteries of our holy religion to one another. The Fathers acted as the judges, praising and rewarding the cleverest and conceding victory to the party that had given the best replies. These conferences did much to advance the speedy learning of the Christian Doctrine, and they were so successful, that other churches tried to establish this same system.

The priests' evangelization work was going peacefully in Guåhan, as the light of the Gospel spread everywhere, owing in part to the services rendered by Don Damián de Esplana. This captain resolved to go to the northern islands to do that which he had done in Guåhan, restoring tranquility and trust. But the seditious barbarians, who were stilled only by the fear within them, decided to take advantage of his absence to get rid of the missionaries and the few Spaniards that had stayed in the island of Guåhan. The inhabitants of Chuchugo [Chochogo'], Mapaz, and other mountain-dwelling residents led this conspiracy. Nevertheless, their intentions could not remain concealed for long without spreading across the island, especially after the murder of a Marianas' native who was killed because he favored the Spaniards. This brutal act angered the commandant, and he was determined to punish its authors. The conspirators entrenched themselves in the mountains, and it would have been impossible to dislodge them if they had defended themselves with courage. Once one of the bravest among them fell dead with the first charge made by the vanguard commanded by [Pampango] Don José de Tapia,[590] the rest fled.

---

590. The Spanish would not have been able to defeat CHamoru resistance without native CHamoru soldiers, who joined the mission's armed contingent, which, for its part, had more soldiers of Philippine than

Finding no resistance, the Spanish reconnoitered the mountains and punished the guilty.

With this, a solid peace was finally established at the request of the intimidated barbarians. One of Agadña's main *chamorris*, whom the Spanish captain had befriended, headed the negotiations. He first went to find the Chuchugo villagers, (page 126) and delivered the following parley:

> *You do not know what you are doing by insisting on your enmity towards the Spanish, whose only purpose is to do us good and show us the way to heaven. Why do you refuse the advantages that they offer us? You live in constant apprehension and evidently expose yourselves to perish by attacking them, secure in your numerical superiority. Experience should make you realize that it is impossible to resist their courage and the deadly weapons that they carry. Be their friends and live in peace with them: this is the only means you have to be happy.*

These reasons convinced them and encouraged them to make a true peace, so that these terrible mountain-dwellers, who until then had been fierce and indomitable, became the Spaniards' allies and friends.

This expedition, and many others that Don Damián later led, greatly helped the progress of religion. The number of believers increased each day, and since the old churches could no longer hold the multitudes that attended the divine mysteries, larger ones had to be built. During this period, the [Church] of Ritidian [Litekyan] was founded, and seminaries to educate the youth of both sexes were established in other parts. The ardor and commitment of these boys and girls were a great consolation [to the missionaries], for they were so desirous to learn that they would often cut their sleeping time in order to have more time to study. They gathered twice a day at the church to pray, recite the catechism, and sing religious hymns, and it was admirable how these naturally libertine young men and women who were accustomed to lives of vice, subjected themselves to the discipline and regularity of these things.

Brother Pedro Díaz,[591] a religious man of proven virtue, was in charge of

---

of European origin. Native soldiers were more readily adapted to the terrain and served as valuable interpreters and mediators, both necessary elements that proved vital for the conquest and colonization of the islands. One of them was Don José de Tapia, who was one of the best Pampango captains to have participated in the CHamoru revolt of 1684 (Augusto V. Viana, "Filipino Natives in Seventeenth Century Marianas: Their Role in the Establishment of the Spanish Mission in the Islands," *Micronesian Journal of the Humanities and Social Sciences*, 3 (1-2), 2004, p. 22.

591. Brother Díaz was born in the village of Calero in Talavera de la Reina (Toledo) in 1651. He joined the Society of Jesus on April 24, 1673 in the Toledo College of Oropesa (Toledo) (AHCJC, Carpeta "EI.b-9/5/4. "Mártires de la Provincia de Filipinas de la Compañía de Jesús. Manila, 12 de mayo de 1903," contenida

these schools. His zeal made him want to destroy a house of corruption [*i mang-guma' uritao*] where ten-to-twelve (page 127) urritaos lived with one woman in such iniquity that they scandalized the Christians. To do it, he resolved to convert the woman, which he accomplished. She therefore left the urritaos and retreated into the seminary of Ritidian [Litekyan] to practice piety and Christian virtues. Angered by the decision of the woman that was the object of their passion, the urritaos resolved to capture her, and with this purpose, three of the most intrepid ones entered the place where the young ladies slept, and caused a great disorder. Brother Díaz was dismayed when he learned about this, and animated by a great zeal, he went to see these young men and scolded them for what they had done. These libertines, angry that a foreigner would dare reproach them their vices and violence, attacked Brother Díaz and a companion that was with him, called Ildefonso de León, and animating each other to take vengeance, they smashed their heads, killing them both and mistreating their corpses.[592] They did not stop there, and as if they were possessed by the devil, they ran to the church and to the missionaries' house, where they killed Nicolás de Espinosa, set fire to the building, destroyed the two schools, and stole the sacred ornaments and anything that they could find.

However, God showed these impious men to respect his sacred objects and refrain from profaning them as others like them had done before, for when one of the urritaos picked up a chalice that was used in the Holy Mass, his hand was burned and he had to drop it immediately. Nevertheless, greed made him grab for it again, and his hand swelled horribly, becoming reddened as if by blood. Another prodigious thing happened that consoled the Christians. Some of the inhabitants of Tarragui [Tarague or Talågi] (page 128), angered at the assassination of Brother Díaz, went to Ritidian [Litekyan] to punish the impious murderers. During their stay that night, they saw three bright stars shining over where the church had been built. These good neophytes did not doubt for a moment that the three stars were a sign that Brother Díaz and his two companions were already in heaven covered in glory. The murderers, learning that they were being pursued, had retired to the island of Zarpana, the usual asylum for their kind, avoiding the punishment that their crimes deserved.

en EI/b-9/5/1-7. "Martirios y varones ilustres." AHCJC.). For more on Brother Pedro Díaz, see H. de la Costa, *The Jesuits in the Philippines...*, p. 457; and P. Murillo Velarde, *Historia de la provincia de Filipinas*, ff. 336r-337v. See also the letter written by Father Gerardo Bouwens to General Giovanni Paolo Oliva (1676) (ARSI, Litterae Annuae Philipp. 1663–1734 (etiam de Insuli Marianis), Tomo 13, f. 129r; ARSI, "Primus Catalogus Anni Personarum Anni 1675." Philippinae Cat. Trien. 1649–1696, Tomo 2-II, f. 395r.

    592. Francisco Gayoso's letter to provincial Javier Riquelme, Manila, September 13, 1676 (transcribed in R. Lévesque, *History of Micronesia*, vol. 6, 1995, pp. 552–553).

Brother Díaz was from Talavera de la Reina in Spain. His mother, who was a virtuous woman, raised him piously, and often told him: *"My son, it is better to die rather than to offend God."* Thus, he always kept the purity of his customs. After having completed his first studies at the College of Oropesa, he went to Salamanca with the intention of graduating, for he aspired to take on ecclesiastical dignities. However, the death of one of his companions, who, murdered with a dagger perished unconfessed, deeply affected him, and he resolved to abandon the world and its empty human vanities. The sermons of the fervent fathers Thyrsus González and Juan Gabriel Guillén, who at that time were holding a mission at Salamanca, convinced him, and he joined the Society of Jesus on April 24, 1673, with the intention of consecrating himself to the conversion of infidels. He began his novitiate with extraordinary fervor, and asked so insistently to be sent to the Marianas that he was destined there at last. During the trip, his virtue made him stand out. His virtue was tested even more at the mission, for besides the work of an apostolic life, the lack of food forced him to eat roots, and the terrible pain and swelling (page 129) that the heat of the country caused in his legs contributed to his suffering.

This young man's loss was felt throughout the whole mission. Here is what one of his companions wrote to Father Francisco García,[593] who wrote a compendium of the life of this virtuous Christian. The letter is from May 25, 1676:

> *I cannot give you the news, without holding back tears,* he said, *regarding the loss we have just experienced in the figure of Brother Díaz. He was particularly suited to the labors of the apostolic life, capable of rendering great services to this mission thanks to his prudence, his zeal for the salvation of souls, and for all the other virtues that he possessed so patently. The perfection with which he spoke the language of the country seemed miraculous, for in less than a year he had mastered it more than any of the older missionaries who had been in the islands far longer. He was martyred for protecting chastity, and with this God Our Lord rewarded him for the vow that he had taken in his youth. His love for this virtue was so high, that he wished to inspire it in everybody, procuring as well as he could that God not be offended in this matter, and to this end he had worked tirelessly in these islands, where licentious customs were so horridly deep-seated.*

---

593. Francisco García, S.J., *Vida y martirio del V. P. Diego Luis de Sanvitores* (Madrid, 1683). This work was translated to Italian in 1686.

Thus ended the letter to Father García.

This precious death was followed closely by that of Father Antonio Maria de San Basilio,[594] one of these islands' most zealous and virtuous missionaries. A man of prayer who was deeply united to God, he was entirely dead to self and detached from all of the material things of this world. The inner peace that he enjoyed shone forth in his face, and it caused such an impression on those who looked upon him that his gaze was enough to dissipate their sorrows, as many have assured. His sweetness and modesty charmed everyone, but his mortification shocked those who knew the extremes to which he took it, for he looked upon his body as if it were his greatest enemy (page 130) and he did not let one day pass without mortifying his flesh. Under his garments, he wore a cross covered in spikes, besides the cilice [hairshirt] that he never took off, and he used three disciplines each day.[595] He also slept on the hard floor; and ate mostly insipid roots, although on rare occasions, he had rice, and on even rarer occasions, some fish.

He traversed the entire island of Guåhan many times, not without fatigue and suffering. Sometimes he walked barefoot over the scorching sand, or he climbed nearly inaccessible mountains in order to succor the sick and baptize children. Moreover, even though he exposed his life every time among those barbarians who did not acknowledge any law other than their own passions, he went day and night wherever the necessities of his ministry called him. Once while climbing a steep cliff that bordered the sea, he had to cling to some swordgrass so as not to fall, but the roots came loose and his body tumbled to the sea. Yet, having invoked God's assistance at that very moment, those who accompanied him were amazed to see that a wave deposited him safely on the shore. This Servant of God was capable of the most heroic actions, and he was often seen sucking the most revolting putrefied sores of the sick in order to clean them, for the most repugnant among the ill were the ones he most solicitously served.

One day, he found an asthmatic whose leg was half eaten by a horrible cancer, and carrying him upon his shoulders, he took him to his own room to care for him. He fed him each day and served him with great charity. He

---

594. Father Antonio de San Basilio was born in Catania (Sicily) in 1643. He joined the Society of Jesus on January 11, 1659 and died while being a missionary to the Indians in the Mariana Islands (ARSI, "Primus Catalogus Anni Personarum Anni 1675." Philippinae Cat. Trien. 1649–1696, Tomo 2-II, f. 392r).

595. In the practice of mortifying the flesh followed by many Catholics at the time, particularly among certain religious orders such as the Society of Jesus, the "discipline" refers to self-flagellation, using whips, lashes, cords or rods.

instructed him in the mysteries of our holy religion, and had the consolation of holding him in his arms during his death, with the sentiments of a foreordained saint.

From his early youth he had been tenderly devoted (page 131) to the Holy Virgin, and considering her as his own mother and advocate, he talked to her with the utmost confidence. In order to ascertain the sentiments that animated him towards this mother of mercy, it would be enough to know the letter that he wrote a few days before his death, and which was found after it.

> *Despite being a great sinner,* he said, *I write this letter to the Holy Virgin Mary, mother of God, queen of heaven and earth, so that she may procure for me a final victory over my passions and the grace of living in perfect union with her Holy Son.*

> *Holy Virgin,* read the letter, *although I am the most wretched of men; a slave to my passions; full of pride in the midst of my miseries; poor and without merit because of my sins; a child of wrath, worthy of the sorrows suffered by demons and those condemned to hell; although I am, in sum, vexed by the weight of my sins and the most abject of all creatures to my own eyes, and I should abandon myself to despair, I have such great trust in you being the refuge of sinners, that I dare kneel at your feet to ask you to accept this letter. Since you are full of tender mercy, listen to my cries and allow me to manifest to you my sorrows and my desire of making myself pleasing to God, who is infinitely merciful. You see that I am in a deplorable state, so please find it worthy to redeem me. I cannot, Holiest Virgin, find the words with which to express all that I feel and my heart's desires. Here are twenty-two wounds that are as many mouths that speak in my favor and shall make my troubles known to you. The desire I have of serving you is such, that I would rather die a thousand times than displease you in the slightest, whether by thought, word, or deed. You know that I ardently desire to give my life for the love of your Holy Son and for the salvation of these abandoned peoples. I beg of you, then,* (page 132), *Holy Virgin, for a perfect abnegation of self, a complete conformity to the will of your Divine Son, and the inner light to know myself and to know my God. Grant me also a great horror of all sin, a sincere love of the cross, a purity that will make me agreeable in your eyes and an ardent zeal for the glory and salvation of souls. I offer you myself completely, all my potencies and senses, and I am willing to spill*

*all of my blood for your love and that of your Holy Son. Look upon
me, then, Mother, with merciful eyes, and do not abandon me now in
the hour of my death.*

The Holy Virgin favored him frequently with extraordinary gifts, demonstrating how agreeable his devotion was to her. Once when the Servant of
God was watching over Luis de Vera Picaso, who was very ill, he knelt at the
foot of the bed and begged God for the health of this fervent catechist. Don
Luis, who was watching him, saw a lady of wondrous beauty next to the priest.
He was tremendously surprised by this vision, but the efficacy of the holy man's
prayer amazed him even more, for he attributed to it his full recovery and his
life, which the Lord conserved for him then so that he could later give it in His
service, as we have already described.

Father Basilio frequently procured health for the sick, more by his
prayers than by the services and medicines that he proferred, so that he ordinarily had the consolation of healing their body and their soul at the same time.
He went on a mission one day with ten Spaniards who served him as catechists.
They were stricken by hunger and fatigue, for they had nothing to eat; and
the Father, more afflicted by his companions' necessities than his own, asked a
Marianas' native that he found [on the way] for something to eat (page 133).
He gave them three small taro roots. The Father took them, blessed them, and
distributed them among his companions, so that each got a very small portion,
and since they were all completely satisfied, they regarded this as a miracle
performed by God because of the merits of his Servant.

Since at this time he was superior of the mission, given Father Coomans's
absence, he was forced to procure life's necessities for his confreres and their
assistants. One day, when there were no provisions in the house, he sent a Marianas' native named Quemado to the town of Upi to bring a certain amount of
yams, a crop that is a common food in this country, paying him in advance to
stimulate him to accomplish his task better. Since the man took a long time
coming back, the Father went looking for him personally [January 5, 1676].
He arrived at the village late that night and early the next morning went to find
Quemado, who gave him a paltry bunch of yams instead of those that he could
have bought. "My friend," the Father told him with his usual sweetness, "you
have not fulfilled the task that I gave you, for these roots are worthless." As he
said this, he bent down to count them. Suddenly, angered by what the Father
had just told him, or perhaps, blinded by greed, the barbarian struck such a

blow upon the head of the Servant of God that it threw him at his feet, and he did not stop hitting him until the Father was dead, aided in this deed by a son of his who was present.[596]

Thus died this great Servant of God, who had always had an ardent desire of giving his life for Jesus Christ. We can judge his sentiments by the letter that he wrote to Father Diego Valdés [?–1693], rector of the college of Alcala, where he resided before going to the Indies:

> *I arrived at these islands last year, and I have remained in them with a happiness interrupted only by the sorrow caused by Father San Vitores' death. I hope that with the grace of God I profit from his instruction and holy counsel, as have the other fathers who had the joy of living with him. Despite his loss, I am consoled nonetheless by the idea that this Servant of God will serve as a guide and model for me, obtaining from the Lord the grace of dying in the exercise of my apostolic duties. We are not lacking in opportunities, whether it is by our dire lack of life's most basic necessities, or by the constant alarm in which we live thanks to the attacks that we are exposed to every day at the hands of the barbarians. These are compelled by their greed for the simple objects that we have brought from Spain, such as bells, knives, and other trinkets that they value more than gold, and that have been the cause behind the murder of some of our secular companions, for their attackers thought that they were carrying objects such as these. If they had been intimidated [early on], we would not have reached this extreme: they would come to Church, listen to the explanations of the Christian Doctrine, and they would not be as indomitable as they are now, that they reagrd us as weak.*
>
> *With these bagatelles, we procure what we need for the sustenance of the auxiliaries that accompany us as well as our own. We suffer a lot, but we are content, for no human consolation is comparable to the joy we feel. For Reverend Father, you must believe me when I tell you that we go as the apostles went, without purse and pack, and often without shoes, whether because we have none, or because these are made of palm leaves and become useless in the swampy terrain or the stretches of sea that we must cross. Roots are our usual food and the basis of*

596. For a brief sketch of Father San Basilio's life, see the letter written by Lorenzo Bustillo to General Giovanni Paolo Oliva on January 17, 1676, in which he relates the unfortunate death of this Jesuit (ARSI, Philippin. Necrologia 1605-1731, Tomo 20, ff. 312r-314v). See also P. Murillo Velarde, *Historia de la provincia de Filipinas*, f. 337v.

everything we consume, for here (page 135) we lack bread and wine; but since man does not live on bread alone, we are not concerned by this, for God showers upon us such an abundance of consolations for our souls that I can assure you, we desire no others. What happiness we feel, Father, when we traverse the mountains in search of children to baptize, and after having walked all day and found none, we finally succumb to exhaustion precisely upon the place where they had been hidden; and with what pleasure do we communicate to them the grace of God by means of baptism.

I am gladdened by the idea that the barbarians, who are so skilled with their lances, wish to hurl them at me and make points with the bones of my arms and my legs; for indeed this is what they do, and they add such poison to them, that if a small bit penetrates the skin it produces death.

The language of these peoples is learned easily, and it is not difficult to pronounce, for they do not join too many consonants in a single syllable. You can know their customs reading the descriptions written about them by Father San Vitores in his letters. The harvest for heaven is abundant, for even if instructing and baptizing the children was all that needed to be done, this task alone would require a hundred workers. God grants us the grace of traversing this land of infidels and guiding so many of them to heaven through the blood of Jesus Christ, for by saving them we also save ourselves. I am Your Reverence's humble and obedient servant in Christ. Antonio María de San Basilio.

The church of Tarragui [Tarague or Talågi] vividly lamented the loss of this holy man. The most zealous Christians went to the mountain of Upi to avenge the Father's death, but since Quemado had fled, they contented themselves with burning down this wretch's house and recovering San Basilio's body, which they buried with great respect in the Church of San Miguel de Tarragui [Tarague or Talågi], which had been founded by the missionary himself.

BOOK SEVEN

At the beginning of 1676, the mission was in a very difficult situation. The barbarians became more fierce and insolent each day. They had killed Father San Basilio and Brother Díaz, and insulted Father Gerardo Bouwens and wounded his companions. The number of Spaniards thus diminished each day, and to add to the misfortunes, the boat that the governor used to keep the savages at bay was lost. It was at sea with Father [Francisco] Gayoso[597] and part of the garrison when a furious storm that carried it all the way to the Philippines surprised it and threw it against the rocks.

This accident was what finally led General Don Damián de Esplana to resolve to leave the islands, something that he had been considering for quite some time. At first, he had distinguished himself for his acts of valor, but when he saw the turn that things were taking, and fearing that he would finally succumb to the barbarians' efforts, he decided to retire to Manila, and the help that arrived on June 10 of 1676 was not enough to change his mind.

The vacuum left by Don Damián's departure was hard to fill. [The mission] needed an able and experienced man who could repress the insolence of the barbarians (page 137) and keep the soldiers in line, so that they fulfilled their duties and were not allowed to engage in vice or sloth, and there was no one in the islands capable of this. The captain of the [*San Antonio de Padua*] galleon that had brought the mentioned reinforcements was interrogated. He was Don Antonio Nieto, a religious man who sympathized with the missionaries, and he recommended Don Francisco de Irisarri y Vivar [1676–1678][598], the only officer on board, as the only man fit for the task at hand. However, this man seemed unwilling to stay in islands that offered such few incentives.

Staying here," Don Antonio told him, "*is not about making a fortune or obtaining those vain honors that men ambition. On the contrary, it is about promoting the glory of God working to solidly*

---

597. Father Gayoso was in the Mariana Islands from 1674 to 1676. He survived this shipwreck and was later sent to the Jesuit province of China (ARSI, "Catalogus Brevis Personarum Provinciae Philippinarum. Anno 1675," Philipp. 4, f. 67r; "Catalogus Brevis Personarum Provinciae Philippinarum Anno 1681," Philipp. 4, f. 77r). See also F. X. Hezel, *From Conquest to Colonization*, p. 89.

598. Navarra-born Irisarri was the first to be designated interim governor of the Mariana Islands. He arrived on Wednesday, June 2, 1676 aboard the galleon *San Antonio de Padua*, accompanied by five Jesuits (four of them Fathers; the fifth, a Brother) and seventy-four soldiers. Although he had experience in war, "he made things worse than they were before," according to Father Luis de Morales, provoking new Chamorro rebellions (Annual Letter of 1676–1677, transcribed in R. Lévesque, *History of Micronesia*, vol. 6, 1995, p. 588–593).

establish religion in these lands of infidels. Moreover, if we are willing
to sacrifice our lives every day serving the princes of the earth, what
should we not be willing to do when an occasion as propitious as this
one presents itself to serve God?

These reasons impressed Irisarri's spirit so much that he accepted the
duty of assisting the Fathers in their apostolic work, but the results obtained
did not correspond with the good intentions of this new governor, for such was
the title given to him. For although he had experience in warfare, the sweetness
of his character and his lack of firmness made things worse than they were
before.[599] The barbarians immediately felt the change, and they began to scoff
at the missionaries and insult the Christians, preventing them from going to
Mass and to the catechisms and instructions offered for them. Their insolence
went so far that they would sometimes beat and abuse them. Thinking that
they should be punished, Irisarri led a few successful expeditions, but the only
consequence was that Aguarín [Agua'lin], one of the most prestigious *chamorris*
(page 138), instigated the barbarians to rise up against the Spanish. This was a
man of sly genius, who had attained credit among his countrymen through his
eloquence and the tone of authority with which he spoke at their assemblies.
He began by encouraging distrust towards the Spaniards, trying especially to
win over those who had taken part in the deaths of the murdered missionaries
by persuading them that the Spaniards would never forgive the crimes that
they had committed, that they were simply waiting for a favorable occasion
to avenge their compatriots; and that the only solution was to prepare to free
themselves from the yoke of tyranny that the Spaniards had imposed on them.

What are you thinking of, compatriots, he would ask them, *that*
you do not see that you will be lost if you continue to suffer the presence
of these cruel enemies that enslave us? They are gaining supporters every
day, reinforced with the help that comes each year from their country.
You already know the means that they use to exterminate us since they
settled among us, for not content with killing our little ones with the
poisoned water that they pour over their heads, they take the children
that we have left with the pretext of instructing and educating them,
and they make them repudiate our customs and distance them from

---

599. On this governor's lack of enthusiasm for the Marianas' mission, see also R. Lévesque, *History of*
Micronesia, vol. 7, 1996, p. 287.

us. Our daughters abandon us to marry Spanish soldiers, and thus we cannot give them to urritaos, which would be more advantageous for us. These foreigners have already killed many of our compatriots under various pretexts, and their intent is to exterminate all of us, and they will do it too if we do not stop them on time. They keep us from pursuing our usual works and diversions on the grounds that we must go to their conference and instructions, (page 139) which they give in order to trick us and seduce us. Why should we suffer such an indignant servitude? We are free; let us, then, conserve this freedom that nature has given us and that all of our ancestors have enjoyed. What would they say if they saw us turned into the slaves of a handful of Europeans that are mighty only because we are scared of them? It would be easy for us to exterminate them, for we are much more numerous than they are. In any case, is a glorious death not preferable to the ignominious lives that we lead?

These reasons impressed spirits that were already prejudiced and ill disposed. The towns of Farisay, Orote [Urotte], Fuña, Sumay, and Agusan following Aguarín's lead, in order to expel the *Guirragos,*[600] which was the name that they gave the Spanish. Since Aguarín knew that all of the previous attempts had so far been unsuccessful, he believed that it was necessary to take all sorts of precautions, and in order to take the Spaniards by surprise, he ordered his supporters to keep this inviolable plan secret.

Let us not frighten the missionaries, he told them, *let them operate freely, so that they become confident and take no precautions against us; this way, we will execute our plan more surely, and get rid once and for all of these enemies of our liberty and peace.*

It is customary in these islands to celebrate the feasts of churches' patron saints with great show. The missionaries come, and all of the parishes come in procession, so that the gatherings are always large. Aguarín chose one of these solemn occasions to carry out his evil intention of killing the missionaries and the other Spaniards. He asked his supporters to go to Tepungan [Te'pungan] on the day of St. Rose [of Viterbo; September 4], patron saint of this parish. And to make sure that his plan would not be ruined by the presence of a large group of Spaniards, (page 140) he also made sure to divide them by calling their atten-

---

600. Juan de Santa Cruz, *Vida y martirio del padre Luis de Medina,* f. 2. See also F. García, S.J., *The Life and Martyrdom of the Venerable Father Diego Luis de San Vitores,* book III, chapter XI, p. 223.

tion to [the village of] Airan. To this effect, he sent some of his men to burn down the church of that parish, which they obediently accomplished, burning down the missionaries' house and the seminaries as well. Part of the garrison stationed in Agadña went to Airan, but it was too late. The barbarians sought to take advantage of their absence from the city and stood before Agadña, but the Spaniards who had remained there seemed so resolute and certain, that they did not dare attack.

Heading a group of the best soldiers, the governor ran to Tepungan [Te'pungan], fearing that the barbarians were going to try something, and he thus foiled Aguarín's plans, for the latter did not dare do anything after the governor arrived. However, his people, who were armed, could not disguise their intentions, and their gestures and movements revealed what was going on inside them, so that, suspecting something, some of them were interrogated, but they said nothing. Not knowing what action to take, for although it was risky to trust them, it was riskier to exasperate them by treating them as traitors and enemies, it was resolved that the most prudent course of action was to pretend that all was well, while being on guard against those people. The only extra precaution taken was that Father Sebastián de Monroy was asked to abandon the town of Orote, because of the suspicions that there might be an attempt on his life there.

The barbarians were disgusted when they learned of this resolution: "*What cause have we given,*" those hypocrites told the missionaries, "*for you to treat us as enemies and punish us as if we were guilty of some crime? We will not suffer the removal of this Father whom we love so much, who teaches us the way to heaven and instructs us in our duties.*" And they insisted so much on retaining Father Monroy that it was decided that it was more convenient (page 141) to leave him there, so as to avoid irritating them, than to give them reason to take up arms. This zealous missionary thus returned to Orote, but to be safe from the violence of the ill-intentioned and traitorous natives, the governor had eight soldiers accompany him, with the orders of keeping him safe and marching together to Agadña the moment they observed the smallest movement on the barbarians' part.

Thus, having retained Father Monroy, the inhabitants of Orote were at peace for some time, but this tranquility was only a trick, for they secretly planned on how to kill him and his companions, and after many deliberations, they resolved to do it while he was saying Mass. They would assault him and his Spaniards all at once and finish all of them off. In order to carry out their infamous attempt, they chose the first Sunday of September, when they would easily conceal themselves in the crowd that went that day to church. However, the plan did not work, because the Father, perhaps by divine inspiration, said Mass earlier than usual, so that the conspirators were late and they did not dare do anything. Angry at having been foiled, they resolved to remove the children from the seminary, and to this effect, they asked the Father to allow them to play at a nearby valley as he sometimes did. As soon as they left the house, the children were ambushed and carried away by a group of barbarians.

The Father deplored this betrayal, and seeing that his complaints were ignored, he decided to leave for Agadña, following the orders that he had been given. A *chamorris* named Cheref whom he believed was a friend, spoke to him under the pretext of returning his children to the seminary, but it was a ploy to entertain him and give the dwellers of the neighboring mountain (page 142) time to join them, so that together, they could attack the Spaniards. Realizing that something was being plotted, the Father set out for Agadña with the eight soldiers that accompanied him to avoid danger. They went to Sumay with the hopes of boarding a boat, but finding no canoe there, they were forced to continue by land. They then realized that a large multitude of barbarians was after them, yelling wildly. Convinced that it was impossible to escape, he exhorted his companions to sacrifice their lives for Jesus Christ, and he prepared them to receive death.

At this point, there were only seven left, for one had been sent ahead to Agadña to give notice of the danger that his companions were in. The barbarians tried to cut ahead of the group in order to surround it, but the soldiers, standing firm, fired upon them, stopping them and producing some casualties. This led them to use treason and deceit to achieve their purpose, as was their custom. Cheref, who was with them, came forth to speak with the Father, but before he reached him, he turned to his men, yelling, "*You do wrong in attacking people who have done us no evil, and there is no reason to come to this extreme, persecuting them with such brutal fury.*" He then went up to the Father, and assuring him that he was trying to save them, showed them a canoe that was ready to be used, and invited him and his companions to board it, offering to accompany them personally to Agadña.

Believing that he was dealing with a friend, the Father did what the savage asked them, but once he and the seven soldiers were inside the canoe, the traitor tipped it, and they all fell in with their muskets, which were rendered useless by the water. The barbarians, who were waiting for this exact moment, furiously assaulted the Spaniards (page 143) and, having dropped his mask, Cheref animated his companions and struck the Spaniards with a musket that he had recovered from the poor unfortunates who were trying to save themselves by swimming to shore. Alone with water up to his neck, Father Monroy exhorted his companions to sacrifice their lives [willingly] for God, for whom they all would have the joy of dying. He was able to fend off some of the barbarians' blows with a cane that he carried and that served as a walking stick in his voyages, until one of them broke his arm and another wounded his neck with his lance. *"Why do you want to kill me?"* the father asked. Without waiting for a reply, he added, *"May God have mercy on your soul!"* He had barely finished pronouncing these words when the infuriated barbarians threw themselves upon him and finished him off with lance blows and punches.[601]

Thus died Father Sebastian de Monroy at the age of 26.[602] He was a religious man of great innocence and an ardent zeal for the glory of God and the salvation of souls. He was already ordained when, through the mission given in Seville during Lent in 1672 by Thyrsus González and Juan Gabriel Guillén, the famed missionaries that we have mentioned before, God inspired in him the [missionary] vocation. Monroy decided to consecrate himself to the mission in the Mariana Islands, and he joined the Society on June 17 of that same year with this purpose in mind. Since the beginning of his novitiate, his detachment from the world and from the calls of flesh and blood was outstanding, for he abstained from having any contact with his parents. His superiors made him write to them before embarking to the Indies. He did, and alarmed, his father traveled to Cadiz immediately, where he begged his son with all of the tenderness and love that a father can muster, not to expose himself to certain death. The novice listened with (page 144) tranquility, and then spoke to him with an inspired tone: *"Oh, my father, you do wrong in opposing the will of God our Lord!"* These words alone penetrated to the bottom of the man's heart, and by a marvelous effect of grace, changed his mind completely. *"Well, since it is God's*

---

601. See also the letter written by Father Coomans to Father Libertus de Pape from Hagåtña, on May 30, 1679 (transcribed in R. Lévesque, *History of Micronesia*, vol. 7, 1996, pp. 47–54); and the Annual Letter of 1678–1679 (transcribed in R. Lévesque, *History of Micronesia*, vol. 7, 1996, pp. 55–78).

602. Father Monroy was not killed for being a missionary only. One of the soldiers that accompanied him had tried to marry a recently baptized girl against her parents' will. They were supposedly angry at her conversion to Christianity. Thus, feelings were running high in the town (Gabriel de Aranda, S.J., *Vida, y gloriosa muerte del padre Sebastián de Monroy. Religioso de la Compañía de Jesús, que murió dilatando la fe alanceado de los bárbaros en las islas Marianas* (Sevilla, 1690). Although we do not know the details, it would not be surprising if the soldier in question had forced himself on this woman, for this was sadly a common practice. In 1690, Father confirmed that various women (married and single), from the districts of Hagåtña and Humåtak were systematically "solicited" by soldiers. Many resisted, and in recognition of their attitude, they were accepted in the congregation of the Santísimo Nombre de María ("Relación del estado y progresos de la misión y cristiandad de las islas Marianas desde mayo de 1690 hasta 1691." AHCJC, *Documentos Manuscritos Historia de las Filipinas* (FILPAS), no. 64, ff. 55v-55r).

will," he said to him, "*go where God calls you, my son. I no longer want to oppose his designs. There is nothing more glorious than to win souls for Jesus Christ, and I assure you that I would consecrate myself to such a holy ministry if there were no indissoluble ties stopping me.*"

Such an admirable and sudden change filled the holy novice with joy, confirming him in his vocation. He departed with his heart full of confidence, and he arrived at the Marianas on June 16, 1674. As soon as he had taken his vows he was sent to Orote, where there was much work to be done. The people of that town were brutal and hard to govern, and what he suffered there in the two years that he remained among them was unimaginable, but God blessed him in such a way that he was able to raise a church and two seminaries, one for girls and one for boys. He put so much effort in his pupils' education that soon they were the best instructed in the entire island. No life was more austere or poorer than his was, in his hut, without a bed, a chair, or any other piece of furniture, where the floor served all of these purposes; his nourishment consisted of insipid roots, and he only drank water, which he had to fetch at a great distance. As if this were not enough, he burdened his body with cilices and mistreated it with frequent disciplines [flagellations]. His companions' amazement at such a mortified life never ceased, and it was enough to judge the suffering that he endured by looking at the marks left on his skin by mosquitos, whose sting was so deep and painful that it was a real torment to be exposed to them. Yet (page 145) he never even tried to brush them off, and suffered them with such patience that he seemed oblivious to the sting of these annoying insects.

Through the practice of these heroic virtues, God was preparing him for the glorious death that indeed took him. The most notable among those who perished with him were Nicolás Rodríguez Carvajal, lieutenant governor of the Marianas; and Juan de los Reyes Carvajal, a valiant soldier and, what was rare among those of his profession, a fervent Christian. This man, after having served the Catholic monarch for ten years in Puerto Rico, resolved to go to these islands to consecrate himself to God's service through this difficult mission, and executed his plan despite the obstacles that his friends and the officers of his corps placed before him.

Juan de los Reyes was a very charitable Filipino, who was as zealous for the conversion of these islanders as any of the most fervent missionaries were. He had had the joy of coming to the Mariana Islands with Father San Vitores, and of having been his usual companion in his apostolic trips and labors. The

Verdadero Retrato del V.P. SEBASTIAN
DE MONROY, de la Compañia de JESVS. Español
y natural de la Villa del Arahal, Arçobispado de Sevi
lla, muerto a lançadas en defensa de la Fè, por los Infie
les, en Orote Pueblo de las Islas Marianas, el dia 6
de Setiembre, año de 1676. su edad 28. años.

{Figure 12: Sebastián de Monroy's martyrdom (1676). Source: Father Sebastián de Monroy (Gabriel de Aranda, SJ, *Vida y gloriosa muerte del Venerable Padre Sebastián de Monroy de la Compañía de Jesús, que murió en las islas Marianas*. Sevilla, Imprenta de Tomás López de Haro, 1690).}

other five who shared these men's fate were from New Spain, and they had rendered great services to this mission. Their names were: Alonso de Aguilar, José López, Antonio Perea, Antonio de Vera and Santiago de Rutía.[603]

Satisfied with having sacrificed these eight victims to their furor, the barbarians celebrated as happily as if they had obtained a great victory, and so that nothing was lacking in their triumph, they returned to Orote and burned the church and the seminaries that Father Monroy had built to educate the young. Aguarín had this supposed victory publicized everywhere, and used it to obtain the support of other towns that also decided to take up arms against the Spanish.

> *What do you fear, my friends?*, he asked them. *We have humili-*
> *ated our* (page 146) *enemies, who are not as formidable as we believe.*
> *We have killed the bravest among them. Those that remain are but a*
> *small number. It is up to us only to get rid of them without combat*
> *and without risking our lives. Nothing is easier than starving them*
> *to death by keeping our compatriots from providing them with food.*
> *Soon we will be left without enemies, and we will be free of their pre-*
> *tensions, of their wanting to introduce their strange and extravagant*
> *customs among us under the pretext of bringing us happiness, when*
> *what they want in reality is to subject us to a shameful slavery. Would*
> *[our ancestors] approve of these laws, which were unknown to them?*
> *We will not approve them, either. Let us defend our liberty; let us*
> *protect it as something that ought to be dearer to us than life itself.*

Aguarín spent the months of September and October going about the land encouraging the villagers with promises, gifts, or threats to join his alliance. Many towns were kind enough to reject his gifts and scoff at his threats. Don Antonio de Ayihi [?–1701],[604] one of the island's most important *chamorris*,

---

603. For more on these soldiers' lives, see Father Monroy's obituary, written by Father Bouwens (RAH, 9/2677, transcribed in R. Lévesque, *History of Micronesia*, vol. 6, 1995, pp. 359–66).

604. In 1685, Father Luis de Morales asked the king to grant Don Antonio de Ayihi and his descendants a medal for his services to the Crown in the Marianas ("Memorial de 1685," transcribed in R. Lévesque, *History of Micronesia*, vol. 8, p. 413). In 1686, the monarch granted this request, sending a medal, "for the oldest son or whoever is chosen, to honor and demonstrate that his services pleased me." The king also ordered Viceroy Don Tomás A. de la Cerda y Aragón to send four silver medals to the Philippines and three to the Marianas, so that their respective governors granted them to those who were worthy of receiving them (AGN, Mexico, Instituciones Coloniales. Gobierno Virreinal. Californias–017. Vol. 26, Exp. 92, f. 251v). For the eulogy written upon his death (on April 15, 1701) by the Jesuits, see "Elogio de don Antonio Ayihi, maestre de campo de los naturales de Marianas. Difunto el 15 de abril de 1701 años" (ARSI, Philippin. Necrologia 1605–1731, Tomo 20, ff. 355r-362r; RAH 9/2678, doc. no. 11; transcribed in R. Lévesque, *History of Micronesia*, vol. 10, 1997, pp. 265–278). For more on Ayihi's life, see Augusto V. de Viana, "Filipino Natives in Seventeenth-Century Marianas: Their Role in the Establishment of the Spanish Mission in the

distinguished himself in these circumstances, for even when the conspiracy was almost universal, he remained faithful and devoted to the Spanish. Since he had acquired great authority among his compatriots, the people in the towns where he had ascendancy did not allow the enemies to pass through their lands. He watched Aguarín's every move, for no matter how able that chief was, he could not keep them concealed from the vigilance of this fervent Christian. Don Antonio would immediately give notice of them to the Spanish, and not content with this important service, he provided them with foodstuffs and exhorted his compatriots to succor and assist them. Aguarín (page 147) regarded him as a traitor, and the most dangerous enemy of his nation. He spoke of him with indignation, and looked for ways to avenge himself upon him, but Don Antonio, who was brave but also prudent, did not fall in the traps set out for him. The Spaniards, interested in conserving such a generous friend, begged him to keep himself from danger: he listened to them, and retreated to the town of Ayran from where he continued serving them with the ardor and loyalty of a true friend.

By his counsel, the walls of the Agadña fort that were in need of repair were reinforced, which disconcerted the barbarians and foiled Aguarín's plans. This chief, who had previously inspected the beach, sent five hundred men to seize some houses that were next to the plaza to facilitate the attack that he was planning, but the men found that the houses had been razed to the ground. They were thus forced to wait for Aguarín's arrival in Agadña. He and his army came into sight on October 15, 1676, and with his troops staying out of fire range, they shouted, as was these barbarians' custom, and fired a volley of rocks, something that these islanders accomplish from afar and with admirable strength and dexterity. The governor waited for them to come closer to fire upon them, but seeing that they were not moving, he placed himself at the head of a group of men and led an incursion. From the very first charge, the entire multitude of barbarians fled and dispersed. Aguarín regrouped them and returned the next day, intending to attack the Spaniards a second time. But the governor, who wanted to surprise them and attract them to the plains, did [not] let himself or his men be seen, and remained behind the palisade without mak-

Islands," *Micronesian: Journal of the Humanities and Social Sciences*, 3:1–2 (2004): 19–26; De Viana, *In the Far Islands: The Role of Natives from the Philippines in the Conquest, Colonization, and Repopulation of the Mariana Islands, 1668–1903*. Manila: University of Santo Tomas Press, 2004; De Viana, "The Pampangos in the Mariana Mission, 1668-1684," *Micronesian: Journal of the Humanities and Social Sciences*, 4(1), p. 16; Alexandre Coello de la Rosa, *Gathering Souls: Jesuit Missions and Missionaries in Oceania (1668-1945)*. Brill Research Perspectives in Jesuit Studies, Leiden: Brill, 2019.

ing a sound, which disconcerted the barbarians who, fearful of falling into some sort of snare, retreated instead (page 148). Aguarín sent the most daring of his men to examine the enemy trenches to make sure, and some of them climbed the palisade and jumped inside the court but finding the sentries asleep, they did not dare do anything [but] pull out some stakes from the palisade and take them to Aguarín. Thereon, the chief advanced with the rest of his men, but by then the Spanish were ready to receive them and, not regarding themselves strong enough to attack, the barbarians retreated. They returned a few days later in greater number and ready to lay a siege to the Spaniards and starve them to death, since they had no means of defeating them by use of force.

There were various combats during the siege: the barbarians suffered great losses, which alarmed them, but what most terrified them was to see that the Spaniards seemed oblivious to their blows, for after three months of siege none of them had died or been wounded, despite the almost continual shower of stones that rained upon the fort. Aguarín decided to try one last effort to take over the fort. He had at his command two armies composed of the bravest people in the islands, one at sea and one at land. He made the land troops advance ahead through the plains, and he had those at sea spread out all across the beach to take on the Spaniards from that flank, while another part hid in the mountains to attack the fort in case the governor led a charge as he was accustomed to do. The plan was well designed, and if Aguarín's orders had been followed, the Spaniards would have been lost. On his part, the governor had taken all of the precautionary measures counseled by prudence. Since he needed to secure his incursions (page 149) through the plains, he had scattered the road that came down from the mountain with poisoned bone points. He had also placed muskets behind the palisade with their ends sticking out, a stratagem that worked out well, for when the barbarians attacked, even though the shots fired back at them were not in fact too many, the sight of the muskets pointing at them struck fear in them and they retreated and fled.

Meanwhile, the governor went out onto the plain, attacking the barbarians who responded by firing at him and his men a hail of stones using their slings. Those at sea fired the most, and wanting to dislodge them from that position, the Spaniards went towards them, but whenever they came close, the barbarians dived under the water, so that the Spaniards were forced to use a piece of field artillery against them.

Aguarín was not discouraged when this first attempt was frustrated, and he repeated the attack eight days later, determined to win or die. The Spaniards, who from the beginning of the siege had implored heaven for succor and placed themselves entirely in God's hands, renewed their invocations, holding public prayer ceremonies and a general communion. On his part, the governor, who was no less pious than he was brave, encouraged everyone with his words and his example.

Since he had been warned ahead of the savages' attack, he ordered that a flag be placed on the seaboard wall within musket fire range, having no doubts that the most daring among the barbarians would strive to seize it. He also had poisoned bone points strewn across the beach, even below the water, and he reinforced that side with soldiers, but ordered them to remain unseen, and not to fire a single shot until he gave the signal.

These orders were executed word for word. On January 24, 1677, an innumerable multitude of barbarians attacked the fort. Thirty canoes advanced towards (page 150) the beach to take the flag. At that moment, the governor gave the signal and the soldiers that were in hiding began firing, as the Spaniards rushed out shouting with all their strength, *Victory, St. Michael, victory!*; and charged at their enemies from all sides. As if a celestial army had fallen upon them, many barbarians were killed and the rest were terrified. They still tried to starve the Spanish out, but seeing that they accomplished nothing, they finally lifted the siege and retreated.

During this siege, which lasted more than six months, the missionaries exhorted the soldiers to fulfill their duties, encouraging them to place their trust in God and to procure the blessings of heaven through their patience and the sanctity of their customs [**end of Father Luis de Morales's manuscript. Henceforth, the text is translated from the Spanish version of Charles Le Gobien's** *Histoire des Isles Mariannes nouvellement converties à la religion chrétienne,* **published as** *Historia de las islas Marianas.* **Estudio y edición de Alexandre Coello de la Rosa. Madrid: Ediciones Polifemo, 2013.**] They had the consolation of seeing that all the soldiers were willing to give their lives for Jesus Christ (page 267) and to sacrifice themselves in the defense of religion. The Marianas women who had married Spaniards showed a sincere love for their husbands, whom they cared for diligently during the siege. Even the children who attended the seminaries did not want to abandon the Fathers for one instant, despite their parents' multiple threats and promises. The Spaniards had managed to build a larger and more comfortable church during the siege, and they dedicated it to the Holy Virgin. In this church, the Office of Holy Week was held with such piety and fervor that the Fathers were reassured in regards to the misfortunes of the past. Peace was restored in the island, and the missionaries carried out their labors and zealously exercised their different tasks, such as baptizing children, assisting the sick, and urging the Christians to fulfill their duties.

Don Juan de Vargas Hurtado, who was on his way to take possession of the Philippine government, arrived at this time at Umatac [1678]. He disembarked[605] to personally examine the port, around which he had formulated some project. He (page 268) befriended the missionaries and assured them of his protection. After having dealt with the affairs related to the conservation and government of these islands he appointed as governor Don Juan Antonio de Salas, in accordance with the king's orders, and left thirty soldiers with the new governor before continuing on to the Philippines. Ardently wanting to distinguish himself against the enemies of the Christian name, the new governor embarked upon a campaign against Aguarín and his supporters, who opposed the advancement of the Gospel with all their might. He left on June 27 and traveled all night, but finding that they would not reach Tarragui [Tarague or Talågi] on time, which was their target, he fell on Apoto [Apotguan], the town where Aguarín lived and where other evil bandits took refuge, fearfully becoming greater in number and more violent each day. Although the natives were ready to defend themselves, having learned of the Spanish expedition in advance, the governor attacked them with such force that he killed some of them, and sacked and burned the town of Apoto, whereupon he returned to Agadña covered in glory. A second incursion against the villages of Tuparao, Fuña, Orote and Sumay, which he burned without losing (page 269) a single man, brought fame to his arms, and procured peace and tranquility to an island where the enemies of religion had perturbed them for three years.

The houses of pandering and prostitution [*i mangguma' uritao*] were a great obstacle for the conversion of the islanders. Urritaos had reestablished them in various places and lived in the most horrible licentiousness ever imagined. The governor resolved to destroy them, and he worked at this with success, while the missionaries applied themselves to recovering the Christians that had been perverted during the revolts, and to regenerating the customs of those who had given themselves over to the most shameful conducts. Don Antonio de Ayihi, the Spaniards' faithful friend, was among the latter. He had left his wife and taken another, scandalizing the Christians. The missionaries had tried several times to break this baleful relationship, but Don Antonio,

605. A note at the margin of Le Gobien's book reads, "on June 18, 1678" next to this line of text.

blinded by passion, had become hardened to their reprimands. They turned to God, and prayed for the conversion of a man who was so useful for the progress of religion. God heard them. Don Antonio (page 270) tore himself away from the person that had obfuscated his judgment, and by the grace of Jesus Christ, he was converted and returned to his first wife. It cost him a great deal to make this sacrifice. For, besides giving up a person to whom his heart was much attached, he drew upon him the scorn and ridicule of his compatriots, which he suffered with a virtue and patience worthy of the first centuries.

Such a renowned conversion had happy results. Many Christians who were in similar relationships renounced them in the light of Don Antonio's example, and led more orderly lives. There was a newfound fervor in all the churches, and religion would have been finally established without obstacles, if some evil natives who were bent on impeding the conversion of their compatriots had been exterminated. The communities of Agofan [Agotguan] and Pigpug [Pipok], the fiercest, proudest, and most unruly of the entire island, protected these men. They were begged up to seven times to expel these conspirators. They excused their refusal under several pretexts. Threatened with the use of force, they scoffed at the threats and got ready to defend themselves. Those of Agofan dug a deep trench with a stone fortification at the entrance (page 271) to their village, and covered the roads with poisoned bone points. These precautions would have been useful if the barbarians had had any courage, but they were invaded by fear the moment they saw the Spaniards, and they fled without opposing the least resistance. They were again offered peace and invited to return home. Nevertheless, this proud people preferred to move to the island of Zarpana rather than give the Spaniards the least satisfaction.

The attack on Pigpug was more difficult. This village was on an almost inaccessible mountain. In order to reach it, one had to pass through gorges that a few men alone could have defended against a whole army. Knowing the Spaniards' intentions, the people of Pigpug had built traps everywhere. The governor would have been unable to avoid them if he had not encountered a Marianas' native who guided him through different roads until they reached the last gorge below the village of Pigpug. This was the most dangerous place. The barbarians had dug deep trenches and had positioned themselves in the highest place to crush the Spaniards who were daring enough to come to (page 272) this gorge. Whether it was because the governor did not see the danger, or because he let himself get carried away by the ardor with which he wanted to punish these rebels, he made his troops go forth. At first, the enemies made no move, so that the Spanish would get deeper into danger. However, when they saw that the Spaniards would no longer be able to retreat, they surrounded them and fired over them an incredible hail of stones and arrows from above with such strength and constancy capable of disconcerting anybody. Peril increased by the moment. Shields broke, and many soldiers were out of combat because of the wounds sustained, when the governor, determined to win or die, placed himself at the head of his troops and started to climb the sharp and difficult rock face with an audacity and intrepidness that frightened the barbarians. His soldiers, encouraged by his example, followed him, reached the heights, expelled the enemies, pursued them, and set fire to the villages of Pigpug [Pipok] and Tala-fofo [Talo'fo'fo], reducing them to ashes.

Don Antonio Ayihi and Don Alonso Soon[g], two of the bravest men in the island, distinguished themselves in this attack. The first had gone before the governor, shielding him from the blows aimed against him; and the second had animated the soldiers and shown them how to avoid the barbarians' arrows. The conspirators regarded Don Alonso as one of their most dangerous enemies, and hated him even more than they hated the Spanish. This *chamorris* had attained such authority over his compatriots, that the moment that they were told, "*Soon[g] has said it; this is what Soon[g] wants*," they obeyed without resistance and without uttering a single word.

These victories succeeded in pacifying the island of Guåhan. All of the villages that were sympathetic towards the Spaniards openly began to take their side and persecute the mutineers. They were quick to capture the evildoers who had murdered the missionaries. A Christian lady that owned the Sydia manor set the example[606]. She had the men who had killed Father Ezquerra and his companions arrested, and placed them in the governor's hands. He deemed it necessary to carry out an exemplary punishment. The instigators of the war of Orote and the murderers of Father Monroy shared the same fate. A (page 274) *chamorris* named Don Ignacio Hineti was named Grand Provost of the island, and he proved his zeal with his valor in acts that were worthy of a Christian hero. He persecuted all evildoers so keenly that they were forced to leave the island of Guåhan and retreat to Zarpana, which was their usual place of refuge.

---

606. The district of Sydia was in the southern part of Guam, where Father Ezquerra was killed. According to Father F. García, "this noble matron was devoted to Christianity and counseled her people, who were already tired of the constant unrest, to purchase peace and friendship from the Spaniards by means of the delinquents' punishment" (F. García, S.J., *The Life and Martyrdom of the Venerable Father Diego Luis de San Vitores*, book III, chapter I, p. 492).

Book Eight

(page 275)

The Mariana Islands' mission flourished thanks to the assistance that it received from Europe, and to the work of Father Manuel de Solórzano,[607] who succeeded Father Bartolomé Besco (or Vesco)[608] as mission superior when the latter died, drained of his vitality by toil and fatigue. Father Solórzano was an educated man of consummate virtue. He had come to the Marianas with the hopes of converting its people and spilling his blood for Jesus Christ. He experienced great suffering in this mission, but he had (page 276) the consolation of doing much good in it, especially after he obtained the support of Don Joseph de Quiroga y Losada,[609] whom God used in these later times to make the glory of His name shine forth in this corner of the world. Here is what I have learned about this man [Quiroga] through trustworthy persons who knew him in Spain, and who lived with him in the Mariana Islands[610], where he is still working with untiring zeal for the advancement of religion.

This officer, who comes from one of the oldest and most illustrious houses of Galicia, left the service after having distinguished himself in the war of Flanders, returning to his country to meditate seriously about the salvation of his soul. He read closely the lives of the Fathers of the desert, and, enchanted by the way that these marvelous men lived, he decided to imitate them and consecrate himself entirely to the service of God. With this intention, he retreated to the hermitage of St. Cecilia, which is a mere two miles from the city of Salamanca, where his relative Don Francisco de Seijas y Losada served as

---

607. Father Manuel de Solórzano was born on December 25, 1649, in Frenegal (Extremadura), where today he is considered one of its most illustrious men. He was the son of Don Cristóbal Ramón Solórzano and Doña Maria Adame y Escobar, who belonged to the local nobility. He studied at the Jesuit school in Fregenal, inaugurated in 1600 by Father Alonso de Paz. He joined the Society on March 21, 1666, in the college of San Luis in Seville. He graduated in humanities (1668-72) and philosophy (1672-75). Once in the Marianas, he was appointed superior of the mission from 1680 to June of 1684. He died on July 23, 1684, victim of a CHamoru attack (ARSI, "Primus Catalogus Anni Personarum Anni 1684." Philippinae Cat. Trien. 1649-1696, Vol. 2-II, f. 425v; Oyola and López Casquete, "Localización de las reliquias del jesuita frexnense…", p. 5). Father Gerardo Bouwens wrote an edifying letter for Father Solórzano on May 20, 1685 (ARSI, Philippin. Necrologia 1605-1731, Vol. 20, ff. 353r-354v.).

608. Father Besco (1614-1680) was mission superior from 1676 until his death in 1680 (ARSI, "Catalogus Personarum Provinciae Philippinaruym Ann. 1680," Philipp. 4, f. 74v). See also F. X. Hezel, *From Conquest to Colonization,* p. 89; F. García, SJ, *The Life and Martyrdom of the Venerable Father Diego Luis de San Vitores,* book III, chapter VIII, p. 325.

609. For more on sergeant major Quiroga, see A. Coello, "Colonialismo y santidad en las islas Marianas: los soldados de Gedeón," pp. 32–34.

610. Notice the first person usage, the first in the entire text. Although the first person perspective allows readers to have a sense that they are getting a direct account of the events from a primary source, Morales did not use it, while LeGobien did.

bishop,[611] before dying some time later as Archbishop of (page 277) Santiago de Compostela.[612] Since Quiroga did not find sufficient solitude in this hermitage, he retreated to the neighboring mountain of Las Velillas, where he built himself a small cell in a rocky crevice. There he dedicated himself to prayer and meditating on the eternal truths, and he only went out to beg for the bits of food that he ate, and to consult his spiritual director. This was Father Thyrsus González de Santalla, the famous missionary and brilliant guide of souls, who is today the General of the Society, as I have stated earlier. In those days, the martyrdom of the Apostle of the Marianas, Father Diego Luis de San Vitores, had just become known in Spain. Father González, who had had a very close relationship with this holy missionary, informed Quiroga of this event. Knowing that the mission needed a man of solid piety and experience in the military arts in order to keep the barbarians in line and to sustain the interests of religion, and (page 278) finding these two qualities in his penitent, Father González thought that Quiroga would render more glory to God if he went to those islands than if he stayed in his solitary retreat, and he proposed this enterprise to him after having entrusted it to our Lord. Quiroga, on his part, consulted God, and he felt so strongly compelled to execute what his Director had proposed, that he had no doubt that this inspiration came from God.

After several interviews with Father González about his new vocation and about the Lord's designs for his person, he renounced the hermit's frock and life, and went to the Mariana Islands, arriving in June of 1679. Tired of living among barbarians who never allowed for true tranquility, the governor, Don [Juan] Antonio de Salas, had been thinking of retiring, which he did within a year. Quiroga was asked to take his place. At first, he humbly declined, but since the missionaries told him that he should accept this job for the glory of God and the welfare of the mission, he accepted until the Court corroborated his appointment.[613]

The new governor's piety and disinterest soon led to a (page 279) transformation of the mission's affairs. Detached from what other men search for

---

611. Don Francisco de Seijas y Losada occupied the bishopric of Valladolid (1664–1668) and Salamanca (1670–1681) until he was appointed Archbishop of Santiago de Compostela (1681–1684). See Vidal Guitarte Izquierdo, *Episcopologio Español (1500–1699): Españoles obispos en España, América, Filipinas y otros países* (Rome: Iglesia Nacional Española, 1994), p. 199, num. 1,245, cited in *The Hierarchy of the Catholic Church: Current and Historical Information about its Bishops and Dioceses.* [Available online] <http://www.catholic-hierarchy.org/bishop/bseijaslo.html> [Consulted March 2, 2024].

612. This is Galicia's main city, famous for the great number of pilgrims that go to it every year to visit the relics of the apostle St. James (note by C. Le Gobien, *Histoire,* book VIII, p. 277).

613. A. Coello, "Colonialismo y santidad en las islas Marianas: los soldados de Gedeón," p. 21.

with great passion, he devoted himself only to that which could contribute to solidly establish[ing] religion in the islands. There were still some unruly natives who disturbed the missionaries and insulted the Christians. Quiroga invited them to live in peace and to join their compatriots. These mutineers became more insolent and daring. Since they were fortified in their mountain peaks and hidden in their caves, they thought that they had nothing to fear from the Spanish. Quiroga soon let it be known that no place was inaccessible to him. He set up camp with his troops in the midst of their mountains, and from there sent incursions to discover the places where the barbarians took refuge. Some of them were captured and then punished as disturbers of the peace, and the rest were so persecuted that, not knowing where to hide anymore, they admitted defeat and sued for peace. They acted in good faith. Soon afterwards, when other conspirators wanted to stir up new trouble, these men handed them over to the Spaniards.

(page 280) Among the mutineers, there was a Christian from Makassar.[614] He had survived the shipwreck of the Concepcion that had taken place in these coasts, as I have already said. This wretch was so perverted by the barbarians and he had adopted their customs so well, that he lived like them, without rules or religion. The missionaries had tried to win him over, but far from submitting to their admonishments, he had taken up arms against the Spanish leading a faction of rebels. When he saw himself in the Spaniards' hands, he recovered his old self, and acknowledging his faults in good faith, he asked God for forgiveness with such manifest sorrow and such an abundance of tears, that Quiroga was moved. He forgave him, and was never given a reason to regret this, for the conversion of this Makassar man was sincere, and his life from then on was irreproachable.

The governor acquired great authority over these islanders because of his good and constant conduct. It was difficult to gather the people for instruction because of the great distance between their homes and the (page 281) multitude of villages that were spread out across the mountains and the most inaccessible places. To remedy this problem, he divided the island of Guåhan in six sectors or *barrios*, three in the north and three in the south, and the following year, he added a seventh, in the island's center, to facilitate communication with the rest. The population from small, spread out hamlets and homes was gathered in a large village in each barrio, and a church was built that could serve to bring the people together. This project met with success. The villages were peopled

---

614. The Kingdom of Makassar is in the island of Celebes [today, Sulawesi], the largest of the Moluccas (Note by C. Le Gobien, *Histoire*, book VIII, p. 280).

more easily than was expected; and the missionaries had the consolation of
seeing the islanders, who had until recently been trying to kill the preachers
of the Gospel and exterminate the Spaniards, raising out of their own free will
temples to the true God, and building houses for their missionaries with more
than admirable fervor.[615]

A hurricane that struck the northern side of the island on November 11,
1680 interrupted this work. Although these islands suffer storms every year,
one as violent as this one was rarely seen.[616] (page 282) Lasting two days, the
hurricane caused a terrible disruption. Almost all of the houses in the island
were in ruins, the canoes were destroyed, and the trees and crops devastated. To
add to this misfortune, the sea, pushed by the impetuous winds, rose so high
that people had to flee to the mountains to not be swallowed by the violence of
the waters that reached the plains and ruined what the winds had spared.

Only God's special protection kept the governor from perishing in this
occasion. He had gone to a neighboring islet to cut wood for the construction
of the church of Umatac. The hurricane nearly caught him there: if he had not
left a moment earlier with his troops, they would have all been engulfed by the
waves that overran this entire island, and even washed part of it away.

This hurricane was very useful for augmenting the population of the
new settlements. Since most houses had been lost, it was easy to convince
the islanders to leave their old dwelling places and move to the areas that
were assigned to them. The work that had been (page 283) interrupted was
renewed, and the churches of [Purísima Concepción of] Pago [Pågu], Inapsan
[Hinapsan], and the rest of the island's barrios were finished. The Spaniards
took charge of the construction of the church at the port of Umatac, where the
ships that came from Spain used to dock. It was dedicated to St. Dyonisius the
Areopagite, in consideration of the Madame Duchess of Aveiro,[617] who had a
special devotion for this saint. This lady, who greatly distinguished herself in
the Spanish court for her piety, her wealth, and her illustrious birth, always
had an ardent zeal for the conversion of infidels. Not content with sending
missionaries to carry the light of the Gospel to particular nations, taking care
of all their expenses, she also extended her liberality to almost all missions,

615. A. Coello, "Colonialismo y santidad en las islas Marianas: los soldados de Gedeón," pp. 21–22.

616. This hurricane destroyed the church at Umatac (A. Coello, "Colonialismo y santidad en las islas Marianas: los soldados de Gedeón," p. 21).

617. Doña María de Guadalupe, Duchess of Aveiro, from Portugal, was the widow of Manuel Ponce de León, Duke of Arcos and Maqueda, from Spain, who had died on November 28, 1693 (Note by C. Le Gobien, *Histoire*, book VIII, p. 283).

and especially to that of the Mariana Islands, which she especially sustained and protected.

Fervor spread among the new Christians. The older folk who had (page 284) until then remained indifferent to the Christian Doctrine applied themselves with the same ardor as the young. Families prayed regularly every morning and every night. Instead of profane or impure songs, spiritual hymns were heard everywhere. People gathered during feast days and Sundays to receive the Holy Mysteries and the instruction that was given to them. Let us see the order that was observed in the church of Agadña, which served as the model for all the other churches. Service was begun with prayer, which was interspersed with spiritual hymns. Mass was celebrated, followed by an explanation of the Christian Doctrine. Absolutely everybody assisted. After dinner, Vespers were sung, and a second instruction was given for those who could not attend the morning explanation. On Wednesdays and Saturdays, the children gathered to recite their catechism[s], and to learn them according to their level. Thursdays were dedicated to the conversion of urritaos, who corrupted the young and encouraged libertinism. Fervent exhortations were held to inspire the fear of God in them and (page 285) to persuade them to abandon the shameful trade that ensnared them. Women gathered in the church on Saturdays to pray the rosary together in honor of the Holy Virgin. Every morning they went to mass, and every afternoon to prayer; the more virtuous of them took communion every month. The most reputable *chamorris* of the nation esteemed and hailed the Christian religion. Their attachment to Christianity grew, and they tried to apply its maxims. Wives were submissive to their husbands; girls were modest and discreet in front of boys, and respectful and obedient to their parents.

These happy beginnings aroused great hopes. After the example of St. Francis Xavier, an officer was appointed in every church to take care of it and act as its syndic and superintendent. His work consisted of watching over the morals of the faithful, instructing the ignorant, baptizing the children in cases of necessity, and assisting the sick and making sure that they received the sacraments. The government for its part established a captain in every village who would govern (page 286) the people and report on the affairs of each barrio.[618] These orders, so wisely given, were very useful in the advancement of religion and the preservation of good customs.

---

618. These captains or *alcaldes mayores* were military overseers that supervised the agricultural production of the district [or *partido*] (M. G. Driver, "Notes and Documents. Quiroga's letter to King Phillip V, 26 May 1720," p. 101).

The number of Christians increased each day, and the missionaries had reason to thank God for the continuous blessings that he showered upon their work. Their only displeasure was seeing that the island of Zarpana was used as a retreat and refuge by evildoers who came to Guåhan every once in a while to pervert their compatriots and foment the spirit of rebellion. The instigators of the recent trouble and the murderers of the missionaries had retired to this island, and did not cease encouraging their partisans to take up arms against the Spaniards.

Don Joseph de Quiroga, determined to put an end to an evil that could cause disturbing breakouts, set out with part of his troops for the island of Zarpana. This commander's arrival disconcerted the inhabitants. Panic beset them, but Quiroga tried to gain their trust by telling them,

> *I do not come here as your enemy, to make war against you* (page 287), *even though you have given us no reason to be happy with your conduct. I come to punish the instigators of the most recent revolt. They have taken refuge here among you and you have admitted them. I order you to hand them to me so that I can give them an exemplary punishment.*

This declaration reassured the islanders. Some men who were sympathetic towards the Spanish were charged with searching across the island for the places where these rascals were hiding, and capturing them. On the very first day, several arrests were made, including that of the famous Aguarín, head of the league against the Spanish and main instigator of the last revolts. During the next few days, many others were captured. They were all tried and executed [1680] with great show in order to strike fear into the barbarians.[619]

The missionaries visited the island, baptized the children that had been born after the war, and reinforced the faith of those who had had the fortune of converting in the midst of so many revolts. More than sixty families who had left Guåhan were taken back, and after ten days in Zarpana, the Spaniards returned to Agadña. The inhabitants of Zarpana, pleased with the governor, wanted to show him their (page 288) gratitude. They looked for the murderers who had remained concealed during his stay, and discovered some of them, including Matapang, who had so cruelly murdered Father San Vitores, the apostle of these islands. Their purpose was to hand him to the governor, but

---

619. Annual Letter of 1680–1681 (R. Lévesque, *History of Micronesia*, vol. 7, 1996, p. 462); P. Murillo Velarde, S.J., *Historia de la provincia de Filipinas*, f. 341v.

this villain defended himself so tenaciously that during his capture, they had been forced to strike him with their lances and he died before they could reach the Spaniards.

This expedition greatly increased Don Joseph de Quiroga's fame and strengthened the peace. It seemed that tranquility would finally be enjoyed, when an unfortunate accident ruined the mission of Inapsan. This was one of the new settlements. Father Gerardo Bouwens was working there with great success, and he was about to celebrate the first mass in the newly finished church, when an arsonist set fire to it in two places. The flames propagated from one extreme to the other so fast, that there was no time to identify him. The fire consumed everything. The house of the missionaries shared the same fate, and Father Bouwens thought that he was going to die there with his companion. After the accident (page 289), whether because they felt guilty, or fearing the punishment that would befall them, the residents of Inapsan were so frightened that they got into their canoes and fled without anybody being able to stop them.

This fire alarmed the missionaries, for they feared that the barbarians were plotting yet another conspiracy. During some time, there was nothing but suspicions and mistrust. Nevertheless, this cloud was finally lifted, and calm was restored. The residents of Inapsan who had retreated to the island of Zarpana were invited to return, and although they promised to return, they later refused to do anything of the sort. The governor, indignant at their obstinacy, went to that island to punish them. Although most of the inhabitants of Zarpana favored the Spanish, those from Inapsan found some who protected them and took up arms in their favor, advancing along the coast to keep the Spaniards from disembarking. The governor, unafraid of this multitude of barbarians, was the first to set foot on land and fire at them, putting them in flight and dominating the shore. This vigorous action caused concern (page 290) among the mutineers, who took to hiding in mountains and caves where pursuit was not judged prudent. Their houses and their canoes were burnt before returning to Guåhan.

The mutineers, freed from their fear of the Spaniards, began making incursions among their neighbors, insulting the Christians, pillaging their lands, burning their homes, and even killing the elderly and those who were incapable of resistance. The governor would not suffer such violence. He quickly returned to [the island of] Zarpana, determined to punish the troublemakers. As soon as these men learned of his arrival, they again fled to the mountains and took cover behind fortifications made of tree trunks and piles of rocks. Since the crags that surrounded them seemed inaccessible, they felt safe. Indeed, they would have been if their enemy had been less brave and daring than Quiroga, for far from feeling discouraged, he decided to attack them from the most difficult side. He set out at nightfall to keep his march hidden from the enemy (page 291) and to fall upon them at daybreak. He crossed terrible precipices and climbed steep cliffs with tremendous difficulty. His labors would have been useless, for he would have been unable to find the barbarians, if he had not fallen by chance upon one of these wretches, who showed him where his companions were hiding. This place was indeed very steep. When the barbarians saw the Spaniards, they let out terrible screams, out of fear of being forced out, or perhaps to animate each other. A volley of arrows followed these howls. The Spaniards responded. They were forced to fight in a gorge, with the rocks thrown by the enemies nearly crushing them, and indeed many of them were seriously wounded. Nevertheless, seeing that they had to win or die, they climbed the rocks and pursued the leaders of the revolt, convinced that if they could get rid of them, the rest would offer no resistance. They were mistaken. The barbarians became even more furious, and fought like people who sought their own deaths. They Spaniards killed the most tenacious of them, and the rest dispersed.

Nothing (page 292) was more pitiable than to see old folks and women with babies in their arms running across horrible precipices, sometimes falling to their death and sometimes letting go of their children to be able to escape more swiftly. Not knowing where to hide, these unfortunate wretches offered to place themselves at the victors' disposal, under the condition that their lives be pardoned. The compassionate Quiroga granted their request and returned to Agadña, where he was received among hails, and where the ship that was expected to come from New Spain arrived soon after.[620]

This vessel brought more troops, together with Don Antonio de Saravia [y Villar],[621] whom the king of Spain had appointed governor general of the Mariana Islands and adjacent lands, with ample powers and absolute independence from the viceroy of Mexico and the Philippine governor. Saravia was a man of merit who had distinguished himself in Sicily and Catalonia as regiment

---

620. Governor Saravia arrived at Guam on June 13, 1681 from Acapulco in the galleon *San Antonio* with his servants Juan Moreno and Antonio Sotera ("Expediente de información y licencia de pasajero a Indias de Antonio Saravia, gobernador y capitán general de las islas Marianas," AGI, Contratación, 5443, N. 1, R. 5, ff. 1-10v).

621. Saravia had served in the Spanish army in Sicily for 30 years. Upon his arrival, Sergeant Major Quiroga left for Manila, where he was appointed superintendent of the Hospital Real (AGI, Filipinas 13, f. 97, c. in M. G. Driver, *Cross, Sword and Silver*, p. 20; M. G. Driver, "Notes and Documents. Quiroga's Letter to King Phillip V, 26 May 1720," p. 99).

commander. He was no less wise than he was virtuous, and if his health had been stronger, no other man would have been more ideal for (page 293) the job. As soon as he took possession of the islands' government, he filled the vacancies in the army and made sure that the soldiers were busy, for idleness easily corrupted them. He had a four-bastion fortress built in Agadña, to be safe from any surprises by the islanders or by insolent foreigners.[622] He appointed governors in the island's main villages, and justices of the peace and police officers to ensure a strict discipline. He named *chamorris* who had demonstrated the most loyalty to the Spanish to these offices, and placed the famous Antonio Ayihi in charge of the whole nation as general lieutenant, to reward him for the services that he had rendered to the state and to the Christian religion in difficult times.

Since the Spaniards' first years in these islands, the people had submitted themselves to their rule, and the islanders who had converted to Christianity had requested the protection of the king of Spain, sending envoys to speak with the governor of the Philippines and the viceroy of Mexico. This protection had been granted, and the king considered them his subjects (page 294). Don Antonio [de Saravia], who was a man of law, considered it necessary to confirm this sovereignty by an authentic act. He conferred with the principal islanders, and convened a general assembly of the whole nation in Agadña for September 8, Feast of the Nativity of the Holy Virgin. All of the island's *chamorris* were present, along with an infinite multitude of people. They consented to swear an oath of allegiance to the king of Spain and recognize him as the lord and sovereign of all these islands. To make the act more solemn, Saravia himself took this oath in the Church of Agadña. Here is the oath's text:

> *We the governors and principal officers of the cities, villages and other places of the Island of San Juan, called Guåhan, capital of the Mariana Islands, being assembled in the Church of the Society of Jesus named Santísimo Nombre de María, promise freely and voluntarily in the presence of the Holy Trinity, Father, Son, and Holy Spirit, and the blessed Virgin Mary, vowing on the Four Gospels, with all the solemnity that we can, to remain faithful subjects to our king and legitimate Lord Charles III, Monarch of Spain[623] and the Indies, and*

622. This was the Fort of Santa Maria de Guadalupe (1683), the first of stone and mortar built in the Marianas, which substituted the wooden palisade built in 1671 (Governor Saravia's letter to Charles II, San Juan [Guåhan], June 6, 1683 (AGI, Filipinas 11). See also Manuel de Solórzano, S.J., "Descripción de las islas Marianas" (1683) (Biblioteca del Palacio Real, II/2866, f. 132r); Annual Letter of 1682–83 [also written by Manuel de Solórzano], transcribed in R. Lévesque, *History of Micronesia*, vol. 8, 1996, p. 20. A strong typhoon destroyed the Fort of Santa María de Guadalupe in 1693 (M. G. Driver, *The Spanish Governors*, p. 12).

623. The original Spanish text said "las Españas," or *Spains*, for this plural form of Spain was that

*to obey his orders the same way that his other vassals and subjects
obey him, submitting ourselves to the just and catholic laws that his
Majesty shall deem necessary for us. In testimony of which those present have signed and sealed with our seal. Done in the Church of the
Santísimo Nombre de María on September 8, Feast of the Nativity of
the Holy Virgin, in the year 1681.*

After the ceremony, the officers left the church amidst the people's
hails. Musket and artillery volley salutes were interrupted only by concerts of
violins and voices. The governor, who was the host of this feast, treated the
chamorris magnificently and was very generous towards the people, leaving
them all very pleased.

From that moment on, the Marianas' natives began adopting Spanish
customs and learning their ways. They were taught to dress properly and make
clothing, to plant Indian wheat,[624] to make bread, and to eat meat. Artisans
(page 296) were sent to various villages to teach them how to spin thread, sew,
make cloth, prepare animal pelts, work iron, carve stone, build in the European
manner, and other trades necessary for life that were totally unknown to them.
The children who were educated in the seminaries became very skilled in all
these arts, and served as teachers for their compatriots. They were taught to
read, write, sing, and play instruments. The young were very helpful, especially
regarding the instruction of the elderly. Each student was assigned four to five
elderly natives to instruct, and the students taught them prayers and catechism,
and even assisted them in death if a missionary was not present. Twelve-and-
thirteen-year-old children were seen holding the crucifix for the moribund and
helping them make acts of contrition and of faith, with the same fervor as that
of the missionaries.

The girls showed signs of an equally great zeal among persons of their
own sex. Their modesty was (page 297) delightful and nothing was more wonderful than to see them guard their innocence in the midst of such a corrupt
nation. An insolent soldier addressed one of these fifteen-or-sixteen-year-old
girls: *"Go away, you rascal,"* she said to him, *"do you not know that God is present?
How dare you offend Him?"* Another Spaniard, carried away by a similar passion,
fell for an older girl, and he tempted her with every means that he could to win
her. The girl never even listened to him or showed him the least pleasantry. The
Spaniard, rejected, disdainfully said to her, *"I do not want anything from you,*

officially used, to imply the variety of kingdoms contained within and ruled by the Spanish Crown.

624. Like most Europeans of his time, C. Le Gobien referred to corn as Indian wheat.

you miserable wench. You are nothing but an Indian who does not even deserve my attention." "*It is true,*" this generous girl replied, "*I am nothing but an Indian, but I have the honor of being a Christian, and I will keep my faith and my innocence at the cost of my own life.*" Some newly arrived Europeans, seeing a group of young women, pursued them with the same criminal intentions as the two men mentioned above. "*Hey! We are Christians!*" the girls said as they fled from them, "*we are Christians! Why do you pursue us, if we would rather die than offend God?*"

(page 298) The customs of the Christian women were no less edifying. Since they had been inspired with the virtues of modesty and shame, and they had been taught to lead a life of retirement, they did so with great innocence. The infidels, enchanted by their conduct and virtue, begged the fathers to make all of their women as modest and dutiful as the Christian women: "*Convert to Christianity with your wives,*" the Fathers told them; "*marry according to the rites of the Church, and your women, by the virtue of the sacrament, will become modest and virtuous.*"

Until then, the islanders had been unreasonable concerning their dead. Stubborn in their chimerical notions, they buried them near their homes, so that the soul of the deceased could visit them more easily and rest in their homes. These burials had been accompanied by lugubrious chants and extravagant ceremonies. They renounced these vain superstitions and, docile to the teachings that they received, they took their dead to be buried in consecrated land several leagues away and with the rites of the Church.

(page 299) Regarding marriage, change was much harder. The custom of these people, as we have said, was to have only one wife, but they were free to leave her for another when they no longer liked the one they had taken. The indissolubility of marriage seemed a terrible yoke to them, and it was one of the laws of Christianity that they found hardest to obey. Women especially, used to dominating and abandoning their husbands when they saw fit, could not subject themselves to a maxim that was so contrary to their natural fickleness, and they considered the law that forced them to live with their first husband until death an absolute tyranny. However, it was necessary for them to submit to this and to renounce the supposed rights of liberty and independence that they enjoyed. The regulations prescribed by the Council of Trent concerning marriage were published throughout the island, and everybody was obligated to conform to them.[625]

---

625. See A. Coello, "Bígamos transoceánicos: reconciliación y abuso y perversión de la "santidad" del matrimonio en las Islas Marianas, siglo XVIII." *Colonial Latin American Historical Review (CLAHR)*, vol. 16,

God blessed the missionaries' toils and cares, for in this time an extraordinary change was produced (page 300) throughout the whole island. The people had become docile, they attended instruction with an ardor that had never been seen before, and, convinced of the sanctity and truth of the religion that they were being taught, they heartily embraced it. The harvest was abundant and the evangelical laborers barely sufficed for its gathering. Churches were built everywhere, seminaries were raised, and the whole island of Guåhan professed the Christian religion, to the consolation of the missionaries. However, they were not satisfied with this. Their thoughts turned to the conversion of the remaining islands. The missionaries had left them due to the [constant] revolts that did not allow them to return. Father Peter Coomans, who had just founded the churches of Pago [Pågu] and Mapupun [or Mapupong], went to the island of Zarpana. He visited the whole island, baptizing various children and instructing a large number of infidels who listened to him meekly. He designated a place where a church was to be built, which was later consecrated to San Francis Borgia, and after regulating all the affairs of this mission and having left three fervent workers there, he went to the northern islands with some officers who had been ordered by the governor to accompany him.

(page 301) These officers reported to Don Antonio de Saravia the state of affairs that they had found in these islands, and it was based on this report that he decided to go and pave the way for the preachers of the Gospel himself. With this intention, he set out with Don Joseph de Quiroga, who had returned from the Philippines where he had gone to deal with an affair related to the mission. Immediately after embarking, a furious tempest surprised them, separating the small fleet. Quiroga resisted for two days and two nights with extraordinary effort, but seeing that he could not resist the rage of the sea much longer, nor the violence of the impetuous winds that pushed him towards the Philippines, he stopped at the island of Zarpana, where the governor had also sought refuge. This accident thwarted the enterprise, which was postponed. Saravia, who wanted to extend the kingdom of Jesus Christ even more zealously than the king's dominions, wanted to follow through his original intentions, but his health was weak, and it could not resist so many fatigues. He fell ill and was taken away four days later: he died on November 3, 1683, with true sentiments of piety, leaving behind (page 302) a mission that was more flourishing than when he had found it. His virtue and his conduct had won him the esteem and affection of the missionaries and the new Christians, who were profoundly saddened by the loss of such an upright man.

Don Damián de Esplana, who had already governed these islands, suc-ceeded him [note at the margins: on August 23, 1683]. He had arrived earlier from Manila in a frigate sent by Don Juan de Vargas Hurtado, the Philip-pine Governor, with considerable reinforcements.[626] Don Damián upheld his predecessor's intentions, and decided to visit the northern islands visited some years earlier by San Vitores and his companions to bring the light of the Gospel to them.[627]

Don Joseph de Quiroga was placed in charge of this important undertak-ing. He left Agadña on March 22, 1684 with twenty canoes and a frigate. A favorable wind took them to the island of Zarpana within a few hours, but sud-denly the winds died, and the small fleet was forced to put into port at Tatgua, where another twenty canoes (page 303) joined them a few days later. Quiroga, stronger with this support, took to sea again and followed such a felicitous route that the next morning Tinian was right before them. Alarm spread through-out the island at the sight of such an unexpected fleet. Women and children fled to the mountains where they hid in the woods. The *chamorris* gathered tumultuously, and seeing themselves incapable of resisting, decided to sue for peace. They advanced towards the Spaniards, but when they saw them ready to disembark, Caiza, who at another time had been the missionaries' friend and protector, entered the water up to his waist, and addressed Quiroga, asking him to forget the past and forgive a people who had been fooled into taking up arms against the Spanish through the trickery of the inhabitants of Saipan.

Quiroga, whose only intention was to establish peace in all of these islands, concluded it on the spot with Caiza. The people of Tinian ratified it by presenting the Spaniards with gifts. They did not stop there, for to show them their faith, they added some canoes to the Spanish fleet, (page 304) which continued on to Saipan without having disembarked in Tinian. It went straight to Catanhitda. This port is in the west, in a deep bay surrounded by woods. The barbarians, who had observed the Spanish route, had gathered, determined to defend themselves to the death. They divided their forces. Some went into the sea to meet the fleet in combat. Others waited on the shore to keep the Spaniards from disembarking. The ardor with which they wanted to fight the Spanish was

626. A. Coello, "Colonialismo y santidad en las islas Marianas: los soldados de Gedeón," p. 25.

627. L. de Morales, S.J., "Relaciones del estado y progresos de la misión de las islas Marianas desde junio de 1681 hasta el 25 de abril de 1684" (RAH, Fondos Jesuitas, Tomo 19, Signatura: 9-3593/26, f. 3v). There also copies of this text in AHCJC, FILPAS, 52, f. 362v and in the Newberry Library (Chicago), Ayer Collection, Ms. 1391.

so great, that instead of waiting for them to reach land, they went out into the waist-deep water, screaming terrible yells and firing arrows and stones.

Quiroga, who animated his people more with his example than with words, jumped into a canoe and made his way between enemy vessels, reaching the shore with his pistol in his hands and killing one of the main *chamorris* who had come forth to strike him with his javelin. Terrified by the death of one of their chiefs and by the blows that came from everywhere (page 305), the barbarians were discouraged, and, unable to resist the Spanish attack, fled and dispersed in a moment. Those at sea remained in combat longer. Their canoes had surrounded the frigate, and they had attacked it rapidly with the intention of taking it. However, they found so much resistance that they could not carry out their plan. A unit of Marianas rowers armed with pikes and scimitars distinguished itself in this battle with valiant acts that greatly contributed to the journey's success.

The following days were spent going through the entire island and engaging in skirmishes with the dismayed barbarians. Nevertheless, the Spanish could reduce them only after burning down several villages, including Araiao, hometown of the island's main *chamorris*.[628] This man, who called himself the Great Father to claim authority over his compatriots, wanted to spare himself the disgust of seeing the Spaniards take possession of his country. The moment that they appeared, he abandoned the village and retreated to the islands further (page 306) north, to avoid the grief of seeing what would become of his father-land [*patria*]. The Saipanese made a few more efforts, but they finally admitted defeat and made peace with the Spaniards thanks to the inhabitants of Opian, who were the first to do so and set the example. Since this is this archipelago's largest and most populated island after Guåhan, Quiroga decided to fix his residence there, building a fort according to the plan that the governor had devised to conserve these islands.

This commander, pleased with this brief and successful expedition, sent part of his troops to conquer the nine northernmost Mariana Islands, called Gani islands.[629] These islands submitted without difficulty to the officer in charge, and only one opposed some resistance. Finding the inhabitants willing to profit from his instruction, Father Coomans announced the kingdom of heaven (page 307) to them and baptized many who had already been taught

---

628. His name was *Radabao* (Note by C. Le Gobien, *Histoire*, book VIII, p. 305).

629. Spaniards call them Volcano Islands, because there are some volcanoes in these islands. There are two in the island of Pag[a]n and one in Asonson. A volcano is a pit full of fire (Note by C. Le Gobien, *Histoire*, book VIII, p. 306).

by Father Tomás Cardeñoso, who had remained in these islands for two years.

After this conquest, the only desire was to solidly establish religion across the archipelago. The missionaries worked with a success that was proportional to their zeal. Churches were built everywhere where the faithful could gather, seminaries were built to teach the young; and the discovery of the southern islands was being planned so as to bring the Gospel to them, when a strange revolution frustrated all of the projects that were being entertained, wrecked the new church, and reduced the Spaniards to a most dire situation. Let us see how it all happened.

The *chamorris* of the island of Guåhan had embraced the Christian religion and submitted themselves to the Spanish, but some had only done this for their own interests, all the while considering the duties of Christian life an unbearable yoke and sighing for their lost liberties without daring to voice their true feelings. The failure of past revolts, the punishment meted to their instigators, particularly the sad lot that befell Aguarín (page 308), chief of the last revolt, made them timid and circumspect. Thus, they were content with lamenting themselves in secret and deploring the loss of their freedom, as they patiently awaited a favorable occasion to unleash their hatred and resentment. This was the state of affairs when they learned of the Spanish attempts to dominate the northern islands. Their disgust and restlessness increased, for they regarded those islands as a refuge where they could retreat to if the Spaniards pushed them to that extreme. They deliberated for a long time on what course of action to take under these vexing circumstances, and finally decided to take up arms and carry out one last effort to recover their liberty.

Don Antonio Yura of the village of Apurguan [or Apotguan] was the leader of the revolt. For a while, he had been confiding in his closest associates, and had stirred in them the spirit of hatred and acrimony towards the Spanish. He believed that the time had come to put their thoughts into action.

> *It is useless, my friends,* he told his supporters, *for us to lament the loss of our freedom and to whimper under the sad yoke that these foreigners have placed on us. We are forever lost* (page 309) *if we allow them to dominate all of our neighbors. For, where will we find a place of refuge? The Spaniards are planning to conquer the islands of the North. Their troops have split and the better part of them has already set out. They have left only the invalids in their fortress. It is easy to surprise them, and thus get rid of them all. The opportunity is favorable to us. Follow me, my friends, let us acquire immortal glory*

and become the liberators of our country.

Yura's speech had the effect that he sought: animated, his supporters managed to include in the revolt some *chamorris* from Ritidian [Litekyan] and Pago [Pågu]. They swore secrecy and inviolable loyalty, and began preparations to carry out their plan. Some wanted to win more people over to their cause, but fearing that their intent would be discovered and that the Spanish would be forewarned, Yura wanted to carry out the undertaking as soon as possible. He called the main conspirators to a meeting on the night of July 22 to 23 of 1684. Sixty men went.

> *My friends,* he told them, *there is no time to lose. If we delay the execution of our plan* (page 310), *we will be infallibly discovered. Tomorrow is a favorable day. We must win or die. We have nothing to fear but disorder and confusion. To avoid this, I am of the opinion that we should each have specific tasks: choose what you think is best. On my part, I am determined to kill the governor, with the help of two or three friends. If each of us does his duty, victory will be ours.*

The assembled party applauded Yura's project and each assumed a task. Some were to kill the sentinels while others were to seize the fortress and kill the soldiers. Some were charged with killing the missionaries and occupying their house. The conspirators armed themselves with swords, daggers and large knives, and set out the next day for Agadña under the pretext of attending Sunday Mass.[630] When Mass was over, they left the Church. When Yura saw Governor Don Damián de Esplana walking in the plaza unarmed and defenseless, he went to attack him with three of his friends, striking him with various dagger blows. The rest of the conspirators spread out everywhere. Some killed the sentries (page 311) while others entered neighboring homes or ran across the plaza killing forty or fifty soldiers and wounding many others. Finally, others entered the missionaries' house yelling with all their might that the governor was dead and that nobody would be left alive. With their daggers, they first killed Father Manuel de Solórzano, a man of ardent zeal and distinguished merit, and Brother Balthasar du Bois,[631] who had excellent qualities. They then

---

630. The letter that Vice-provincial Gerardo Bouwens wrote to the King on May 15, 1685, specifies that Governor Esplana was accompanied by 40 soldiers (AHCJC, FILPAS, 52, f. 371r). See also R. Lévesque, *History of Micronesia*, vol. 8, 1996, p. 191.

631. Balthasar du Bois was born in Belgium on March 15, 1654. He joined the Society of Jesus on January 29, 1675, and arrived at the Marianas on June 1, 1680. He took care of domestic chores in the mission until his death on July 23, 1684. (ARSI, "Primus Catalogus Anni Personarum Anni 1684." Philippinae Cat.

gravely wounded Father Diego Zarzosa[632] and Father Gerardo Bouwens, then Vice-provincial and Superior of the mission; and they would have killed him if Brother Antonio de los Reyes had not come to his aid and seized him from the hands of the barbarians.[633]

The worst treatment was received by Father Pedro Pavón,[634] a holy old man from the Philippines who was eighty years old and had been consecrated to the service of this mission for many years.[635] The barbarians struck them so many times with the points of (page 312) their weapons that not a single part of their bodies was left unscathed. The blood that flowed everywhere seemed to animate these furious men, when a servant of the governor came to the door screaming with all his might that the governor was not dead and that he required his confessor. This was Father Diego Zarzosa, who, despite being gravely wounded himself, went to hear the governor's confession. These yells scared the barbarians, but when they learned of Yura's death, their disconcertment was complete. This wretch, who had encouraged the others to sedition, was killed at the very beginning of the bloody affair. He had just wounded the governor when two soldiers ran, alarmed by the noise, and jumping on him left him dead at their feet.[636]

In this horrible confusion, God made his power shine forth in a miraculous way. Father Antonio Cerezo[637] was saying mass when he learned of what

---

Trien. 1649-1696, Vol. 2-II, f. 429r). On May 20, 1685, Father Gerardo Bouwens wrote Baltasar du Bois' edifying letter (ARSI, Philippin. Necrologia 1605-1731, Tomo 20, ff. 340r-340v).

632. Father Diego de Zarzosa was born in Antequera (Malaga, Andalusia) on August 15, 1648. He joined the Society on November 2, 1662. He was a graduate at the time of his death, which occured five years after he first set foot in the Marianas (ARSI, "Primus Catalogus Anni Personarum Anni 1684." Philippinae Cat. Trien. 1649-1696, Vol. 2-II, f. 425v). See also Hezel, SJ. *From Conquest to Colonization*, p. 89.

633. For more on the so-called Second Great [CHamoru] War (1683-1685), see A. Coello, "Colonialismo y santidad en las islas Marianas: los soldados de Gedeón," pp. 25-37.

634. Brother Pedro Pavón was born in 1655 in Pasarón de la Vera (Plasencia, Extremadura). He arrived in the Marianas on June 1, 1680, and was in charge of the mission's domestic chores until his death on March 24, 1686 (ARSI, "Primus Catalogus Anni Personarum Anni 1684." Philippinae Cat. Trien. 1649-1696, Tomo 2-II, f. 429r). Father Paul V. Daly's 1949 translation of Morales/Le Gobien's *History* obviates the existence of Father Pedro Pavón, and seems to confound him with oblate Felipe Sansón, for it says, "Only Brother Peter Panon (his name was Phillip Sanson) was left at their mercy. He was a holy man of 84 years old, who for a long time had dedicated himself to this mission. The barbarians struck him and Father Bouwens with the points and edges of their weapons repeatedly. [...]" (book XVIII, p. 18).

635. He is referring to Felipe Sonsón (¿-1686) (Note by C. Le Gobien, *Histoire*, Libro VIII, p. 311), who was in the Marianas since June 1, 1680. Although he was an oblate, the missionaries regarded Felipe Sonsón as a fellow Jesuit. This is confirmed in the "edifying letter" that refers to him as Brother (RAH 9/2677, transcribed in R. Lévesque, *History of Micronesia*, vol. 8, 1996, pp. 555–573).

636. M. Cuculino, S.J., *Relación desde junio de 1684 a mayo de 1685* (AHCJC, FILCAR, E.I, a-18/1. ff. 33–36). See also L. Ibáñez y García, *Historia de las Islas Marianas*, p. 60; Carano & Sánchez, *A Complete History of Guam*, p. 81; P. C. Sánchez, *Guahan / Guam*, p. 45; and G. Barratt, *The Chamorros of the Mariana Islands*, p. 154.

637. His given name was Augustine Kersehbauer, but he was known as Antonio (or Antón) Cerezo. Born

was happening. However, thinking that the Sacrifice should not be interrupted, he continued, immutable. A soldier came at the end of the Mass to take communion, and the Father gave it to him. Immediately, a group of barbarians that was (page 313) chasing the Spaniards entered the church. The Father, who was holding the holy Chalice in his hands, advanced towards them with a security that terrorized them. The angriest among them fled as if an invisible hand had slapped them, and the others turned away with an amazement noticed by everybody. The Father remained in the church, where he heard the confession of another soldier who was gravely wounded, and gave him communion. At that moment, a Spaniard erupted into the church hoping to save himself from the furor of the barbarians that pursued him. The Father, who was still holding the Chalice in his hands, went to stand before these furious men as he had done the first time. Upon seeing him, the barbarians stopped and remained motionless, like statues, for a considerable time. They then retreated in a confusion and terror that was not at all natural.[638]

When the tumult ceased, Father Cerezo went to the house, where he found most of the missionaries gravely wounded, Brother Dubois [du Bois] dead, and Father de Solórzano dying. This was a man whom God had (page 314) blessed with extraordinary graces since his youth. He had wanted to join the Society of Jesus, but his parents could not allow themselves to lose a son whom they considered the mainstay of their Andalusian family; with no regard for his entreaties and his tears, they constantly opposed his purpose. Tired of this resistance, Solórzano fled his parents' home and presented himself to the Jesuit novitiate. Nevertheless, upon learning that he had left his parents' house without their consent, the priests denied him entrance. The young man was not discouraged, and he remained steadfast in his resolution, with such firmness and such valor, that his father, who was a man of God, was moved by his son's perseverance, and finally granted what he had so long desired.[639]

---

on March 3, 1643, he joined the Society on September 20, 1661. He was a graduate and had professed the four vows (15/8/1679) (ARSI, "Primus Catalogus Anni Personarum Anni 1684." Philippinae Cat. Trien. 1649–1696, Tomo 2-II, f. 423v). In 1688, he was sent to the College of Manila (Provincial Joseph Sánchez's letter to Procurator Antonio Jaramillo, Manila, April 29, 1690 (RAH, Fondo Cortes, 567, Legajo 9/2669/20, f. 1v); ARSI, "Catalogus Brevis Personarum Provinciae Philippinarum Anno 1689," Philipp. 4, f. 86r).

638. AHN, Sección Clero/Jesuitas. Legajo 93, Expediente 2, f. 2v. See also M. Cuculino, *Relación* (1685), in AHCJC, FILCAR, E.I, a-18/1, f. 37; Annual Letter of 1684-85, written by Father Matías Cuculino [Matthias Kuklein], S.J. (transcribed in R. Lévesque, *History of Micronesia*, vol. 8, 1996, p. 245). On the idea of the presence of the divine, or immanence, in physical representations of it, see William B. Taylor, "Images and Immanence in Colonial Mexico," in *Shrines and Miraculous Images: Religious Life in Mexico Before la Reforma*, Albuquerque, NM: University of New Mexico Press, 2010, pp. 15–62).

639. This comment must be understood in the context of Father Morales's critique of Andalusian provincials, some of whom prevented their best men from departing as missionaries to the Indies.

Since he entered the house of the Lord, he made great progress on his path to God through the most solid practice of virtues, and he advanced so much in such a short time, that he was considered a saint. He showed great abilities in the sciences thanks to his dedication and the vivacity of his spirit, and his superiors (page 315) had already destined him to his first employments in the province when God called him to the apostolic life.

Moved by the martyrdom of Father San Vitores and by the admirable conversions that this new apostle had carried out in the Marianas, Solórzano felt urged to consecrate himself to the service of this difficult mission. He opened his heart to his superior regarding his vocation, but since they had other plans for him, he did not find them well disposed to support his desire. He turned to God, and redoubled his prayers and his penances, and he addressed Father Giovanni Paolo Oliva, then the General of the Society.[640] The General listened to him and, convinced that his vocation came from God, granted him permission to go to the Marianas. Overjoyed that his desires were to be satisfied, he immediately departed for Cadiz. Let us see what he wrote at that time to a friend of his who was in Cordoba. His letter is dated July 6, 1675:

> *I have finally been granted what I have long desired. I am so* (page 316) *rejoiced by this that I cannot express myself. I ask you for your blessing, dear Father* [General *Giovanni Paolo Oliva*], *and I beg you to obtain from God the grace of using the mercy that he has granted me without my merit for his greater glory. Although there are many missionaries in the fleet, only three priests and a novice are destined to the Mariana Islands.[641] The rest are going to Mexico or California.[642] Many envy our lot, for the harvest* (page 317) *is abundant, and there is much work in our mission. The islands that I*

---

640. He died in Rome on November 26, 1681, at the age of 82 (Note by C. Le Gobien, *Histoire*, book VIII, p. 315). Born in 1600 and a member of the Society since 1616, Giovanni Paolo Oliva [or Juan Pablo Oliva, in Spanish] was the eleventh superior general of the Society of Jesus, from 1664 to 1681. During his generalship, relations with other orders improved, and more overseas missions were founded, particularly in East Asia.

641. These were Fathers Manuel de Solórzano, Bartolomé Besco, Antonio Jaramillo and Miguel Rubio (Francis X. Hezel, S.J., *From Conquest to Colonization*, p. 89).

642. This is one of the largest islands of the world, extending from 24° southern latitude, to c. 46, along the coast of New Andalusia and New Mexico. It is only separated by a sea arm that the Spanish call Mar Bermejo, or the Red Sea of Cortés, because this captain was the first to discover the island of California. Only the coasts of this great island are known today. Some believe that it is in fact a peninsula joined to the continent in its northern reach. The Spaniards have taken diverse measures to settle it and to bring the faith to its peoples, but they have not succeeded. Fathers Juan Maria de Salvatierra and Francisco Maria Picolo, renowned missionaries of the Society of Jesus, founded a new mission there two years ago, as we know from a letter written by Father Salvatierra from California on November 27, 1697 (Note by C. Le Gobien, *Histoire*, book VIII, p. 316).

*am going to are not far from Japan and China. How happy we would
be, my dear Father, if we could find a way to enter this first kingdom,
closed to the Gospel for such a long time! May God grant me the grace
to give my life to such a glorious enterprise! Ask it of Our Lord for me.*

Father Solórzano arrived happily at New Spain. There, his merit and his
virtue were soon renowned, and the Archbishop of Mexico, who was then acting Viceroy, showed him signs of esteem and distinction, and let him go most
unwillingly, for he was loath to lose a man capable of moving men's hearts and
minds.[643] This talent was useful, for this zealous missionary on the voyage from
Acapulco to the Marianas. He had embarked along with a group of villains who
were being sent to the Philippines to establish new settlements there.[644] Appalled
(page 318) by seeing themselves exiled to a faraway land, these wretched men
decided to rid themselves of the missionaries and kill the captain and his officers
in order to seize the ship and traverse the seas as filibusters.[645] They were about
to carry out this detestable enterprise, when Father Solórzano, who had gained
the trust of one of its accomplices, learned of it: he prevented them from carrying it out, and arrived at the Marianas without incident. Since workers were
sorely needed, he barely had time to rest. He was employed with great success
in the conversion of the barbarians. After four years of continuous work, which
God showered with blessings, he was appointed Vice-provincial and Superior of
the mission. He then dedicated himself with all the ardor of his zeal to advancing (page 319) God's work. He had churches built in every barrio of the island
of Guåhan, founding seminaries to teach the young, sending missionaries to
Zarpana, and causing the mission to flourish more than it ever had before. He
was assisted greatly by Father Antonio Maria Jaramillo, who visited these islands
during this time and contributed to establish order in them.[646]

Father Solórzano joined an extraordinary mortification and an almost

---

643. This was Don Payo Enríquez de Rivera Manrique, OSA [Order of St. Augustine], Archbishop of
Mexico (1668–1681) and Viceroy of New Spain (1673–1680).

644. Spaniards in Mexico who did not deserve to be condemned to the gallows were sent to the Philippines to establish colonies (Note by C. Le Gobien, *Histoire*, book VIII, p. 317).

645. Filibusters are adventurers from various nations, who for the last forty or fifty years have settled
in the islands of Tortuga and Santo Domingo, in America, from where they traverse the seas and wage a
permanent war against the Spanish, whom they torment with their persecutions. The name *filibustero* comes
from the English word filibuster, which means privateer, because the first of these adventurers were English
(Note by C. Le Gobien, *Histoire*, book VIII, p. 318). According to the Oxford English Dictionary, the root
of the English word "filibuster" is "freebooter," which comes from the Danish term *vrijbuiter*: vrij ("free") +
buit ("booty") + -er ("agent").

646. This father, who has worked for so long in the conversion of these barbarians, is today distinguished
in Spain by his virtue and merit (Note by C. Le Gobien, *Histoire*, book VIII, p. 319).

continual union with our Lord to ardent zeal. He had often begged for the grace of spilling his blood for His love. God seemed to grant his desire when the conspirators of Apurguan, resolved to exterminate religion and expel the Spaniards, erupted full of fury into the house of the missionaries in Agadña, as we have already described. After having wounded Father Bouwens, they attacked Father Solórzano, striking him several blows with their knives. They cut a finger off, and then his hand, and (page 320) stabbed him in the head four or five times. The holy man fell half dead and covered in blood. A Marianas' native who served the missionaries, and had particular obligations to Father Solórzano, went to him. It seemed that he would assist him, but according to his compatriots, this traitor sunk his knife into the holy man's throat. The Father let out a great sigh and, elevating his eyes to heaven, begged Our Lord to forgive this wretch. He then died with sentiments of love full of tenderness and gratitude towards God.[647]

Brother Balthasar du Bois had died a few hours earlier. He had been admitted in the Society at Tournay [Tournai, Belgium] at the age of twenty-one. He had sanctified himself at an early age by fulfilling his exercises of piety with great dedication and fidelity. His love for suffering led him to consecrate himself to the Marianas mission. He spent five years of continuous labors there, and had the consolation of working in the construction of several churches, doing all of the carpentry himself. He was one of the first victims [of the above-mentioned attack] immolated by the barbarians in their furor. He was struck with such (page 321) rage, that his skull was crushed. He died on July 23, 1684, after having confessed and received the extreme unction.

---

647. On June 23, 1684, the natives killed Father Solórzano, wounding him on the head and nearly severing it from his body; they also cut off his right hand. Soon after, on June 9, 1685, in Manila, Father Zarzosa certified the authenticity of Father Solórzano's skull, with the authorization of Vice-provincial Bouwens and Father Pedro Oriol, rector of the College of San Ignacio in Manila. This means that the martyr's body was exhumed to obtain this relic. The skulls of the dead held a special significance for the Marianas' natives, but Catholics also regarded them as powerful relics (R. Lévesque, *History of Micronesia*, vol. 9, 1997, pp. 53–54). See also ("Carta dirigida al padre Francisco García por un jesuita [Bouwens] de la misión en Marianas con noticias sobre la reducción de las islas Marianas durante los gobiernos de don Damián de Esplana y Joseph de Quiroga, y en particular de la rebelión de julio de 1684," Guåhan, May 20, 1685 (RAH, Fondos Jesuitas, Tomo 57, Signatura: 9-3631/43, 3v); "Relación del gobernador de Filipinas, don Gabriel de Curuzeálegui y Arriola, al rey Carlos II del alzamiento que hicieron los indios de las islas Marianas el año de 1684" (AGI, Filipinas 12, Ramo 18, f. 1v); and P. Murillo Velarde, *Historia de la provincia de Filipinas,* ff. 364v-365r. Afterwards, the skull and a collection of private letters went to the missionary's nephew, Don Juan Casquete de Prado Solórzano, and the relics were in his family's custody for two more generations until 1984, when Josefa Jaraquemada Tous de Monsalve deposited them in the Jesuit School of Villafranca de los Barros, where it remains to this day (José Quintero Carrasco, *Historia de Fregenal de la Sierra* (third edition), [1983] 1996, pp. 325-27; M. López Casquete and A. Oyola Fabián, "Localización de las reliquias del jesuita frexnense Manuel Solórzano y Escobar (1649-1648), evangelizador de las islas Marianas", in F. Lorenzana de la Puente (eds.), *España, El Atlántico y el Pacífico: Y otros estudios sobre Extremadura,* 95-108. Llerena: Sociedad Extremeña de Historia, 2014, pp. 95-108).

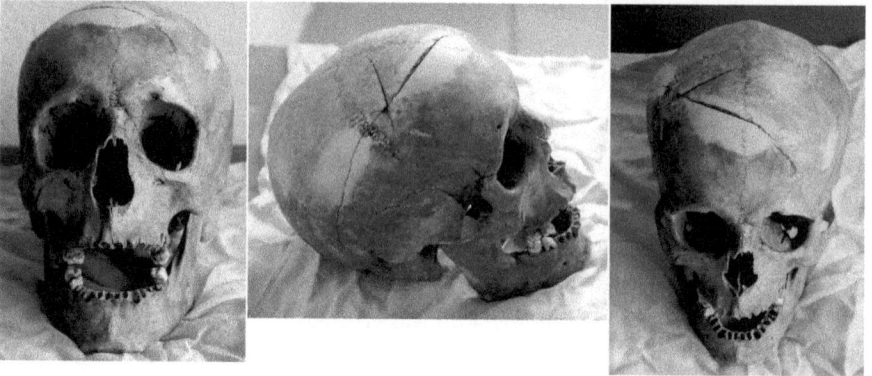

{Figures 13, 14, 15, 16 and 17: Relics of Manuel de Solórzano}

BOOK NINE

(page 322)

When the Spaniards recovered from the surprise and the consternation in which Yura's conspiracy had thrown them, they retreated into the fortress with the missionaries and they prepared for a vigorous defense. As soon as they had entered, the sentinels gave the alarm that barbarians were coming down the neighboring mountain. They ran to their arms and were ready to repulse them, but to everybody's surprise, it was the valiant Sergeant Major Don Ignacio Hineti who came with his people to succor the (page 323) Spanish.[648] The rebels had tried to win his support and have him take their side, but this fervent Christian, far from listening to them, had declared that he would pursue them relentlessly and become their greatest enemy. He kept his word, for during the duration of this war he alone caused them more harm than all the Spaniards together.

The conspirators, seeing that their purpose had failed due to their leader's death, did not give up. They convinced the people of Guåhan and the neighboring islands to take up arms against the Spaniards. They spread the news that they had killed the governor and murdered the garrison [and] that it was necessary to recover their liberty and free themselves from foreign tyranny forever. A *chamorris* from Ritidian [Litekyan] placed himself at the head of the rebels. He was full of resentment because Father Teófilo de Angelis[649] had kept him from giving his daughter to those perverted young men who lived in scandalous disorder and libertinage. Indignant at the priest's action, this *chamorris* tried to convince his compatriots to murder Father Angelis. The best among them felt esteem and affection for this holy man, and they were horri-

---

648. AHN, Sección Clero/Jesuitas. Legajo 93, Expediente 2, f. 2v. See also the letter written by Father Bouwens to the King on May 15, 1685 (AHCJC, FILPAS, 52, f. 372v-373r); Cuculino, *Relación* (1685), in AHCJC, FILCAR, E.I, a-18/1, f. 38. For more on the "faithful Indian" Don Ignacio de Hineti, see A. Coello, "Colonialismo y santidad en las islas Marianas: los soldados de Gedeón," pp. 17–44.

649. Teófilo de Angelis was born in Siena (Tuscany, Italy) on January 15, 1652. He joined the Society of Jesus on November 6, 1673. At the time of his death, in 1684, he was a graduate (ARSI, "Primus Catalogus Anni Personarum Anni 1684." Philippinae Cat. Trien. 1649-1696, Vol. 2-II, f. 427v). On December 3, 1677 and January 22, 1678, Angelis wrote two letters to the General of the Society from his school in Naples, asking him to send him to any "uncultivated vineyard" in the Indies, manifesting how strongly he felt the missionary "call" (ARSI, "Epistolae Indipetae," Vol. 748 (Italia), no. 203; no. 219). On May 20, 1685, Father Gerardo Bouwens wrote Brother Teófilo's edifying letter (ARSI, Philippin. Necrologia 1605-1731, Vol. 20, ff. 341r-342v).

fied at such a detestable project. Instead of (page 324) sharing the *chamorris'* exaltation, they advised the Father to leave for the island of Zarpana to escape the rebels' fury and to inform Don Joseph de Quiroga of the revolt that was taking place in Guåhan. The *chamorris* found out, and he had the priest followed by two urritaos. This missionary had already embarked and was waiting for the immediate return of his pilot, who had absented himself, when these two villains jumped into his canoe, tied a rope around his neck, and hung him from the vessel's mast. Since the man of God did not expire quickly enough to suit them, they cut the rope, threw the Father to the floor, beat him to death, stripped his body, and threw it overboard.

Such was the death of Father Angelis, whom God had seemed to call to the Society of Jesus only to make a martyr of him. He came from a distinguished Tuscan family. He had embraced the ecclesiastical state with the intention of rising to the most eminent dignities of the Church, and he was already a Subdeacon when God opened his eyes to the dangers to which he was exposing his soul. Docile to the inner voice that urged him to leave (page 325) the world and join the Jesuits, he took the precautions that were necessary to carry out an undertaking that thwarted the grand plans that his family had for him. He left for Rome and from there went to Naples, where he entered the novitiate. It was in this house that God communicated to him an abundance of extraordinary graces and inspired in him the desire for martyrdom. This fervent novice decided at that moment that he would go to the Indies and dedicate himself to the conversion of infidels. His feeble health and almost continuous illnesses seemed an invincible obstacle for his desires; but he always trusted that God, who inspired this sentiment, would give him the means and the strength to carry out such a great enterprise. He was so certain, that one day, after vomiting blood and suffering such a terrible oppression in his chest that he could barely breathe, he told those who pitied him, "*You see me without strength, my Brothers, in a state that inspires your compassion, but you will be surprised when I go to the Indies and you hear it said that I have been martyrized.*" He had not finished his studies when he asked permission to (page 326) dedicate himself to this unfortunate mission. His requests were so insistent that, despite the weakness of his health, his superiors felt in their consciences that they were obligated to send him to the missions.

After having had the honor of kissing Pope Innocent XI[650] and having received his blessing, he went to Genoa, where he embarked. During the trip,

---

650. Born Benedetto Odescalchi in Lombardy (Italy) on May 16, 1611, Cardinal Odescalchi was Pope from September 21, 1676 to his death in August of 1689.

his only thought was that of martyrdom; and as if he had had some premonition on the way in which his sacrifice was to be consummated, he had written the following letter to a friend of his on June 3, 1678:

> *I am in good health thanks to our Lord, and I hope to embark next week for Cadiz with my companions, who are many. For, besides the nine Fathers from Bohemia who are already here, we are waiting for the arrival of six others from Austria and one from Milan. I beg you, my Reverend Father, to recommend me to Our Lord's sacred wounds, to the Holy Virgin, St. Joseph, my protector, St. Michael, and St. Ignatius and the great Apostle of the Indies, St. Francis Xavier, so that, animated by a truly apostolic spirit, I seek only (page 327) God, and my only desire is that of spilling my blood for Jesus Christ after having suffered much and worked hard for the salvation of souls. I know that the sea will not agree with me and that I will suffer in it. I might even die in it from weakness and infirmity. This thought, far from scaring me, brings me joy because I feel willing, by the grace of God, to die and suffer all that our Lord sees fit. Happy, if I accomplish His holy will. However, I have a secret presentiment that I will not die, at least, not so soon, and that instead I will live and preach the wonderful work of our Lord to the confines of the world.[651] I sigh for those islands where obedience sends me. What a joy this is for me, dear Father, if after a thousand sorrows and a thousand labors I am hung for the love of Jesus Christ! However, it is a grace far too great for a wretch like me.*

He reached Cadiz with his companions, but missed the fleet for New Spain by one day. His trip (page 328) to the Indies was postponed for a year, and he was sent to Salamanca to finish his studies. After having given examples of a rare virtue and evidence of his great capacity in this famous university, he returned to Cadiz and embarked with various other Jesuits on the boat called Nazareth. The fleet was about to leave the harbor when a ship, crowding sails, headed straight against the Nazareth. This vessel moved back in order to allow the other ship passage, but it struck a rock that lies at the entrance of the Cadiz port so forcefully that it split. The ship's cargo, which was very valuable, was lost, but the missionaries and the crew were saved. Father Angelis returned to Cadiz with his companions. Perfectly resigned to the will of God,

---

651. *Non moriar, sed vivam et narrabo opera Domini Pg 117* (Note by C.by C. Le Gobien, *Histoire*, book IX, p. 327).

he was consoled to the misfortune and delay of his trip, which was postponed because of this accident, to the following year, when a corvette[652] came in the middle of the night to take them to the fleet that was already at sea. They were divided among the ships that wanted to receive them. Since they had (page 329) embarked in a hurry, they had taken nothing but their breviaries and crucifixes, and thus they found themselves without beds, without provisions, and with none of the refreshments that are normally taken on such long journeys. They slept on the open deck and lived on what they were given as alms. Besides these discomforts, which were common to Father Angelis and his companions, he was also afflicted by a cruel seasickness. When he reached land, he recovered. He stayed in the novitiate of Tepotzotlán[653] in the kingdom of Mexico and did not arrive at the Marianas until June 13, 1681.

Nothing was more surprising than the way he lived. It is incomprehensible how, with such a frail health, he was able to practice such penances. He only ate one meal a day, insipid roots more appropriate to make him ill than to nourish and strengthen him. After sleeping for four hours, he took a rude discipline and spent the rest of the night before the Blessed Sacrament. He wore a hair shirt and cilice, and used an iron cincture with (page 330) spikes that bit into his flesh. He sustained all of these penances with a profound humility and in intimate union with God. He felt such a great love for the person of the Savior, and such tender devotion for His Sacred Wounds, that he let out tears whenever he thought of His passion. He learned the language of the Marianas so well, that nobody spoke it better than him. He composed a book called *El Espejo la Confesión* (The Mirror of Confession) in this language, which is used in the islands with success. He had begun work on various treaties to use with these people, but God, who wanted to reward his servant, interrupted his work and granted him the grace of dying in the defense of His holy name, as we have narrated.

This precious death would have been followed by that of all the missionaries that worked in the churches in the island of Guåhan, if God had not conserved them by a special providence. They had the custom of gathering every year in Agadña to confer together on the affairs of the mission and the means of advancing God's work. Since this meeting was normally held (page

---

652. This is a very light boat that is used to take information to the fleets when they are at sea (Note by C. Le Gobien, *Histoire*, book IX p. 328).

653. Tepotzotlán was the Jesuit's novitiate seminary in viceroyal Mexico, where the young men of many regions of the kingdom went to study and join the Society of Jesus. Many of the Mexican Jesuits that distinguished themselves in various activities in the Society, whether as teachers or as missionaries in the northern regions, received their instruction in this novitiate. (Rafael Heliodoro Valle, "Jesuitas de Tepotzotlán," *BICC*, Tomo IX, nos. 1, 2, 3, 1953, pp. 159–263).

331) eight days before the feast of St. Ignatius, founder of the Society of Jesus, many of the missionaries were on their way to Agadña when the revolt broke out. Fathers Cardeñoso, Bustillo, and [Basilio le] Roulx[654] arrived that very day, and Johann] Tilpe[655] and [Juan de] Ahumada[656] arrived on the following day, thanks to the help of Don Ignacio Hineti, who kept them from the peril of being assassinated in Pago [Pågu] by the conspirators. Thus, only Father Angelis was sacrificed by the vengeance of the *chamorris* of Ritidian [Litekyan].

This barbarian, who placed himself at the head of the rebels after Yura's death, sent his supporters to the island of Zarpana to encourage its people to join the revolt and kill Fathers [Agustin] Strobach[657] and [Karl Von] Boranga,[658] who were in charge of this new Christian church. Whether it was because these islanders were convinced that if they stained their hands with the missionaries' blood, Don Joseph de Quiroga would take his vengeance for the death of these apostolic men; or whether it was that they were happy with the Fathers, as they said at that time, the truth is that they did not heed the *chamorris*' pressing requests. Indeed, they went further, for although Father Bouwens, then Superior (page 332) of the mission, had ordered those two missionaries to retreat to Agadña so as not to be exposed to the barbarians' fury, the Christians refused,

---

654. Basilio Le Roulx was born on June 24, 1643. He joined the Society of Jesus on September 29, 1662. He was a graduate and professed of the four vows (2/2/1678) (ARSI, "Primus Catalogus Anni Personarum Anni 1684." Philippinae Cat. Trien. 1649–1696, Tomo 2-II, f. 423r).

655. Johann Tilpe was born in Bohemia (contemporary Czech Republic) on October 10, 1644. He joined the Society on October 14, 1666. He was a graduate and professed of the four vows (31/7/1700) (ARSI, "Primus Catalogus Anni Personarum Anni 1701." Philippinae Cat. Trien. 1701–1755 1768, Tomo 3, f. 2r).

656. Father Juan de Ahumada arrived in the Marianas in 1674 and died in Guåhan in 1687 (F. X. Hezel, *From Conquest to Colonization*, p. 89).

657. Augustin Strobach was born in Jihlava (Moravia, in the Kingdom of Bohemia—contemporary Czech Republic) on March 12, 1646. He joined the Society of Jesus on January 19, 1661. He was a graduate and professed of the four vows (2/2/1682) (ARSI, "Primus Catalogus Anni Personarum Anni 1684." Philippinae Cat. Trien. 1649–1696, Tomo 2-II, f. 424r). For its proximity to the Protestant "heretics," the Spanish authorities were reluctant to send German Jesuits to the Indies. For this reason, Strobach changed his name to Carolus Calvanese de Calva, as well as his origin, said to be Milan (L. Clossey, *Salvation and Globalization*, p. 34). In 1684, Father Emmanuel de Boye (1639–1700) published Father Strobach's *Vita* (transcribed in R. Lévesque, *History of Micronesia*, vol. 8, 1996, pp. 447–461).

658. Father Karl (Carlos) Boranga was born in Vienna on July 8, 1640. He joined the Society of Jesus on Ocobter 5, 1656. He was a graduate and professed of the four vows (2/2/1674) (ARSI, "Primus Catalogus Anni Personarum Anni 1684." Philippinae Cat. Trien. 1649–1696, Tomo 2-II, f. 423r). Like Father Teófilo de Angelis, his missionary vocation was manifested during his novitiate, as is shown in the September 8, 1669 letter that he wrote to the General of the Society from his school in Austria, asking to be sent to the barbarians' missions in the Indies (ARSI, "Epistolae Indipetae," Vol. 755 (Austria), no. 277). And like Father Strobach, he adopted a new name, becoming Juan Bautista Pérez de Calatayud, and claiming to be from Aragon. But he still had to stay in Madrid for a year before being allowed to board a ship to the Mariana Islands (L. Clossey, *Salvation and Globalization*, pp. 34; 151). On May 20, 1685, Father Gerardo Bouwens wrote Father Boranga's edifying letter (ARSI, Philippin. Necrologia 1605–1731, Tomo 20, ff. 338r-339v. See also R. Lévesque, *History of Micronesia*, vol. 8, 1996, pp. 462–476).

and they begged the Fathers so insistently to not abandon them, that the two priests believed it was best to separate. Father Boranga stayed in Zarpana to care for his flock, and Father Strobach departed on July 27 for Agadña following his superior's orders. He reached the port on the following day, but seeing that the church, the missionaries' house, and the two seminaries where the island's youth was educated had been burned down, he thought that the barbarians had seized the fortress and that all was lost. Therefore, he returned to the island of Zarpana on the same boat, after a long pursuit by the enemy.

The conspirators, who had led the revolt of part of Guåhan, returned to attack the Spaniards that had taken to the fortress. Don Ignacio Hineti watched the surrounding areas and defended the church with his troops. He valiantly resisted various attacks, and put the enemy to flight (page 333). However, when the barbarians returned with new troops, Don Ignacio did not consider himself capable of resisting them. He asked the governor for assistance, requesting the presence of ten-to-twelve Spaniards to keep the barbarians at bay. Instead of sending him support, the governor asked him to move under cover of the fortress's cannon fire. The *chamorris*, saddened by this unfortunate order, obeyed, and had to watch how the church, the two seminaries, and the missionaries' house were burned down[659]. The enemies took control of the shore on which the missionaries' house stood. Angry by the loss that they had just suffered, Don Ignacio charged against the enemy with the succor of a company of young men from Aniguag [or Anigua] that had opportunely just joined his forces, and the enemy was again put to flight.

The barbarians gathered more supporters with the intention of carrying out one final assault. They advanced until they were almost at musket fire range. Several volleys were fired at them without success, for these people fight very differently from those of Europe. They stand apart from each other, looking only at their enemies' movements (page 334), observing the smallest acts. Since they are very afraid of firearms, they do not take their eyes off them. As soon as they see that they are about to be fired, they fall to the ground and rise up a moment later with great whoops of joy, as if to celebrate the fact that they were not hit. Don Ignacio watched them. He saw their advance, and decided to spare them part of the walk. He went to meet them with his small troops, charged at them several times, and performed actions of such valor, that the Spaniards

---

659. Master Sergeant of the Marianas militia, Don Ignacio de Hineti, possibly from the *mana'chang* low class of Chamorro society, remained loyal to the Spanish. However, we do not know why the governor did not offer him military support.

were awed.

This day's failure did not discourage the rebels. They returned to Agadña for a third time, but the rains came at that moment, and the determination shown by the Spaniards made them turn back without having attempted anything. Don Ignacio Hineti's bravery was a great obstacle for their plans. They again tried to win him over to their party. They sent him trusted envoys to convince him that it was shameful that a man of heart like himself should give himself up to foreigners and combat against his own nation and his own blood; that the salvation of his homeland was (page 335) entirely in his hands; that the Northern islands had already freed themselves from the yoke. That Quiroga could do nothing, for he was besieged on all sides in the island of Saipan; that the Spaniards, who needed his help, saw themselves reduced to extremes; that he could not continue to support a ruined party; that if he was sensible to glory and to his own interests, it was necessary to place himself at the side of his country, kill the governor, join his compatriots, and help them exterminate the enemy of their peace, so that they could recover their old liberty. That, on the other hand, if he rejected the offers that they were making him, he should regard himself as a lost man, cursed and burdened by the hatred of the entire nation.

Don Ignacio spurned his compatriots' offers and threats. He replied firmly that he had the honor of being a Christian, and that he would never do anything unworthy of so glorious a name; that his religion was dearer to him than life itself; and that, since he had it thanks to the Spaniards who had opened his eyes and shown him the true God, it was his duty and his honor to help them reduce the rebels and punish the guilty (page 336). Hineti's response disconcerted the barbarians, for they knew that their failure was certain and that the Spaniards would avenge the death of so many illustrious missionaries if Quiroga, aware of their rebellion, fell on them with his troops. Thus, their only recourse was deceit.

A few Filipino soldiers in the fortress had married native Marianas women. The barbarians had the mothers of these recent brides go to the fortress with the pretext of bringing their daughters some refreshments, but their real purpose was to pervert those Filipinos and win them to their side. They went to [Francisco] Masongsong who was in charge. This officer had distinguished himself on several occasions and had ascended, thanks to his valor and his conduct, to Maestro de Campo [lieutenant colonel]. Since he had married a Marianas' native, the enemy sent his mother-in-law. This deceitful woman told him that it was shameful for the Filipinos to be slaves to the Spanish, exposing their lives constantly for people who had conquered their country and reduced their nation to slavery; that (page 337) this was a favorable occasion to shake off their yoke and end a war that could only be disastrous for the Spanish; that it was not fair for her daughter and him to get caught up in the Spaniards' misfortune and die with them; [and] that the means of securing their safety was to kill the governor and join the Marianas' natives, who would consider him their liberator.

Masongsong listened to the woman quietly and seemed to accept the propositions that she made: "*This enterprise of which you speak is momentous,*" he told her. "*It would take time to succeed without risking anything. Return and I will tell you what I have resolved and what measures I shall take.*" As soon as his mother-in-law left, he went to the governor and reported what he had just heard: "That said," he added, "I have only one piece of advice to give you: do not trust the Filipinos at your service, for they might be corrupted. Double your guard and from now on, entrust the safety of your person only to real Spaniards."

The barbarians' artifices alarmed the governor, and their intense (page 338) desire to see him dead caused him new unrest. His fear increased when five Filipinos defected to the enemy. To prevent this kind of desertion, he had Masongsong's mother-in-law arrested when she came back to the fortress. This feeble remedy did not bring the governor peace, for he was worried about the lack of news from Quiroga. He made new attempts to find some news about him. He wrote to him and ordered a Marianas envoy to take the letter without stopping in Zarpana or Tinian. Whether because this unfortunate wretch let himself be corrupted by the rebels, or because he was intimidated by them, he stopped in Zarpana and refused to go further.

Being notified of this, Father Strobach went to meet the messenger and, animated by a holy zeal for the salvation of his brothers and the conservation of this new Christianity, he took the governor's letter with the intent of delivering it to Quiroga himself. He set out with only one man, put into port at Tinian, and sent his pilot ashore to inform himself of the situation there while he remained hidden away in the canoe. Nevertheless, the barbarians discovered him, assaulted him, dragged him on to land, (page 339) and after tying his hands behind his back, took him to the Sungharon *chamorris*. The barbarian asked him some questions, which the Servant of God answered sweetly; but impelled by an ardent thirst and exhausted by the fatigue of the voyage, he asked for a drop of water. The *chamorris* ignored him, and sent him to Marpo, a village some four miles from Sungharon. He clearly saw by how they treated him that he was going to be killed. He thought of nothing other than offering up his death as a sacrifice to God. He recited his breviary, and with his eyes on the crucifix, spoke with our Lord for a long while affectionately and with tenderness. The barbarians, who had gone away, returned to him, tore the crucifix from his hands, and began mistreating him. *"What have I done to you?"* the holy man said to them. *"In what have I offended you?"* The barbarians did not answer, and continued abusing him. *"I have no fear of you killing me,"* he continued, *"for I am innocent. God, who is my protector, will defend me; he will never abandon me."* *"Go ahead, you wretch,"* the barbarians replied to him, *"we do not know him* (page 340) *and we are not worried."* At that moment, Zuijan, one of the most furious of the group, struck him a terrible blow on the head with a stick and he fell to the ground. His companions, encouraged by his example, hit him with the same violence and doubled their blows until he expired.

Thus died Father Agustin Strobach, more illustrious by his virtue than by his birth, even though he came from a good Moravian family. His parents, who were very pious, had given him a happy education. The revolts in Hungary had forced them to send him to study at Olmuts [Olomouc, Moravia, Czech Republic], the capital of the province and famed for its University. There he studied Philosophy and some theology courses, when, touched by the desire to sanctify himself and work for the salvation of souls, he entered the novitiate of the Society of Jesus at Brin [Brno, Czech Republic] on October 15, 1667. Two of his brothers had made this same choice. This gladdened his father, a venerable old man who frequently declared that he would have followed his son's example if he were not bound by ties that he was not allowed to break.

(page 341) Father Strobach spent the first years of his youth in the ordinary duties of the Society. The innocence of his customs and his exact practice of Christian virtues made those around him regard him as a saint even then. He ardently wished to go to the Indies, to have the joy of spilling his blood for Jesus Christ. However, he was sent to the army of Count [Maximilian] of Thun [1638–1701], who had asked the Jesuits to inspire piety and religion in his soldiers. He performed this duty to the satisfaction of the General, who commanded the emperor's troops. On his return, he learned that missionaries were being sent to the Indies and that one of them had taken ill and was not capable of undertaking such a long voyage. Using this opportunity, he presented himself to his superiors and begged them to let him take the sick man's place.

He obtained it and left Prague with his companions for Genoa, where he met Father Teófilo de Angelis and other missionaries. They embarked and went to Spain, spending an entire year waiting to depart. During this long stay, (page 342) the talents that God had given this holy man to guide souls were made manifest, and his virtue was held in such high regard, that he was charged with caring for the novices that were being sent that year to the Philippines, where he was himself destined. However, he did not make it there, for while they were in the Marianas he was so deeply moved by the deplorable state of the natives that he asked to stay there to work in their instruction. Very soon, his vocation for the conversion of these peoples was made evident. He obtained a great harvest in Inarajan, where he first worked, and in Zarpana, where he worked later, and founded the church.

He was a very introspective man, enlightened in the ways of the Lord. He felt such attraction for prayer that he employed all of the time that he had when he was not teaching or serving others in this holy exercise. He treated his body with a surprising rigor. Besides sleeping on a plank and eating only one meal a day, he performed great penances every day and usually wore a spiked cincture or cilice the mere sight of which was horrifying. He never brushed off those small gnats (page 343) called mosquitoes. He suffered them on his face and hands even though their stings were so deep and painful that it was a real martyrdom to be exposed to them. I will let those who write the life of this holy man talk about his sweetness, his obedience and his humility, of which we have so many examples. I will only add that he had such an ardent desire that all men know and love God that he was disconsolate when he saw that there were so many nations enveloped in the darkness of infidelity. Such a holy life was followed by an even holier death, for he had the felicity of spilling his blood for Jesus Christ, just as he had always wanted.

Father Carlos Boranga, his companion, met an equally happy fate. He had stayed in the island of Zarpana to assist the Christians who had begged him not to abandon them. He worked with good results. His sweetness and charity had won the hearts of these barbarians. The rebels, who regarded him as an obstacle, could not stand him. They had him assassinated by their emissaries who then convinced the (page 344) Zarpana natives to take up arms against the Spanish.

This apostolic man was born in Vienna, Austria. Placed under the protection of the Holy Virgin and St. Joseph since his birth, he was educated with great care. Since at that time the Swedes were attacking Germany and threatening Austria and the very capital of the Empire, he was sent to Venice, where he stayed until after the peace of Münster.[660] He devoted himself to his studies, making considerable progress in them, but he progressed even more in virtue. The almost constant miracles that St. Francis Xavier carried out in the Kingdom of Naples, especially in Potamo, in the remotest part of Calabria, where a famous image of this great saint stood, led Boranga to read the life of the Apostle of the Indies. He was so enchanted that he decided to dedicate himself to the conversion of infidels. With this purpose in mind, he became a Jesuit, entering the novitiate on October 7, 1656. After completing his studies and his training for the apostolic life, he begged his superiors to send him to the most difficult and grievous mission in the Indies. He sent (page 345) urgent pleas, but he had an accident that foiled all of his plans. He fell from a relatively high place, and broke a leg. This fall caused him such violent pains, that they all feared its consequences. Seeing himself in such danger, he asked permission to consecrate himself by vow to the missions of the Indies if God restored his health. And God did. His strength and his health were briefly restored, and of his fall, there remained nothing but a slight discomfort that was barely noticeable.

As soon as he could walk, all his thoughts turned to his voyage to the Indies. He went to Spain and was about to embark on the fleet that goes every year to Mexico, when all his plans were disrupted by problems with the Spanish Ministers. The President of the Council of the Indies sent him notice that, since he was not a subject of the Catholic Monarch, he could not allow him to go to the Indies. This blow afflicted Father Boranga, but he did not give up. He thought he could win over the president. He spent an entire year requesting [permission] personally and though the mediation of friends. The emperor's ambassador, who had this (page 346) priest and his family in high regard, became interested in his case and tried several initiatives, but the president, who had made his decision, was inflexible, and Father Boranga had to return to Germany. After all these futile efforts, it would seem that he should no longer think of the Indies. However, God, who called him to this mission, kept urging him internally to consecrate himself to it, despite all the obstacles that presented themselves to thwart this holy enterprise. He was sent to the army of [general field marshal] Kops who commanded the Emperor's troops in Hungary

---

660. The peace treaties of Osnabruck (May 15, 1648) and Münster (October 24, 1648) constituted the well-known Peace of Westphalia, which ended the Thirty Year War in Germany and the Eighty Year War between Spain and the Netherlands.

[1677].[661] In this charitable employment, he found abundant opportunities to exercise his zeal and his patience. He suffered much and practiced such severe penances, that his superiors were forced to moderate them.

During this campaign and the rest, he had so many mediators insisting his cause before the king of Spain that the Council of the Indies finally consented to allow the subjects of the emperor to work in the conversion of infidels. This consent was necessary, moreover, because the missions of the East Indies lacked apostolic workers. Missionaries from Austria were requested (page 347). Father Boranga, shaken by this news, asked God to make His will known to him; and, feeling so impelled to dedicate himself to this mission, he went to meet with his provincial, threw himself at his feet, and begged him to send him. His request was granted. He went to Genoa, where he met Father Angelis, Father Strobach, and various other missionaries of the Society, who named him their superior. They arrived at Cadiz after the fleet had departed to New Spain, which delayed the voyage for another year. During that time he was sent to Seville, where he had occasion to exercise his zeal in the celebrated mission that Father Thyrsus González de Santalla held there for a second time, according to the orders given by the Archbishop of this great city, Don Ambrosio Spínola,[662] one of the most virtuous and diligent prelates of this century.

Father Boranga spent the rest of the year (page 348) in the small city of Jerez, and he did not return to Cadiz until the fleet was ready to set sail for New Spain. He embarked with Fathers Strobach and de Angelis in the ship Nazareth, which sank upon leaving the port, as we have already described. This accident did not surprise the new apostle. Used to encountering obstacles at every step of the way, he was resigned to God's will. The following night, he was taken to the fleet that was already at sea. He was cruelly tormented by seasickness from Cadiz to New Mexico, but he suffered even more in the trip from Mexico to the Marianas. The great heat, which he was unused to, gave him a burning fever that almost took him to the other life. His joy upon seeing that he had finally obtained what he had for so long desired, gave him his health back. He learned the Marianas' language with such ease that he was immediately capable of working in the conversion of these poor islanders. He had the consolation

---

661. For more on the subject of the imperial armes in Hungary and the *General de batalla* Kops, see Álvaro Bernaldo de Quirós, *Historia de las revoluciones de Hungría*, Madrid: Imprenta de Bernardo de Villa-Diego, 1687, pp. 368–374. Suffice it to say, that this general, of German origin, was so brutal in his treatment of captured enemies that Emperor Leopold I had him banished before the war was over.

662. He died in the odour of sanctity in the month of May in 1684. A eulogy was written by Spaniard Don Juan de Loaysa [1633–1709], canon of Seville's Iglesia Metropolitana (Note by C. Le Gobien, *Histoire*, book IX, p. 347).

of teaching and baptizing many of them in the island of Guåhan. This success brought him attention (page 349), and led him to be considered the right man to win the people in the island of Zarpana and bring them to the Christian fold. He was sent there, and he founded the Church of Aguisan, taking care of it until his death, which took place soon after that of his companion, Father Strobach. This would be the place to write the eulogy of this zealous missionary, but I will content myself with citing what Father Nicolás Avancín, a man distinguished in the Society for the offices that he has held and the books that he has given the public, wrote to a friend of his soon after Father Boranga had departed for the Indies:

> *I can speak to you about Father Carlos Boranga more accurately than anybody else, for I have known him since his infancy and I have been his spiritual director for many years. He is a man of great genius, with a talent for thinking up great enterprises and even more talent for carrying them out. His trust in God knows no limits. Nothing is beyond his hope and nothing is capable of striking fear in him. His zeal for the salvation of souls is ardent and his obedience immediate. His detachment is such that mundane honors and vain praise (page 350) horrify him. This is what he was like when I met him in Europe, and I have no doubt that he has continued making even greater progress in virtue since he has been in the Indies.*

The revolt in the island of Zarpana gave Don Damián de Esplana new reasons for concern. The governor, who had formerly been the terror of the barbarians, trembled at the enemies' last attack, and did not dare present himself to repel them.[663] He counted only on Quiroga, but he did not know if the latter was still alive, and he had given up hope of ever receiving any news from him when a fleet of seventy sails appeared on the northern coast. He watched it amidst great bursts of joy, convinced that it was Quiroga coming to help him. But this happy idea vanished immediately. The small fleet broke up and headed for different ports. The governor, more anxious than before, sent spies to the Aban area to fetch information, and his spies only brought him devastating

---

663. The differences between Don Damián de Esplana and superior Father Gerardo Bouwens were well known. In the Jesuit priest's opinion, most of the disasters that befell the mission could have been avoided if the Mariana Islands' government had been placed in the hands of sergeant major Don Joseph de Quiroga instead of Esplana, whom he considered incapable of governing ("Carta dirigida al padre Francisco García por un jesuita de la misión en Marianas…," Guåhan, May 20, 1685) (RAH, Fondos Jesuitas, Tomo 57, Signatura: 9-3631/43, f. 8v; R. Lévesque, *History of Micronesia*, vol. 8, 1996, p. 273). See also the letter written by Father Bouwens to the Philippine Governor, Don Gabriel de Curuzealegui y Arriola, Guåhan, May 31, 1685 (transcribed in R. Lévesque, *History of Micronesia*, vol. 8, 1996, pp. 384–388).

news. They told him that the fleet had come from Zarpana and the northern islands; that (page 351) the Spaniards in the island of Tinian and those who had been sent to conquer the islands of Gani had all been murdered; and that Quiroga could no longer resist the multitude of enemies that besieged him. This news finally overwhelmed the governor, who was already terrified. His fear spread to the soldiers. They thought themselves lost. The missionaries tried to animate them and sustain them with their speeches.

> *Why are you alarmed, my friends?*, they would tell them. *We are not fighting for our lives only, but for our religion and the glory of our God. He has not abandoned us since we have been in these infidel lands. He will not abandon us now. Has He not rendered all of the barbarians' efforts useless up to now? What should we fear if the Lord is fighting for us? Let us pray to Him with confidence; let us place Him on our side by the sanctity of our customs, and let us appease His wrath through salutary penance.*

These exhortations sustained the weak and discouraged garrison. The soldiers prayed assiduously (page 352), placing all of their trust in God, attracting over them the blessings of Heaven through their frequent reception of the Sacraments and the regularity of their lives.

BOOK TEN

(page 353)

During this period of trouble and revolt, Quiroga and his companions were living peacefully in the island of Saipan, unaware of the revolt in Guåhan and of the events that were taking place elsewhere. The barbarians took care to conceal everything from him, and acted with all the dissimulation of which a cunning people is capable. Quiroga would have been taken by surprise if he had been less competent and circumspect. For, after God, it was due to his conduct and the wise precautions that he took (page 354) that the Christian religion was preserved in these infidel lands.

As soon as the revolt in Guåhan broke out, the conspiracy was almost general. The leaders of the uprising had partisans in the other islands; [and] they did not fail to inform them of what was happening, and of the advantages that they had over the Spanish. It was not hard to set in motion spirits that were naturally anxious and unaccustomed to dependence. The inhabitants of the island of Tinian were the first to show themselves. They killed seventeen Spaniards that were spread out throughout various places in the island's villages, burned the frigate that served as transport for Quiroga's troops, and resolved to drive him from the islands of the north or make him perish at sea.

With this purpose in mind, they invited the inhabitants of Zarpana and Aguiguan to take up arms and join them on an assault on the island of Saipan, under the pretense of clearing a forest and cultivating new lands. However, the barbarians were betrayed by the extraordinary happiness that they revealed in various occasions, and the abusive way in which they (page 355) treated a Filipino, inspiring such suspicions in Quiroga, that he doubled the sentinels in his camp and gave the order that nobody should be allowed inside. Thinking that their plan had been discovered, the barbarians took their mask off and killed two soldiers that were apart from the others. This crime was the signal for war. The whole island took up arms against the Spaniards, who were soon reduced to dire circumstances.

Even though he found himself confined in a simple palisade with thirty-six men, practically without provisions or ammunition, and with no hopes that succor would come, this general uprising did not frighten Quiroga at all. Since he knew that the barbarians' temperament was prone to insolence before any sign of moderation, he set out with his bravest men. The enemy, numbering eight hundred, advanced to engage in combat. He met them with valor and charged at them with such vigor that they gave way and were about to flee when a group that was ambushed came to their aid. Quiroga saw himself surrounded by enemies who (page 356) were firing poisoned arrows and stones from all sides. Aware that the barbarians were sending a dispatch to attack his camp, Quiroga had his unit retreat there without giving their backs to the enemy, which pursued them for quite some time.

The following days he made various sorties with good results, razing various villages and pushing the barbarians to the neighboring mountain. These incursions, however, did not quiet his anxiety for the troops that he had sent to Tinian and the Gani Islands. He had received no news from them, and he had reason to fear that some bad turn had been done to them, especially since the barbarians spread the news that Guåhan had revolted; that they had strangled the governor, the missionaries, and almost all of the garrison members; and that all of the islands had taken up arms to exterminate the Spaniards.

These rumors scared Quiroga's companions. They saw themselves at the ends of the earth, constrained in the corner of an island far from the world's affairs, surrounded by enemies and reduced to a small number of men; without boats on which to (page 357) flee, without hopes of receiving assistance; and worse still, without provisions and almost without munitions for war. However, Quiroga was not discouraged. *"We should pay no heed,"* he told his companions, *"to the rumors that our enemies spread to scare us. They are cowardly souls that are only trying to surprise and deceive us. Remember, my friends that we are fighting for God and that if He is with us, we have nothing to fear from our enemies' evil or their great numbers."*

On his part, Father Matías Cuculino (1641–1696)[664] exhorted the

---

664. Matías Cuculino was born Mathias Kuklein, on January 17, 1641 in Bohemia (Czech Republic). He joined the Society of Jesus on December 31, 1661. He was a graduate and professed the four vows (August 15, 1679) (ARSI, "Primus Catalogus Anni Personarum Anni 1684." Philippinae Cat. Trien. 1649–1696, Tomo 2-II, f. 423v). On January 27, 1678, he wrote a letter to the General of the Society asking to be sent to the missions of New Spain and the Philippines (ARSI, "Epistolae Indipetae," Vol. 756 (Bohemia), no. 133). On May 20, 1685, he wrote a letter to Father Emmanuel de Boye, the Provincial of Bohemia, who had been his teacher in the Clement College of Prague, as well as a *Relación* of the Chamorro revolts of June 1684 to May 1685 (AHCJC, FILCAR, E.I, a-18/1, ff. 33-52). He died in 1696 in the Mariana Islands (F. X. Hezel, S.J., *From Conquest to Colonization*, p. 89). For more on Father Cuculino, see Josef Koláck, S.J., "Matías Cuculino (1641–1696)." *Iesuitica* B 52 (1), 1996.

soldiers to place all their trust in God, and to mollify his wrath through the saintliness of their lives. The soldiers, docile to the opinions of this fervent missionary and to their commander's reprimands, practiced all of the exercises of piety that they were ordered. They prayed assiduously, listening intently to the word of God, zealously receiving the Sacraments. Some even did great penances. Their fervor went so far, that Father Cuculino wrote in his letters that it became necessary to curb them instead of encourage them.

Such a Christian life drew (page 358) the blessings of Heaven over them. Quiroga had set up camp on a neck of land that had the sea at the north and a swamp to the south. His camp was therefore accessible only through two flanks, and it was through both those flanks that the barbarians decided to attack. The troops of Tinian and Aguiguan, together with those of Saipan, approached the camp with screams that would have terrified anybody unused to them. In a moment, a hail of stones and poisoned arrows fell on the camp. The enemy then charged at the palisade trying to pull out the stakes, but they found more resistance than they had imagined. They redoubled their attacks, returning up to three times in assaults that were full of rage and fury. The Spaniards repelled them each time, forcing them to retreat.

The barbarians had two camps, one to the east and one to the west. Although these camps were surrounded by ditches and palisades, and guarded by three to four hundred men, Quiroga attacked them and defeated them more than once. He did not stop there, for when he saw that the barbarians took flight (page 359) as soon as they saw Spanish firearms, he carried out incursions through half the island, pillaging, burning and sacking everything he found. What is most surprising is that none of his men were wounded in these raids nor in the course of the war. He attributed this favor to St. Joseph, patron saint of the island of Saipan, whose protection he had already felt on various occasions. Finding themselves surrounded one day by the barbarians, who were firing their arrows at close range, Quiroga turned to this great saint, and at that moment, he saw, and all of his troops saw it as well, that these arrows broke in midair and fell at their feet without touching them.

It is not surprising that God seemed interested in the conservation of a man who was willing to sacrifice himself and give his life to prevent the slightest sin. Something that he did during this siege is absolute proof of this. Among his troops there was a brave and determined soldier who was also impious, a blasphemer, unscrupulous in the fulfillment of his service, and given to excesses. The [former] Governor, Don Antonio de Saravia, had often been inclined to execute this soldier for the crimes that he committed. However, (page 360) the scarcity of soldiers had persuaded him not to. Instead of profiting from the indulgence that was shown him, this wretch was not a better person or a better Christian in Saipan than he had been in Agadña. He continued scandalizing the troops with his impious acts and his bad example; and forgetting his duty, he insulted the officers vomiting horrible blasphemies. Quiroga, more sensitive to the injury against God than to the insult to his officers, ordered his arrest and had him executed. His soldiers, who were vexed by seeing themselves so few amidst so many enemies, were surprised at this punishment. Quiroga, who read the emotion on their faces, spoke to them in these terms, with a firmness worthy of a Christian hero:

> *You are surprised, brothers and companions of mine, by what I have just done, because you believe that the service that this rascal rendered us was necessary to resist our enemies. No, my brothers, this evil man was more capable of drawing God's wrath upon us than of rendering us any service. I have had him punished* (page 361) *for his impiety more than for his insults, and I will treat anybody who follows his bad example in the same way. For I am happy to let you know that I do not fear death or the enemies; the only fear that I have in the world, is that of offending God. Thus, I tell you today that I want the law of God to be strictly observed, and that I will punish more severely those who trespass it, than those who neglect their military duties. I will forgive nobody, even if I were left alone to combat the enemies of our holy faith.*

The soldiers, who loved Quiroga as a father and considered him a man of solid virtue, were not worried by this reprimand. On the contrary, they vowed him inviolable loyalty and pointed out that they were willing to follow him and obey his orders.

From that very day, God seemed to want to reward Quiroga's zeal. A woman came to the camp to sell the garrison some refreshments, and Quiroga ordered her detainment and forced her to take him to some *chamorris* that he wanted to capture. His project turned out (page 362) just as he had planned it. He went out at night to keep his march hidden from his enemies, and kidnapped some *chamorris* that he found lying asleep. There were five of them. Quiroga chose the most important of them to take a letter to Governor Don Damián de Esplana, of whom he had not had any news since the uprising: "*If you do not take this letter,"* he told the *chamorris, "and bring me back the answer, be certain that your four relatives will answer for it and die."*

The distressed *chamorris* took the message and left for the island of Guåhan. He went to a friend and told him in secret the motive for his trip and the juncture in which he found himself. *"Do not trouble yourself,"* his friend told him. *"I will take your letter at the risk of my own life; just stay here and keep silent about the secret that you have confided in me."* This Marianas' native was a Spanish ally. He left his home at nightfall and arrived at the fortress at two in the morning. He gave his usual signal, they opened the door, and he went inside and (page 363) gave the governor the letter that had been entrusted to him. The governor read it in the presence of the garrison with inexpressible transports of joy. He wrote a reply then and there, and ordered that the reply be taken to the *chamorris,* who immediately carried it to Quiroga. The governor told him that they were in a dire situation, and that if he did not come to help them, they were in danger of succumbing to the barbarians who had laid a siege on them for four months. Quiroga, pleased to know that Don Damián and the missionaries had not been killed as the barbarians had said, could think of nothing other than leaving the island of Saipan and returning to Agadña with his people. He had so exhausted his enemies with his constant raids and attacks, that they had been forced to sue for peace. Quiroga had just concluded it when he received the governor's reply. However, it was not easy to carry out his orders. He had no boats or canoes to leave Saipan; and even if he had had them, he had every reason to believe that the barbarians would immediately fall upon him with their canoes if (page 364) they found out his plans. Although these difficulties seemed insurmountable, Quiroga succeeded. Here is how he did it.

His camp was very close to the village of Opian, which he had razed several times in his sorties. The inhabitants of this village ardently desired to see themselves free of such a terrible and dangerous neighbor. Quiroga, who did not ignore their sentiments, asked them for canoes under the pretext of going to the island of Tinian. They sent them immediately along with the necessary pilots to maneuver them. They could not appear more obliging. But these barbarians were always deceitful, and they had ordered the pilots to overturn the canoes the moment the Spaniards got on them, and thus make them all perish.

Quiroga, who knew the insidious and cunning character of this nation, was on guard against these barbarians lest they take him by surprise. He had a small campaign tent carried to the neighboring mountain, and fired some volleys there to keep the barbarians nervous on that side. At the same time, he ordered his troops to embark, and left on the night of November 21, 1684 (page 365). And to make sure that the pilots could not carry out any evil purposes, he had them tied up and left only their hands free to move so that they could maneuver their small vessels. These precautions were indeed necessary, for, as soon as the small fleet lost sight of land, these pilots lowered the sails and tried to jump into the sea. They were stopped, and made to continue their task by force. The sea was heavy and furiously rough. They were in danger of perishing more than once. Of the eight canoes that had sailed out, three were shipwrecked by the wind when they got too close to Tinian. The others wanted to go in their aid, but the fear of all being lost made them abandon fifteen Spaniards to the mercy of the barbarians and continue en route. They arrived at Agadña around three in the morning after two days of sailing. A soldier jumped ashore and yelled with all his might to the sentry: "*Tell the governor that Sergeant Major Don Joseph de Quiroga has just arrived and awaits his orders to disembark!*" At this cry, the whole garrison jumped for joy. The governor ordered that weapons be (page 366) taken and ran out to shore to receive their liberator. He embraced him tenderly and took him to the fortress amidst the noise of drums and trumpets, pointing out that he was infinitely grateful for the service that Quiroga rendered to religion and the State.

Their happiness at seeing themselves reunited after so many troublesome adventures increased with the arrival of the fifteen Spaniards that had been shipwrecked off the coast of Tinian. They had been given up for dead, but they had the fortune of being saved. The Tinian natives had received them favorably and sent them on their way unharmed. It seems that these barbarians who had stained their hands with the blood of Father Strobach and seventeen Spaniards a few months earlier, now wanted to atone for their crime and to reconcile with the governor by treating these new guests well.

The troops that had been sent to conquer the islands of Gani were not so fortunate. Let us see what was learned at that moment. When Don Joseph de Quiroga arrived in Saipan and settled there, he had, as we have said, sent part of (page 367) his troops to the northern islands under the command of Don José de Tapia. Tapia did not find any resistance. They were received everywhere as he would have wished, and, pleased with his expedition, he decided to return to Saipan with his troops towards the end of September, when their pilots received notice of their compatriots' revolt and conspired to make them perish at sea. These traitors convened that they would overturn the canoes when a branch was raised in the air, thus throwing all of the Spaniards into the sea. This plan worked out as they intended. When they saw the branch raised, all of the canoes were overturned; and of the twenty-five soldiers that were in this small fleet, only five or six survived. Father Coomans was one of them. He took hold of his pilot, who was not fast enough to upset his canoe, and forced him to stop at the island of Alamagan, where one of the main *chamorris* took him under his protection.

From there, he went to the island of Saipan, where the barbarians assassinated him in the month of July the following year.[665] I would gladly recount the circumstances of this (page 368) death here, but despite all of the work that I have carried out, I have been unable to recover what Father Bouwens wrote about it at that time [and sent] to Europe. I have only learned, from trustworthy people, that the barbarians tied him to a tree, where they beset him with stones and pierced him with arrows.

This was a real loss for the mission, which needed workers of his character. His zeal for the salvation of souls was so great, that he had dedicated himself to the missions despite the asthma that he suffered since infancy. This malady had gotten worse with age, and it became so violent that he had to spend his nights on a chair, so as to avoid the danger of asphyxiation. This is what people who lived with him in Leuven, where he studied theology, said. But what is most surprising is that this bothersome asthma disappeared the moment he arrived in the Mariana Islands, and that, as if by a particular providence of God, he was freed from all other discomforts, according to what he himself wrote more than once to his friends.

Peter Coomans joined the Society of Jesus (page 369) at the age of eighteen.[666] Faithful to the calls of grace, and moved by the infinite loss of infidels, he consecrated himself to the missions after finishing his career in theology. He asked his superiors to send him to the Philippines with the hope of going to Japan and dying there for Jesus Christ. His request was granted, and he set out from Ostend in early 1670 in the ship of Captain Coots, which took him to Corunna, the famous Galician port. He arrived at Madrid in the month of March after making a trip to Compostela, where he visited the relic of the apostle St. James, with his compatriot and dear companion, Father Gerardo Bouwens.

The Council of the Indies' orders kept them there longer than they would

---

665. M. Cuculino, *Relación* (1685), in AHCJC, FILCAR, E.I, a-18/1, ff. 43-44; Father Cooman's edifying letter, written by Father Bouwens, Guåhan, May 31, 1686 (ARSI, Filipinas 13, ff. 218-219, transcribed in R. Lévesque, *History of Micronesia*, vol. 8, 1996, pp. 513–529). See also P. Murillo Velarde, *Historia de la provincia de Filipinas*, f. 364v; L. Ibáñez y García, *Historia de las Islas Marianas*, p. 63.

666. He was born in Antwerp [Belgium] on January 30, 1638 and had become a Jesuit in Mechelen on September 14, 1656 (Note by C. Le Gobien, *Histoire*, book X, p. 369).

have wished. They dedicated themselves to useful occupations. Father Bouwens perfected his mathematical knowledge, of which he had some basic notions; and Father Coomans, who had much talent for oratory and the literary arts (page 370), taught rhetoric in the Imperial College with tremendous success.

The indecision and slowness of the Spanish ministers made the new missionaries nervous, for, although they were subjects of the Catholic king, they were prevented from going to the Philippines. Pleasant words were spoken, but no decisions were made, which made the missionaries fear that after their long wait, they would be sent back to their own country. Father [Prospero] Intorcetta's arrival in Portugal freed them from this fear. This famous missionary[667] had left China two years earlier and was looking for apostolic men to convert this vast and flourishing Empire in which the preachers of the gospel had just been reestablished in their Churches [note on the margin: in 1669] after a long imprisonment. [Father Intorcetta] was on his way to Rome for this purpose.

Father Coomans and his companion (page 371) were determined to go with this glorious confessor of Jesus Christ if the Council of the Indies decided against their desires. However, the Council pronounced itself in their favor, and they obtained what they had been requesting for more than a year. They left on the fleet to New Spain and from there, continued to the Mariana Islands. Since this nascent mission had just lost its most solid supporter, Father Diego Luis de San Vitores, they were invited to stay in these islands and take part in the overwhelming work. They accepted. Father Coomans had the happiness of giving his life for Jesus Christ, and Father Bouwens has worked there for more than thirty years [sic] with an indefatigable zeal, assisted by Don Joseph de Quiroga, who has garnered absolute authority in this country through his virtue and his conduct.

This officer's arrival in the island of Guåhan struck terror and consternation in the barbarians. Far from attacking the Spaniards, they thought only of finding a safe place to avoid his vengeance. Quiroga, wanting to profit from such a favorable situation, went out on campaigns, burning the rebels' houses (page 372) and razing their lands to force them to return to their duties and sue for peace. He was not as successful as he had hoped. He terrorized the barbar-

---

667. Father Prospero Intorcetta died in China in the odor of sanctity on October 3, 1696, after having worked for thirty-seven years in the conversion of its peoples, founding various churches there. I have written the life of this illustrious missionary in *Historia del Edicto del Emperador de la China, a favor de la Religión Cristiana* (Note by C. Le Gobien, *Histoire*, book X, p. 370). For more on Father Intorcetta (1626–1696), see Liam Matthew Brockey, *Journey to the East: the Jesuit Mission to China, 1579–1724*, Cambridge, Mass., Harvard University Press, 2007, pp. 277–79.

ians but did not submit them. The most able retired to the neighboring islands and the others hid in the jungle and in caves, from where they often came out to harass the Spaniards and plunder the Christians. These ferocious men would not let themselves be pacified, as all the efforts made in that direction failed, and for many years, the Spanish were exposed to the insults of these rascals thanks to the misconduct of Don Damián de Esplana.

This governor, jealous of the fame that Quiroga had acquired through his valor and because of the services that he had just rendered the State and religion, could not stand him, and was constantly looking for pretexts to upset him and foil his plans. Even though he owed his salvation and the conservation of these islands to this officer's wise conduct, the governor took evil pleasure in censuring him and disgracing him before the troops by his speeches and jeers. Quiroga, who had come to these islands only to procure the glory of God and (page 373) to advance the affairs of religion, faced the governor's persecution like a true Christian, and without concerning himself with the disservices that were made to him, he continued fulfilling his duty and edifying everybody with his piety and his virtue. However, this almost cost him his life. Let us see how.

Don Damián had gone to the Philippines in early 1688 under the pretext of recovering his health, but in reality, he was going to safeguard the fortune that he had amassed. He left the government of the Mariana Islands to Quiroga, according to the orders given by the king. For a long time, the soldiers had lived in a libertinage and disorder that scandalized all good people. Quiroga tried to have them return to their duties, reestablishing military discipline among them, but used to such a licentiuous life, they could not submit to order. First, they murmured against the laws that Quiroga was attempting to impose on them, and then, going from murmurs to revolt, they took up arms and seized the fortress, locking Quiroga up in a dungeon after pulling him from the missionaries' house where he had gone to see (page 374) a friend. This violent conduct led to the fear that the mutineers would take this to the ultimate consequences. Father Gerardo Bouwens, who was then Superior of the mission, met with them, reminded them of their duties, and begged them to free a man whom they should all regard as these islands' guardian angel. The rebels paid no attention to his prayers and remonstrance; all that he could obtain was a promise that they would make no attempt on Quiroga's life. Although there were many reasons to mistrust the word of a gang of mutineers, there was nothing to do but to be satisfied with it.

The missionaries were pierced by grief upon seeing a man to whom religion and the State owed so much in prison. Since all their efforts had been useless, they redoubled their prayers to God and made it a special mission to ask that the conspirators return to their duties. God heard their prayers and conferred unction and strength to their words. The rebel leader was moved by the fervent exhortations that he heard every day, and decided to convert. He went secretly to Quiroga and, throwing himself at his feet, he burst into tears and begged his (page 375) forgiveness for the affront that he had carried out against him, and he promised to repair the horrible scandal that he had given. Since his return to God was sincere, he spoke to his confidants, won them over, and freed Quiroga amidst all the honors that were owed to this rank and dignity. Some mutineers opposed this course of action, seizing the fortress, but it was all in vain, for the new penitent found the way to repel them, punishing the guiltiest among them and detaining the rest, who were immediately sent to the Philippines, and obtained their liberty only through Quiroga's intercession.

This commander's primary concern, given his zeal for the salvation of souls, was to have religion flourish again across the island of Guåhan, rebuilding the churches that had been destroyed during the war. The church of Pago [Pågu] was completed in early 1689. Its inauguration took place amongst much pomp, and an image of the Holy Virgin famous for many miracles was brought over from Agadña.[668] The church of Umatac was also rebuilt.

There had been no rain in five months. Fruit and crops were ruined, and a cruel famine threatened them all. Quiroga, moved by this poor people's desolation (page 376), attended mass in this church [Umatac], which was consecrated to St. Dyonisius the Areopagite. It was falling apart and unsafe, so he promised to build a new one if he obtained from God, through the intercession of this great saint, a favorable rain that would refresh the goods of the land. His prayers were heard, and soon after, such favorable rains fell that year, that an abundant harvest was gathered.

This was not the only grace that God granted the islanders. The ship that came from New Spain in the month of June brought considerable reinforcements, but it also carried such a bad air, that the entire island was contaminated.

---

668. The church of Pågu, in Guåhan, was built in 1689 after the image of the Blessed Virgin, which had been lost during the uprising of 1684, was found. For a community of Catholic believers, the image of a saint or a virgin constituted a symbol of identity, very much like a coat of arms or a flag. It should not surprise us, then, that this finding served to consecrate the new district [partido] and build a new residence (ARSI, Lorenzo Bustillo, S.J., "Relación del estado y progresos de la misión de las islas Marianas desde mayo pasado de 1689 hasta el de 1690," Supplementum ad Historiam, 1584–1750, Tomo 14, f. 75r-75v).

Nobody escaped the malady that it caused. It was a sort of catarrh accompanied by fevers, bleeding, and other unpleasant symptoms. The missionaries increased their zeal and their charity in this occasion: moved by the ravages caused by such a contagious disease, Father Juan [Johann] Schirmeisen (1657–1719),[669] exhorted the sick to place all their trust in God and drink holy water. They did, and the effect was so wonderful, that all who drank it were cured.

Quiroga was not content with having (page 377) reestablished order and piety in the island of Guåhan. He decided to submit the islands of the North, and make Christianity flourish in them, too. He set out in early July of 1689 with his best troops. At first, they sailed in favorable weather, but the wind changed and the sea became so rough that they had to return and postpone this undertaking for another time. Things were no better in the south. Some years earlier, Admiral Don Francisco Lazcano had discovered a large island in the south that he had called Carolina, in honor of the king of Spain, Charles II.[670] Quiroga, who ardently desired to take the faith to these infidel lands, sent *chamorris* Don Alonso Soon with a few soldiers to make a faithful report on the state of this island, and to examine the customs and temperament of its inhabitants. Soon tried very hard to find it, but he did not succeed.[671]

Don Damián de Esplana, who had returned from the Philippines, formed a similar plan, and the following year ordered Soon to return to the sea to make that discovery. The *chamorris* had no better luck than the first time (page 378). He could not find the Caroline island, and the governor was more easily reconciled with this than with what happened to him in the Philippines. He had obtained permission to make this trip, as we have said, but his impatience led him to set out before the permit reached the Marianas. The Philippine governor, who did not like him, considered this an offense and

---

669. Father Johann Schirmeisen was born in Silesia (a historical region that is now in southwestern Poland) on January 17, 1657. He joined the Society of Jesus on August 21, 1677. In 1701 he was still a spiritual coadjutor (15/8/1687). He died on February 20, 1719 (ARSI, "Primus Catalogus Anni Personarum Anni 1701." Philippinae Cat. Trien. 1701–1755, 1768, Tomo 3, f. 4r; Corrado Caracristi and Johann Mayr, *Il missionario di Rumo P. Giuseppe Bonanni, SJ (1685-1752) e le sue lettere*, Comune di Rumo: Tipografía Quaresima di Cles, p. 43).

670. Since August of 1526, there was knowledge of the Carolinian archipelago, but it was not until 1686 that its "rediscovery" took place. Sailing from Acapulco to Manila on the *Santa Rosa* galleon, General Francisco Lazcano stumbled upon those tiny islands—referred to as the "garbanzo islands"—and christened them as the Caroline Islands in honor of Charles II (Patricio Hidalgo Nuchera, (eds.), *Redescubrimiento de las islas Palaos* Madrid: Miraguano Ediciones & Ediciones Polifemo, 1993).

671. Indeed, in March of 1690, Governor Damián de Esplana had twice sent Don Alonso Soon, principal and sergeant major of the natives of Hågat and Humåtak, with four boats and eight soldiers to look for the Caroline island, "but he could not find it the first or the second time" (ARSI, "Relación del estado y progresos de la misión de las islas Marianas desde mayo pasado de 1689 hasta el de 1690," Supplementum ad Historiam, 1584–1750, Tomo 14, f. 77r).

had him arrested and imprisoned. He remained in prison for four months, in extreme desolation, and he surely would have been there longer if he had not found powerful assistance. This trip cost him very much, but he tried to recoup it. As soon as he arrived in the Mariana Islands, he had a boat built with the purpose of selling it to the Philippines. However, he said that he was building it to submit the northern islands. In effect, he embarked with the elite of his troops and the entire munition necessary for such an important expedition. A favorable wind took him swiftly in view of the island of Zarpana, but since his purpose had nothing to do with making new conquests, he did not stop there; and after various maneuvers, returned to port (page 379) three days after having set out. Such a precipitous return exposed him to everyone's gossip, but he did not care: he sold his ship and sent it to the Philippines with eighty soldiers, who were removed from the Marianas' garrison to satisfy his shameful avarice.[672]

A ship from New Spain that broke against the rocks trying to get to port made up for this loss.[673] More than twenty priests from the Order of St. Francis were saved along with the crew, which numbered more than one hundred men. It seemed that God had sent the governor these reinforcements to reduce the Northern islands. Their conquest would have been easy and certain, if it had been undertaken, but worried only about amassing wealth, [the governor] did not concern himself with advancing religion or the welfare of the state. Since the sale of the ship that he had sent to the Philippines had earned him considerable profit, he was determined to build more ships for the same purpose. He employed all of the people who had been saved from the shipwreck in cutting and preparing the necessary lumber for the construction. Among them was a great number of rascals the likes of whom are sent every year from New Spain

---

672. As Marjorie G. Driver has pointed out, avarice was one of the most constant characteristics of the Marianas' governors. According to Driver, Don Damián de Esplana returned to Guåhan because of his commercial interests related to the Manila galleon. To reduce corruption, the Crown dictated a Royal Decree in 1700 that "condemned the governors of the Marianas for neglecting to further the conquest, for misadministration of the *situado*, for mistreatment of the islanders, for imposing unauthorized royal fees, and for openly engaging in commercial activities by operating a store (*tienda abierta*)" (M. G. Driver, *Cross, Sword, and Silver: The Nascent Spanish Colony in the Mariana Islands*. Mangilao, Guam: MARC & University of Guam, 1993).

673. The *nao almiranta Nuestra Señora del Pilar de Zaragoza* was shipwrecked on June 2, 1690, off the Umatac bay in Cocos islet, losing most of the situado and the Marianas' stipend. According to the Relación written by the new vice-provincial Father Lorenzo Bustillo in San Ignacio de Hagåtña on May 21, 1691, the only provisions saved were "the wine and the flour for the holy Host, and a few other victuals." Nearly two hundred people disembarked, among them a group of Franciscans (15 fathers and 4 lay brothers) whose superior was Friar Miguel Sánchez, and six Jesuit missionaries who remained in Guåhan when the rest continued on their trip to Manila aboard the *nao capitana* Santo Niño de Cebú ("Relación del estado y progresos de la misión y cristiandad de las islas Marianas desde mayo de 1690 hasta 1691." AHCJC, *Documentos Manuscritos Historia de las Filipinas* (FILPAS), 64, ff. 32r-32v; RAH, Fondo Cortes, 567, 2/2669, Doc. 29, f. 1r-1v).

to the Philippines to establish colonies.[674] Seeing themselves overburdened with work and poorly fed, these wretches conceived the detestable project of killing the governor and his officers, sacking everything that they could find in the island, taking the first ship that was built, and leaving towards wherever they saw fit. The missionaries were not to be spared. They were all to be killed, except for one, left alive to absolve those who wanted to be converted. The day set for this bloody enterprise was the feast of St. Rose [September 4], patron saint of one of the island's most important parishes.[675] However, God looked with merciful eyes upon this mission that had suffered for so many years. He touched the heart of one of the conspirators. This man could not stand the pangs of conscience that afflicted him, and he revealed the conspiracy.[676] The culpable men were arrested, and after they confessed their crime [September 10–11] they were condemned to death, the ultimate torment.[677] The Jesuits assisted them at death and had the consolation of inspiring in them sentiments of repentance. They were helped in this (page 381) work of charity by the Franciscan friars. These priests, who were destined to the Philippines, remained in the island of Guåhan for six months. The Jesuits tried to temper their long stay with their cares and their proofs of friendship. Let us see what one of the most important among this apostolic troop from the Philippines wrote to Father Lorenzo Bustillos, then Vice-provincial and Superior of the mission of the Marianas:

MY REVEREND FATHER,

It had been some time since the ship sent here by the governor of the Mariana Islands, Don Damián de Esplana, had arrived, when I happened by chance upon the city of Macuvayan, which is three

674. In his 1685 *Memorial*, Father Luis de Morales asked the Council of the Indies not to allow the Philippine Governors to send exiled soldiers or subjects to the Marianas. On August 10, 1685, the Council of the Indies dispatched the necessary orders, but their frequent non-application rendered them useless (transcribed in R. Lévesque, *History of Micronesia*, vol. 8, 1996, pp. 411; 417–19).

675. He is referring to Hågat (Agat), whose church was completed in 1681. The parish or *partido* of Hågat (1689–94) was founded during the CHamoru War (1684-1685), when the Spaniards decided to build a church and town to concentrate the Chamorro population from Hagåtña's hinterland in one place (N. Ross Crumrine, "Praying and Feasting: Modern Guamanian Fiestas," *Anthropos*, 77, 1982, pp. 89–90).

676. "Relación del estado y progresos de la misión y cristiandad...." (AHCJC, FILPAS, 64, ff. 33r-38r); Diego de Zarzosa's letter to procurator Antonio Jaramillo, Någu, May 5, 1691 (transcribed in R. Lévesque, *History of Micronesia*, vol. 9, 1997, pp. 447–448). See also P. Murillo Velarde, *Historia de la provincia de Filipinas*, f. 365v.

677. "Relación del estado y progresos de la misión y cristiandad..." (AHCJC, FILPAS, 64, f. 41v); Pedro Cano's letter, to Antonio Jaramillo, Manila, June 22, 1691 (RAH, Fondo Cortes, 567, 9/2669, Doc. 37, f. 1r). See also R. Lévesque, *History of Micronesia*, vol. 9, 1997, p. 472; and P. Murillo Velarde, *Historia de la provincia de Filipinas*, f. 365v.

leagues from the city of Manila, and there I received the letter that you did me the honor of sending. I was happy to learn that you were in perfect health, as well as all the Fathers that work with you in the mission of the Mariana Islands. Allow me to greet you from here and express my gratitude and the sentiments of esteem (page 382) *and friendship that I harbor towards them. I pray to our Lord that he may shower on them His grace and blessings, for the conservation and growth of this Christianity. I do not doubt that I will be satisfied, for those missionaries work with great zeal and success in the conversion of souls. I have witnessed it during the six months that we remained there, my brothers and I in Umatac, in the house of the Society where I have been marvelously edified by their conduct. I have seen everything that these evangelical workers go through in these abandoned islands, the contradictions that they encounter, the persecution that they suffer, and the almost constant dangers to which they were exposed. I assure you, my Reverend Father, that I carry these apostolic men in my heart and that I will do so for the rest of my life, regarding them as models on whose imitation I should form myself, thanks to the example that you have given us and the charity with which you received us, you and Father Miguel de Aparicio. I am so imbued with it, that I cannot find the words to express my feelings and my gratitude. I have often told our* (page 383) *Fathers and laymen, albeit with a stammering tongue, for I am like a child who has not learned how to speak yet, and form only poorly articulated sounds, I have told them that I have seen these zealous missionaries hurrying through the countryside to administer the sacraments in the most horrible weather; walking on the muddiest, often impracticable paths; and embarking with evident peril to their lives, to fulfill the same duties of charity, without ever speaking of how much they suffer because of the lack of the most basic provisions, or mentioning the discomfort caused by mosquitoes and other insects, which are abundant in these islands. But what I cannot admire enough is the peace and charity that reigns among them; their zeal for the salvation of souls; their poverty, both in their dress and in their food; their example, admired by many but imitated by few; their kindness and their charity towards these poor Marianas' natives whom they instruct with sweetness and with patience; their continual preaching; and the success with which God has rewarded their labor. In sum, I cannot better express the sentiments that* (page 384) *I harbor for these apostolic men, nor thank you enough for the graces that*

*you have given us. I have tried to fulfill this last duty in the letter that
I wrote last year to Father Miguel de Aparicio by way of Acapulco.
I beg you not to forget me in your prayers, and to be sure that I am,*

MY REVEREND FATHER,

*your humble and loyal chaplain, Brother Antonio de la
Concepción y de Urrea.*

From the Philippines, April 8, 1692.

The kind and cordial way with which the Jesuits received these apostolic
men did not stop the Society's enemies from divulging that they had behaved
very badly (page 383b) towards them. Although such a gross calumny falls by
itself, and there is countless evidence of the contrary, I have thought that repro-
ducing this letter, which I have faithfully translated from the Spanish original,
is sufficient vindication for the missionaries of the Marianas.

After so many revolts and seditions, it was hoped that they would finally
enjoy the serenity and sweetness of peace, but God afflicted the Mariana Islands
with the most terrible hurricane that has ever been known. It started on the
nightfall of November 20, 1693 with an impetuous wind mixed with thunder
and lightning, accompanied by such a prodigious rain that it seemed more like
a deluge. The wind, which increased unabated, passed from north to south, and
whipped up the sea in such an extraordinary manner that it seemed that the
island of Guåhan would disappear. The waves crashed against one another and
rose as high as mountains, breaking [beyond] their usual limits, and running
over the plains with such speed that they carried along everything in sight: trees,
churches, cities, everything was washed away by the waves. Even the fortress
of Agadña could not resist (page 384b) such violence. It was overturned and
covered by the water overed. The inhabitants of this unfortunate city, like those
of Umatac, thought that they were all going to die. All you could hear were the
cries and groans of those who were drowning or struggling against the waves
during the horror of that dark night.

The missionaries were saved only through God's particular protection.
For a long time, the sea and the wind played with Father Diego Zarzosa, while
he clung to a plank, until they tossed him at last upon a neighboring hilltop.
Father Paravicin[678] held on to a tall stake, and there he was beaten by the waves

---

678. Father Francesco Pallavicino, an Italian or Bohemian missionary who lived in the Marianas from
1688 to 1695 (F. X. Hezel, S.J., *From Conquest to Colonization*, p. 89). Simona Binková, "Contribución

until, unable to stand their violence, he was about to succumb when help came to him. Father Miguel de Aparicio was in no less danger. He was the principal of the children's seminary. The water surprised him there with Brother Melchor de la Cruz and two young seminarians who had not followed their companions. Locked inside a house, they had no hope of survival, when the wind blew the roof off. They threw themselves on some planks of wood that (page 385) had come loose, and after rowing about for some time, they had the luck of being carried to the ruins of the fortress, where they remained until the hurricane passed. The hurricane ceased the following morning, after having caused horrible destruction, for in the whole island not a single house was left standing. Most of the plains as well as the ports were filled with stones and sand left by the sea. Desolation was general, but the most surprising thing was that the neighboring islands barely suffered any damages, for all of the fury of this tempest fell on Guåhan alone.

As soon as the hurricane passed, reparation of the damage that it caused was begun. The land was cultivated again to stave off hunger, and houses and churches were erected. The most important *chamorris* distinguished themselves for their zeal and their piety on this occasion. Don Antonio Ayihi undertook the reconstruction of the church of Agadña; Don Alonso Soon, that of Agat; Don Ignacio Fuguon and Don Pedro Tao those of Fuña and Pago [Pågu]. The governor took charge of the churches of Merizo and Umatac. He chose this last site to (page 386) build his house, taking the garrison that used to be in Agadña. This change was not favorable for the governor, for soon after he fell into a sort of torpor that degenerated into hydropsy and took him to the grave. All of the projects of grandeur and fortune that had occupied him for so many years vanished in a moment. He had time to do penance and, after much suffering, received the sacraments of the Church and finally passed away on August 16, 1694.[679]

Don Joseph de Quiroga succeeded him. A man of his character was needed to reestablish the affairs of a mission that had been desolated by his predecessor's misconduct.[680] He began by reestablishing military discipline among

de los jesuitas de la provincial de Bohemia para la evangelización de las islas Filipinas y Marianas y para su conocimiento en Europa central", in Simona Binková, Markéta Křížová et al. (eds.), *Ir más allá… Fuentes bohemicales para el estudio comparativo de la expansión colonial española en la temprana Edad Moderna*, Praha: Universidad Carolina de Praga & Editorial Karolinum, 2016, pp. 40-41).

679. A. Coello, "Colonialismo y santidad en las islas Marianas: los soldados de Gedeón," p. 41.

680. The dismal state of the mission was not just the governor's fault. On June 20, 1694, the patache that was destined to bring Guåhan the assistance that it so badly needed, burned down in the Philippine port of

the troops, which were made to return to Agadña. This earned him the soldiers' good will.[681] Since his only interest was to spread Christianity in these infidel lands, he went to the island of Zarpana at the beginning of October, 1694.[682] The islanders were not scared by the arrival of a man whose good intentions were well known, (page 387) and they brought him refreshments and placed themselves at his command, bringing their children. Father Basilio le Roulx, who accompanied Quiroga, had the consolation of baptizing one hundred and fifty children, some of whom were sent to Agadña to be educated in the seminary. Happy with the rapid success of this expedition, Quiroga prepared during winter for the conquest of the northern islands, which he had been unable to submit for ten years. He departed with his troops on July 11, 1695, in a frigate and twenty canoes.[683] A heavy wind separated the small fleet, with the canoes stopping in Zarpana while the governor's frigate continued en route until it came in sight of Tinian. The barbarians, alarmed, waited at the coast while they embarked their women and children for the island of Aguiguan to keep them safe.[684] Quiroga advanced, trying to reach the beach, but since he was unable to do so because the frigate took in too much water, he reunited with his fleet and sailed towards Saipan. These islanders gathered and decided to fight the Spaniards and keep them from disembarking. However, they were fired upon so powerfully that (page 388) they disappeared in a moment. [The Spaniards] went about the island and crossed it without encountering any resistance, for the barbarians, terrified, had retreated to the mountains and hidden in inaccessible places. Some were seized and taken to Quiroga, who explained

---

Cavite loaded with more than 20,000 pesos (Pedro de Silva's letterl to Antonio Matías Jaramillo, Manila, July 20, 1694, in RAH, Fondo Cortes 567, Legajo 9/2669/ 46, f. 1r).

681. By then, the Crown sent the governor 37.800 pesos, which covered the salary of 60 infantrymen (transcribed in R. Lévesque, *History of Micronesia*, vol. 9, 1997, pp. 646–648). However, by then the presidium had 131 Spanish soldiers (the Spanish infantry) and 53 Filipino soldiers (the Pampango infantry) (M. G. Driver, *The Spanish Governors*, p. 14).

682. Annual Letter of 1696 (ARSI, Filipinas 13, ff. 306-325v, transcribed in R. Lévesque, History of Micronesia, vol. 9, 1997, pp. 665–666).

683. According to the *Relación del padre Diego Zarzosa*, written on April 30, 1696, Governor Quiroga organized a new expedition and set out to conquer the thirteen northern islands on a galleot, a frigate, and twenty Marianas canoes with 80 sailors (both Spanish and Philippine natives) (RAH, Fondo Jesuitas, Signatura: 9-3732/Documento 52, ff. 189r-190v); Annual Letter of 1696 (ARSI, Filipinas 13, ff. 306-325v, transcribed in R. Lévesque, *History of Micronesia*, vol. 9, 1997, pp. 666–667).

684. *Relación del padre Diego Zarzosa*, April 30, 1696 (RAH, Fondo Jesuitas, Signatura: 9-3732/Documento 52, f. 189v); "Relación de la conquista y restauración de las islas Marianas. 1696" (ARSI, Litterae Annuae Philipp. 1663-1734 (etiam de Insuli Marianis), Vol. 13, f. 306r-307v; Annual Letter of 1696 (ARSI, Filipinas 13, ff. 306-325v, transcribed in R. Lévesque, History of Micronesia, vol. 9, 1997, p. 667). For more general information, see V. Muñoz Barreda, *La Micronesia española*, p. 96; and D. A. Farrell, *History of the Northern Mariana Islands*, pp. 175–76.

to them that he had not come to expel them from their homes or to avenge
the Spaniards that they had killed, but to live in peace with them and procure
them all kinds of benefits: "*I only ask you one thing,*" he said, "*and that is that you
listen to the preachers of the Gospel and are docile to their instruction.*" The Saipan
natives liked these proposals and promised to do all that he asked.

After having subjugated this island, Quiroga returned to Tinian with the
fleet. He found no one there, for, overcome by panic, they had all withdrawn to
Aguiguan. This island, which is only one league from Tinian, rises up from the
sea like a fortress, tall and craggy. It seems inaccessible, and indeed, it would be
if it were not for some narrow passes. The Tinian natives took this island and
resolved to defend themselves there to the death (page 389). Since forcing them
would not be easy, Quiroga sent trustworthy envoys to invite them to return
to their homes and live in peace [August 12, 1696].[685] However, the barbarians were not docile. They were threatened with having their lands laid waste
and their houses burned, but to no avail. They were obstinately determined
to remain in their trenches. First, the Spanish tried to tire them out through
hunger and thirst, for there are no fountains or springs in this island. When
Quiroga saw that these barbarians were willing to die rather than give in, he
decided to attack them and force them out of their trenches. The undertaking
was bold, not to say foolhardy, and there are reasons to believe that he would
not have decided upon this if he could not count on God's help much more
than on his troops' valor: let us see how he did it.

There were two gorges on each side of the island, the only two places
through which it was accessible. Quiroga divided his small fleet. He sent one
part to take the western pass, while the rest of his troops went with him to
attack the other.[686] (page 390) He approaches the enemy trenches, and they send
him joking messages telling him that peace is being negotiated on the western
side, and that it is nearly concluded. Quiroga, who sees that the pass that he
wants to attack is inaccessible, pretends to believe them and rapidly goes to the
other side, where he finds his troops ready to charge the barbarians. He gives
the order, encouraging them, and at that moment Captain Nicolás Rodríguez

685. In his *Relación* of 1696, Father Zarzosa hailed him as a true apostolic hero (*Relación del padre Diego Zarzosa*, 30/4/1696 (RAH, Fondo Jesuitas, Signatura: 9-3732/Documento 52, f. 189r-190r). See also the Annual Letter of 1696 (ARSI, Filipinas 13, ff. 306-325v, transcribed in R. Lévesque, History of Micronesia, vol. 9, 1997, pp. 667–668).

686. "Relación de la conquista y restauración de las islas Marianas. 1696" (ARSI, Litterae Annuae Philipp. 1663-1734 (etiam de Insuli Marianis), Vol. 13, f. 307r-307v); Annual Letter of 1696 (ARSI, Filipinas 13, ff. 306-325v, transcribed in R. Lévesque, History of Micronesia, vol. 9, 1997, p. 668).

[alias "Conejo," Rabbit], Adjutant José Rodríguez, Philippine Sergeant Major Matías de Guevara, and *chamorris* Don Pedro Ynoc, followed by Adjutant Nicolás de Vega and others, jump on land and advance towards the enemy. The great number of stones that the barbarians rolled down from atop their high trenches dulled these brave men's ardor and forced them to seek refuge under an outcropping to avoid being crushed. Sergeant Juan Pérez Vello and a Philippine captain named Pablo de la Cruz could not stand that these officers' advance had been stopped, and without any regard for the danger to which they exposed themselves, climbed the steep rocks. Their companions followed them, and they climbed together, fueled by fury, until they reached their enemies' trenches (page 391), forcing them out and becoming their masters. Scared, the barbarians lowered their weapons, threw themselves upon the ground and asked for mercy. It was granted, but under the condition that they would all be taken to the island of Guåhan to be instructed in the truths and maxims of the Christian religion. This was done the next day.[687]

This victory established a solid peace and brought happy results for religion. Rumors soon reached the islands of Gani, the name given to the Marianas that are at the upper north, as I have said before. Those islanders did not wait for troops to come: they submitted themselves voluntarily to the Spanish. They were ordered to go to the island of Saipan to be instructed. They obeyed, and applied themselves with such ardor to learning the Christian doctrine and practicing what they were taught, that this Christianity was soon one of the most flourishing of these islands. This was thanks to the untiring labor of Father Gerardo Bouwens, an excellent missionary who despite being more than sixty years old, is still working today in the conversion (page 392) of these peoples with an indefatigable zeal.[688]

After this highly successful expedition, Quiroga returned to Agadña, only to nearly perish at the port. His ship, pushed by an impetuous wind, broke between two rocks that were very close to the surface. The entire crew was in

687. From August of 1695 to April of 1696, more than fifty vessels drained the populations of Tinian and Agrigan as well as the northern islands, relocating them in the main islands to guarantee the survival of the mission. See the letter written by Father Bouwens to Father Edigio Estrie from Saipan, May 28, 1696 (AHCJC, FILCAR, E.I, a-18/1, f. 69; transcribed in R. Lévesque, *History of Micronesia*, vol. 9, 1997, pp. 679–688).

688. However, the need for evangelization was ever-present. In 1697, Father Bouwens was concerned that some natives from Gani who had been relocated to Saipan were returning to their islands. A year later, stationed in Saipan, he informed the Governor of this worrisome trend (ARSI, "Relación de la última reducción de las islas Marianas llamadas de Gani que están al norte de las islas de Saipan y Tinian" (1699), Philipp. Suppl. 1584–1750, Tomo 14, ff. 88r). See also the letter written by Father Bouwens from Guåhan to Father Conrad Janning, in Antwerp, on November 20, 1698 (transcribed in R. Lévesque, *History of Micronesia*, vol. 10, pp. 153–157).

danger. Some drowned, while others survived after much trouble, including the governor and Father Basilio le Roulx, the Viceprovincial and Superior of this mission. The Father was underwater for a long time, and he was taken out half dead, without a pulse and without his senses. He only came to after many remedies were given to him, and was not fully recovered until a few years later. God conserved the lives of these two men so that they could work for his glory and for the growth of our Holy Faith, as they do today with such zeal that one cannot admire enough. The other missionaries work with the same ardor. Churches to the true God are erected everywhere, and the people go to them en masse to receive instruction, and they progress with a rapidity that makes one think that (page 393) all of these islands will soon be Christian. The number of baptisms and conversions multiplies each day.[689] The missionaries are too few. They ask for help to gather the fruit of such an abundant harvest. The time for the conversion of these abandoned peoples seems to have come, as the apostle of these islands, Father Diego Luis de San Vitores, used to say.

A new field for the preaching of the Gospel has been opened. More than thirty islands have been discovered to the south. Nobody knows their extension or size, but they are sure that there is an infinite number of people who do not know Jesus Christ and who have never heard of him. You can read how this discovery was made in the letter appended at the end of this History.[690] All that is needed are saints and fervent missionaries to go in the name of Jesus Christ to take possession of these infidel lands and bring the light of the Gospel to these people, who have been buried for centuries in the deepest darkness of paganism. *Messis quidem multa operari* (folio 394) *vero pauci. Rogate ergo dominum messisut mittst operarios in messem suam.* [The harvest is abundant, but there are few workers; pray, then, so that the owner of this harvest sends workers to his field].

---

689. According to Manuel de Solórzano's *Relación*, in June 1669 the Jesuits had baptized 13,560 natives, and 200 children of those baptized had died sometime after receiving the sacrament ("Descripción de las islas Marianas, costumbres de sus naturales. Una relación del estado en que se hallaban las misiones..." (1683), f. 124r). This obvious exaggeration is in keeping with the policy of requesting financial aid for the outlying Jesuit missions.

690. It refers to the letter Father Paul Klein addressed to General Tirso González, dated in June 10, 1697, informing him about the discovery of the 32 islands down south of the Marianas. These islands were part of the Palaos and the Caroline archipelago. For more information about this, see Coello, *Jesuits at the Margins*, chapter 4.

GOVERNORS OF THE MARIANA ISLANDS (1668–1700)

Military officers

Don Juan de Santa Cruz (June 16, 1668)
Don Juan de Santiago Bozo (May 2, 1678)
Don Damián de Esplana (June 16, 1674)

Governors

Don Francisco de Irisarri (June 10, 1676)
Don Juan Antonio de Salas (June 21,1678)
Don Joseph de Quiroga (June 5, 1680)
Don Antonio Saravia (1681)
Don Damián de Esplana (November 3, 1683)
Don Joseph de Quiroga (1688)
Don Damián de Esplana (June 1690)
Don Joseph de Quiroga (August 16, 1694)
Don Joseph de Madrazo (August 1, 1696)
Don Francisco Madrazo y Asiam (September 15, 1700)

VICEROYS OF NEW SPAIN
(1664–1701)

Don Antonio Sebastián de Toledo, Marquise of Mancera (October 1664–November 1673)

Don Pedro Nuño Colón de Portugal, Marquise of Jamaica (December 8–13, 1673)

Don Payo Enríquez de Rivera, OSA (Archbishop of México) (December 1673–October 1680)

Don Tomás Antonio de la Cerda, conde de Paredes, Marquise of la Laguna (November 1680–November 1686)

Don Melchor Portocarrero, Count of la Monclova (November 1686–September 1688)

Don Gaspar de la Cerda Sandoval Silva, Count of Galve (November 1688–February 1696)

Don Juan de Ortega y Montañes, Archbishop of México and Michoacán (February 1696–December 1696)

Don José Sarmiento y Valladares, Count of Moctezuma and Tula de Allende (December 1696–Novemebr 1701)

GOVERNOR-GENERALS OF THE PHILIPPINE ISLANDS (1668–1701)

Don Juan Manuel de la Peña Bonifaz (1668–1669)
Don Manuel de León (1669–1677)
Interregnum, Audiencia of Manila (1677–1678)
Don Juan de Vargas (1678–1684)
Don Gabriel de Curuzeláegui (1684–1689)
Interregnum, Audiencia of Manila (1689–1690)
Don Fausto Cruzat y Góngora (1690–1701)

Name Index

Subject Index

9 781935 198093